While competitiveness has its place, cooperation............ ore important in many contexts, and individual relationships are the mos. ant elements in our lives. In this book, authors from disciplines as diverse as biology, philosophy and the social sciences consider the nature of cooperation, altruism and prosocial behaviour between individuals, within and between groups, and between nations. Three valuable case studies of international cooperation are also included. The chapters are integrated by a series of useful editorials, which emphasize that a full understanding of cooperation and prosocial behaviour requires us to move between different levels of social complexity.

This unique volume will be of interest to students and researchers in the fields of psychology, biology, philosophy, and the social and political sciences, as well as policy makers, diplomats and the concerned general reader.

Cooperation and prosocial behaviour

Cooperation and Prosocial Behaviour

EDITED BY

ROBERT A. HINDE

Master of St John's College, Cambridge
and
M.R.C. Unit on the Development and Integration of Behaviour,
Madingley, Cambridge

AND

JO GROEBEL

Chair, Department of the Social Psychology of Mass Communication
and Public Relations,
University of Utrecht

Foreword by
Her Royal Highness The Princess Royal

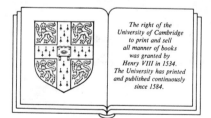

The right of the
University of Cambridge
to print and sell
all manner of books
was granted by
Henry VIII in 1534.
The University has printed
and published continuously
since 1584.

CAMBRIDGE UNIVERSITY PRESS

Cambridge
New York   Port Chester
Melbourne   Sydney

Published by the Press Syndicate of the University of Cambridge
The Pitt Building, Trumpington Street, Cambridge CB2 1RP
40 West 20th Street, New York, NY 10011-4211, USA
10 Stamford Road, Oakleigh, Victoria 3166, Australia

First published 1991

Printed in Great Britain at the University Press, Cambridge

British Library cataloguing in publication data

Cooperation and prosocial behaviour.
1. Social behaviour
I. Hinde, Robert A. (Robert Aubrey), *1923* – II. Groebel, Jo
302.14

Library of Congress cataloguing in publication data

Cooperation and prosocial behaviour / edited by Robert A. Hinde and
J. Groebel : foreword by H.R.H. the Princess Royal.
    p.cm.
ISBN 0–521–39110–5 (hardcover). – ISBN 0–521–39999–8 (paperback)
1. Interpersonal relations.   2. Intergroup relations.
3. International cooperation.   I. Hinde, Robert A.   II. Groebel, Jo.
HM131.C74733   1991
302 – dc20 91–9981 CIP

ISBN 0 521 39110 5 hardback
ISBN 0 521 39999 8 paperback

Contents

List of Contributors   *page* x

Foreword by Her Royal Highness The Princess Royal   xv

Introduction   1
R. A. HINDE AND J. GROEBEL

A. COOPERATION IN ANIMALS AND MAN   9

Editorial   11

1   Help, cooperation and trust in animals   15
    A. H. HARCOURT

2   Culture and cooperation   27
    R. BOYD AND P. J. RICHERSON

**B. THE DEVELOPMENT OF PROSOCIAL
PROPENSITIES**   49

Editorial   51

3   The development and socialization of prosocial
    behavior   54
    P. A. MILLER, J. BERNZWEIG, N. EISENBERG AND R. A.
    FABES

4   Cross-cultural differences in
    assertiveness/competition vs. group
    loyalty/cooperation   78
    H. C. TRIANDIS

5   The development of prosocial behavior in
    large-scale collective societies: China and Japan   89
    H. W. STEVENSON

6   The learning of prosocial behaviour in small-scale
    egalitarian societies: an anthropological view   106
    E. GOODY

C. SITUATIONAL AND PERSONALITY DETERMINANTS OF PROSOCIAL BEHAVIOUR   129

Editorial   131

7   Situational and personality determinants of the quantity and quality of helping   135
J. FULTZ AND R. B. CIALDINI

8   Perceiving the causes of altruism   147
W. C. SWAP

9   Altruism   159
J. HEAL

10   Complications and complexity in the pursuit of justice   173
S. D. CLAYTON AND M. J. LERNER

D. TRUST, COMMITMENT AND COOPERATION   185

Editorial   187

11   The dynamics of interpersonal trust: resolving uncertainty in the face of risk   190
S. D. BOON AND J. G. HOLMES

12   Commitment old and new: social pressure and individual choice in making relationships last   212
M. LUND

13   Cooperation in a microcosm: lessons from laboratory games   224
D. A. GOOD

14   Determinants of instrumental intra-group cooperation   238
J. M. RABBIE

E. COOPERATION BETWEEN GROUPS   263

Editorial   265

15   Changing assumptions about conflict and negotiation   268
J. Z. RUBIN

16   Cooperation between groups   281
H. FEGER

17   The role of UNESCO in the development of international cooperation   301
F. MAYOR

18 U.S.–Soviet cooperation against terrorism:
Common ground 316
I. BELIAEV AND J. MARKS

19 U.S. policy towards the Soviet Union from Carter
to Bush 329
E.-O. CZEMPIEL

Name index 355
Subject index 360

Contributors

I. Beliaev
Liternaturnaya Gazeta
13 Kostiansky Lane
Moscow, U.S.S.R.

J. Bernzweig
Arizona State University
Tempe, Arizona 85287, U.S.A.

S. D. Boon
Department of Psychology
University of Waterloo
Ontario N2L 3GI, Canada

R. Boyd
Department of Anthropology
University of California
Los Angeles, California 90024, U.S.A.

R. B. Cialdini
Department of Psychology
Arizona State University
Tempe, Arizona 85287, U.S.A.

S. D. Clayton
Department of Psychology
Allegheny College
Meadville, Pennsylvania 16335, U.S.A.

E.-O. Czempiel
Inst. Internationale Beziehungen
Johann Wolfgang Goethe-Universität
6000 Frankfurt, Germany

N. Eisenberg
Department of Psychology
Arizona State University
Tempe, Arizona 85287, U.S.A.

R. A. Fabes
Department of Family Resources & Human Development
Arizona State University
Tempe, Arizona 85287, U.S.A.

H. Feger
Institut für Psychologie
Freie Universität Berlin
Habelschweerter Allee 45
1000 Berlin 33, Germany

J. Fultz
Department of Psychology
Northern Illinois University
De Kalb, Illinois 60115, U.S.A.

D. A. Good
Department of Social & Political Sciences
University of Cambridge
Free School Lane
Cambridge CB2 3RQ, U.K.

E. Goody
Department of Anthropology
University of Cambridge
Free School Lane
Cambridge CB2 3RQ, U.K.

J. Groebel
University of Koblenz-Landau, Germany
and Department of the Social Psychology of Mass
Communication and Public Relations,
University of Utrecht,
Heidelberglaan 1, 3508 TC Utrecht, The Netherlands

A. H. Harcourt
Department of Anthropology
University of California
Davis, California 95616, U.S.A.

J. Heal
St John's College
Cambridge CB2 1TP, U.K.

R. A. Hinde
Master's Lodge
St John's College
Cambridge CB2 1TP, U.K.

J. G. Holmes
Department of Psychology
University of Waterloo
Ontario N2L 3GI, Canada

M. J. Lerner
Department of Psychology
University of Waterloo
Ontario N2L 3GI
Canada

M. Lund
Didi Hirsch Community Mental Health Centre
Los Angeles Psychiatric Service
4760 So. Sepulveda Blvd
Culver City, California 90230, U.S.A.

J. Marks
Search for Common Ground
2005 Massachusetts Avenue, N.W.
Lower Level
Washington DC 20036, U.S.A.

F. Mayor
Director General
UNESCO
7 Place de Fontenoy
75700 Paris, France

P. A. Miller
Arts & Science
Arizona State University
West Campus
4701 West Thunderbird
Phoenix, Arizona 85069–7100, U.S.A.

J. M. Rabbie
Faculteit der Sociale Wetenschappen
Rijksuniversiteit te Utrecht
Heidelberglaan 1,
3508 TC Utrecht, The Netherlands

P. J. Richerson
Institute of Ecology
University of California
Davis, California 95616, U.S.A.

J. Z. Rubin
Harvard Law School & Department of Psychology
Tufts University
Medford, Massachusetts 02155, U.S.A.

H. W. Stevenson
Center for Human Growth & Development
University of Michigan
300 North Ingalls
Ann Arbor, Michigan 48109, U.S.A.

W. C. Swap
Tufts University
Medford, Massachusetts 02155, U.S.A.

H. C. Triandis
Department of Psychology
University of Illinois
603 East Daniel Street
Champaign, Illinois 61820, U.S.A.

The pursuit of individual self-fulfilment, accompanied by acceptance of competitiveness as a virtue, is accepted by many as a basis for living. Western values play down the importance of cooperation in group-living and turn a blind eye to the fact that interpersonal relationships are the most important element in the lives of everyone of us. But we must not lose sight of the other side of the picture. Competition may have its place, but cooperation is more important in many contexts. If we value cooperation and mutual tolerance we can achieve a better world for all than undiluted aggressive competition will ever provide. Self-fulfilment is a proper goal, but for most if not all it is to be found in or through interpersonal relationships.

While many additional complexities enter in, what is true of individuals is also true of nations: we need to find ways to build international relationships based on trust and mutual understanding. Hopefully, the world will remain as diverse and colourful a place as it is now - people do not have to become more similar in order to develop mutual trust. To acknowledge differences and at the same time to respect each other's needs and values so long as they are not destructive many not be easy, but it is the only solution for the future of the world. In times of increasing global interdependence nations have to learn mutual respect and to commit themselves to reasoned co-existence - but this will only be the case if nations can learn to respect each other's values and needs.

To achieve these aims, we shall need the help of diverse sorts of people - religious leaders, politicians, business men and industrialists, academics and social workers. As a contribution to that end I welcome this collection of essays by psychologists and biologists on the bases of cooperation, trust and commitment, in the belief that it will help us to steer our societies onto a balanced path.

Anne

Introduction

ROBERT A. HINDE AND JO GROEBEL

The problem

In 1989, the editors of this volume published a collection of papers on 'Aggression and War: their biological and social bases.' Chapters presenting data from genetics, physiology, ethology, psychology, sociology, history and political science showed that *violent and antisocial behaviour* at different levels of social complexity required different principles of explanation and could not be described in monocausal terms. *Prosocial behaviour* is often assumed to be just the other side of the coin to antisocial behaviour. In this volume, therefore, we ask whether the same, i.e. a high complexity, is true of prosocial behaviour. Does its nature also differ at different levels of social complexity?

Briefly, let us summarize the findings for aggression. At the level of interaction between individuals, biological and environmental (including social) influences can contribute to the development of a potential for aggressive behaviour. A major factor which determines whether this predisposition leads to aggression or not is social experience, especially childhood experience within the family.

At higher levels of social complexity, different issues arise. Individual propensities may play an important role but are not sufficient to explain, for instance, the occurrence of war, even if mass media tend to 'personalize' wars, as was the case with the Gulf War, and Saddam Hussein in 1991. With each social level – relationship, group, society, nation – additional influences come into play. And the socio-cultural structure – the system of beliefs, values, myths, customs, institutions, etc. – both influences and is influenced by behaviour at each level. The institution of

1

war, for instance, is in part a consequence of complex historical events which influence the willingness of individuals or groups to act violently. In an actual war situation, however, individual aggression may be absent and the primary psychological process may even be prosocial behaviour and cooperation between soldiers of the same army who simply fulfil the task they are given.

As this example shows, the relationship between antisocial and pro-social behaviour is not as clear-cut as it might seem at first sight. In the book on 'Aggression and War' two chapters (by Norma Feshbach and Arnold Goldstein) were devoted to the reduction of aggression. An attempt to reduce aggression may itself be a prosocial act, e.g., by a teacher, but does not necessarily lead to prosocial behaviour from the (former) aggressive actor, e.g., a student. Furthermore, prosocial and aggressive behaviours may have some elements in common, other elements which are exclusive to one or other, and yet other characteristics which cannot be placed along the same phenomenological dimension.

A common factor for both is the importance of interaction between individuals and of their relationship. Several chapters of this book deal with these social variables: the history of interactions between two indi-viduals determines the nature of their relationship and, in turn, their relationship influences further interactions. These interactions may include antisocial or prosocial behaviours and may of course change from one to the other over time.

Exclusive elements for antisocial and prosocial behaviour are by defi-nition the consequences for the interaction partner, his/her suffering or (relative) well-being. However the distinction is not always clear-cut. For example short-term harm may be a necessary precondition for long-term well-being. Yelling at a child when it approaches a hot fire may be regarded by some as a crucial means to prevent it suffering from the long-term consequences of such behaviour. Thus, time and the process-characteristics play a crucial role in the evaluation of behaviour as pro- or antisocial. An additional possible complication is the perspective of the evaluator. In our example, the child may regard the yelling as aggression, while the parent feels he/she is using a justifiable though aggressive means for a prosocial end. And perspectives may change over time. Again, war is a striking example: a nation is attacked by its neighbour (aggression), another nation may try to help (prosocial behaviour) by attacking the neighbour (aggression), while during the attack soldiers help each other (prosocial behaviour). After some time, the 'helping' nation may be perceived by a fourth nation as aggressive because it seems to have followed 'egoistic' goals. Again, the attribution of the behaviours depends on the position of the evaluator in the whole interaction process.

An example of the difficulty of placing anti- and prosocial behaviour along the same phenomenological dimension comes from studies of the consequences of television violence. Violent programmes are often imitated by child viewers, and can serve as a model for social learning. Parents and teachers have postulated that by replacing violent programmes with prosocial ones these same observation learning qualities can be used to foster prosocial behaviour. One dramatic feature of a violent programme, however, is its arousal effects. The success of films can be traced only partly to their content. The content is usually combined with 'action', with the actors' fast movements and motor activities creating (positively experienced) physiological arousal among the viewers. Most prosocial behaviours lack this arousal quality and so, from a purely dramaturgical point of view, they would not be an adequate replacement for aggressive programmes. Sadly, aggression is often perceived as 'more interesting' than prosocial acts. News makers know this effect and use it, with the consequence that violence has a much higher probability of being reported than prosocial acts. Even scientists may regard aggression as more stimulating, as a more common behaviour, or as a socially more important problem, than prosocial behaviour: until recently, the scientific literature on violence and aggression by far exceeded that on prosocial behaviour, cooperation and altruism.

Even with an increase of studies on such positively evaluated behaviours it is interesting to note that mostly only one level of analysis is taken: either the biological, or the psychological, or the sociological one. Few authors try to link the findings across the different levels of complexity. An exception, to a certain extent, is the book on 'Trust' by Gambetta (1988) which offers a collection of papers from different disciplines including economics, psychology, biology, sociology and philosophy. It is shown that trust (a crucial prerequisite for cooperation – see the editorial to Part D) depends on motivations, on expectations and on the surrounding social and political system: thus, the interaction between social, psychological, biological and environmental factors is a major variable for the analysis of trust.

Many, if not most, of the studies dealing with prosocial behaviour analyse its functions. Do people only act prosocially because they can expect a direct or indirect reward in the long run? As several chapters in this volume demonstrate, it is not so easy as that. Prosocial activities may be based on mixed motives, some of them centred around positive outcomes for the individual, some however more related to survival or well-being of the group, society, or nation. Yet, membership within a particular group can play a crucial role: the member of a group from a collective society may show the highest level of selfless prosocial behaviour towards members of his own group and, at the same time, the lowest imaginable level of empathy towards members of another group. For instance, an individual may even be willing to commit suicide for the sake of his/her own group in order to inflict cruel suffering on those (enemies) he is to 'take with him'. This example shows that it might be naive simply to try to establish collective societies (Chapter 4) as this might neglect the presence of in-group/out-group thinking. In the long run, defining the *global* social system as the common in-group may be a solution, but this would demand many more answers to the problems raised in his volume. One of these problems of course is the continuing, perhaps even increasing, presence of numerous economic and ideological conflicts around the globe. Dealing with prosocial behaviour and cooperation thus also requires us to include chapters on negotiation and conflict resolution as a necessary basis for the development of inter-group trust and empathy.

Definitional issues

Before proceeding, it is as well to take up a few definitional issues. We may start with '*Help*', which the *Concise Oxford Dictionary* defines as 'Provide (person, etc.) with means towards what is needed or sought, be of use or service to'. Helping may involve costs to the helper, but it may also bring rewards: in most cases, probably, it does both.

Cooperation occurs when two individuals help each other to reach or obtain 'what is needed or sought'. The essence of cooperation is that two (or more) individuals assist each other to reach the same end. (Boyd and Richerson (Chapter 2) use 'cooperation in a slightly different sense, to refer to behaviour intended to benefit a group at a cost to the individual).

When two or more individuals help each other to reach different rewards, their partnership is described as involving *Mutualism*. This is related to the biological concept of *Symbiosis*, referring to a situation in which the proximity of two animals is to their mutual advantage.

Leaving aside the basic question of the currency in which benefits or rewards and costs are measured (see the editorial to Part A), where helping involves costs not commensurate with the rewards, it may be referred to as *Altruism*. (In its everyday sense, the term altruism is used for

'Regard for others as a principle of action; unselfishness' (*Concise Oxford Dictionary*).) There is, of course, often considerable difficulty in determining whether or not costs outweigh rewards, and thus whether a given action is 'truly' altruistic, or in discriminating between behaviour that is genuinely altruistic and behaviour that is basically selfish. For that reason, the term *Prosocial behaviour* is often used for all behaviour that benefits others.

Prosocial behaviour or cooperation may involve two or more individuals previously unacquainted with each other: the helping behaviour occurs in a relatively short-term *Interaction*. More usually, perhaps, the individuals concerned have a *Relationship*. A relationship can be defined as involving individuals who interact on a series of occasions so that each interaction is affected by past interactions with the same individual and/or by expectations of future ones. (That there is no sharp dividing line between a long interaction and a relationship is an issue that need not detain us here.) Of course there is more to a relationship than the behaviour shown on successive encounters, but the important point here is that relationships are extended in time and involve individuals known to each other. Relationships therefore have properties that are irrelevant to individual interactions – for instance, a relationship may involve just one type of interaction (e.g. many doctor–patient relationships) or many (e.g. husband–wife). Not surprisingly, therefore, new issues arise in relation to prosocial behaviour.

In a relationship, one individual may help another in the expectation of subsequent reciprocation. But in that case, he must *trust* the other – that is, have a firm belief in his/her reliability, honesty, etc. (*Concise Oxford Dictionary*). A more technical definition of trust has been given by Gambetta: 'Trust . . . is a particular level of the subjective probability with which an agent assesses that another agent or group of agents will perform a particular action, both *before* he can monitor such action (or independently of his capacity ever to monitor it) *and* in a context in which it affects *his own* action'. As Dunn has emphasized, fundamentally, trust is a device for coping with the freedom of others: it presupposes a risk of defection by the other party, and on that basis has been contrasted with 'confidence'. Within a relationship, trust usually implies a belief that the partner has *commitment* to the relationship. Commitment ('engagement or involvement that restricts freedom of action' (*C.O.D.*)) is used in the context of a relationship to imply that one or both parties either accept their relationship as continuing indefinitely or direct their behaviour towards ensuring its continuance or optimizing its properties. Clearly, if one individual is to help another in the expectation of reciprocation, he must not only be committed himself, but also believe in the partner's commitment. He is more likely to sustain such a belief if it is compatible with past experience with the individual concerned.

Relationships are usually set within *groups* – families, communities, and so on. Just as relationships have properties that are not relevant to interactions, so groups have properties not relevant to relationships. For example, the arrangement of the relationships within the group may be an important property of the group (hierarchical, linear, etc.), but is not a property of the individual relationships. A group usually implies perceived interdependence between the individuals involved (see Chapter 14), but *aggregations* may involve diverse individuals who may lack interpersonal relationships with each other. In such cases the coherence of the group is maintained by allegiance to a person, symbol or idea. *'Loyalty'* then takes the place of commitment, but has many of the same properties.

Groups may interact with other groups, and many of the terms used with reference to interpersonal relationships are sometimes applied also to groups, though the phenomena involved may be quite different. Usually this is recognised by a change in terminology: cooperation becomes an *alliance*, for instance, and loyalty may become *patriotism*. Where groups are in conflict, i.e. where incompatible interests clash, the process of trying to solve the conflict in a way acceptable to both parties (i.e. the best possible outcome for both) is called *Negotiation*. If the outcome is indeed accepted by both parties we talk of (non-violent) *Conflict resolution*. Negotiations may involve all kinds of behaviour, even threat and bluff, but its basic assumption is the willingness on both sides to reach a resolution of the conflict in a non-violent way.

Whether we are dealing with two individuals or a whole society, shared beliefs, values, norms, etc. play a crucial part in determining behaviour. Confidence that beliefs and values are shared may be essential for trust and commitment: loyalty and patriotism may depend on agreement with group values. The beliefs, values and norms, together with institutions with their constituent roles, shared by more or less all the individuals concerned, can be described as the *'socio-cultural structure'* of the group.

The successive levels of social complexity that we have mentioned: individuals, interactions, relationships, groups and societies, and also the socio-cultural structure, influence and are influenced by each other. For instance the nature of an interaction is affected both by the individuals taking part and by the relationship in which it is embedded: similar principles apply at other levels of social complexity. In part for that reason, the levels are not to be regarded as entities, but as processes in continuous creation through the agency of these dialectical relations between levels (Hinde, 1987).

Finally, a few words about the organization of the chapters that follow. Two issues underline the division of the material into sections. First, the full understanding of behaviour requires answers to a number of distinct questions: What causes the behaviour? How did the propensity to show the behaviour develop in the individual? And how did the behaviour evolve? The latter includes the course of its evolution and the consequences through which natural selection acted to maintain it in the individual. For the last of these three questions, much of the data comes from animals. The first chapter gives an overview of this material, and the second provides a link to studies of our own species. Chapters 3–6 take up the developmental question, three of the chapters being especially concerned with the influence of the socio-cultural structure on the development of individuals.

The remaining sections are concerned with the nature and causation of cooperation, trust and commitment. Here a second issue influences the arrangement of the material. Because each level of social complexity has properties not relevant to lower ones, the relevant principles are liable to differ between them. Accordingly, Chapters 7–10 are concerned with situational and personality determinants of prosocial behaviour, and with the issue of what is fair. Chapters 11–14 focus on dyads and within group behaviour and Chapters 15–19 on principles of cooperation and trust between groups and nations.

Acknowledgement

We are grateful to Jane Armitage and Eve Rule for help in the preparation of the manuscript, to Ingrid Vatter for her advice over the art layout, and to Heike for her emotional support.

Further reading

Gambetta, D. (1988). Can we trust trust? In D. Gambetta, (ed.), *Trust*. Oxford: Blackwell.

Goldstein, A. P. (1989). Aggression reduction. Some vital steps. In J. Groebel & R. A. Hinde (eds.). *Aggression and War: their biological and social bases*. Cambridge: Cambridge University Press.

Dunn, J. (1988). Trust – political agency. In D. Gambetta (ed.). *Trust*. Oxford: Blackwell.

Feshbach, N. D. (1989). Empathy training and prosocial behaviour. In J. Groebel & R. A. Hinde. (eds.). *Aggression and war: their biological and social bases.* Cambridge: Cambridge University Press.

Groebel, J. & Hinde, R. A. (eds.) (1991). *Aggression and War: their biological & social bases.* Cambridge: Cambridge University Press.

Hinde, R. A. (1987). *Individuals, Relationships & Culture.* Cambridge: Cambridge University Press.

Hinde, R. A. (1991). *The Institution of War.* London: Macmillan.

A.

COOPERATION IN ANIMALS AND MAN

Editorial

Cooperation, trust and altruism are sometimes regarded as amongst the highest human virtues. We start by asking how and where comparable behaviour occurs in animals, and how the pervasiveness of cooperative behaviour in humans could have arisen. The first issue is the subject of Harcourt's chapter, who asks under what conditions animals cooperate, who cooperates with whom, and how does the occurrence of cooperation affect the nature of animal groups.

Before proceeding, it may be as well to clarify three issues concerning the relations between biological and social psychological perspectives. The first concerns the currency in which rewards and costs are to be reckoned. Let us consider first the biological approach. In classical evolutionary biology, since natural selection acts through the success of individuals in leaving offspring in the next generation, rewards and costs were assessed in terms of an individual's reproductive success, or potential for reproduction. It is now recognized that natural selection can favour also acts that enhance the reproductive success of related individuals, even though the actor him/herself incurs costs, if the actor is thereby enhancing the propagation of genes identical with his own. Thus biologists now assess the results of actions in terms of 'inclusive fitness' – that is, the contribution to the individual's own reproductive success measured in terms of descendants in subsequent generations plus that of their relatives devalued according to their degree of relatedness (Hamilton, 1964; Trivers, 1985).

Long-term reproductive success is difficult to measure. However, individuals cannot reproduce unless they survive, and their reproduction is more likely to be successful if they are healthy. Thus biologists, like

11

psychologists, are often content to assess the consequences of an action in terms of its contribution to survival or well-being – though in the bio- logists' case this is regarded only as a first step. Actions that augment the well-being of the actors are usually pleasurable or rewarding to the actor. There is thus a link between the biologists' and the everyday use of 'reward'. However not everything that gives pleasure is conducive to reproductive success.

In general, individuals are likely to repeat actions that yield pleasure and not to repeat actions that bring pain. In laboratory studies of learning, stimuli that alter the probability of the recurrence of a response are referred to as 'rewards' or 'reinforcers'. Of the two terms, 'reinforcer' is usually preferred because it carries no implication of pleasure to the recipient. Social psychologists, concerned with interactions between adult human beings, more often use 'reward'. This is in keeping with general usage, and with the fact that the primary emphasis is on the perception by the recipient of pleasure or value. 'Cost' is the opposite to 'reward': for both biologist and psychologist costs may include the extent to which the performance of an action results in alternative rewards being foregone.

A second issue concerns the role of intentionality in the goals of behaviour. Biologists, assessing benefits in terms of reproductive success in subsequent generations, emphasize that the goals of helping, cooper- ative or altruistic behaviour need not be conscious (see Harcourt, Chapter 1 this volume). By contrast, psychologists interested in the immediate causation of behaviour, usually imply the opposite. Thus Miller et al. (Chapter 3) are concerned with voluntary behaviours intended to benefit others (but see p. 149).

The third issue concerns a problem confronting both evolutionary biologists and psychologists, though they arrive at it by somewhat differ- ent routes – namely, the extent to which apparently altruistic behaviour does in fact benefit the altruist. Most psychologists would say that it does not: behaviour that benefits the actor is by definition not altruistic. However any act is likely to bring both rewards and costs to the actor: thus others argue that an act is properly called altruistic if the costs outweigh the benefits, or are perceived to do so by the actor. This issue is taken up in Chapters 7–9.

Given that biologists tend to reckon rewards and costs in terms of inclusive fitness, they need to explain why altruistic behaviour occurs at all. How could natural selection have favoured actions that reduce the individual's welfare or chances of surviving and reproducing? They there- fore suggest that altruism is more apparent than real, and may in fact benefit the altruist in two ways. One possibility depends on the concept of inclusive fitness (see above). Imagine a gene that predisposed an individual to give up his/her life for others. One copy of that gene would disappear when the altruist died. Nevertheless, if the altruist's act led to the preservation of identical genes in other individuals, the sacrifice might be favoured by natural selection. This would be more likely to be the case if the altruist's sacrifice saved the life of individuals who, because they were closely related to him, were likely also to carry the same genes. In general, if the act saved the life of more than two offspring, more than two siblings or more than eight cousins (etc.), the altruistic gene could still increase in frequency. This is the basis of 'kin selection theory': apparently altruistic acts may be selected for because they increase the prevalence in the next generation of the genes that give rise to them.

This sociobiological argument may seem remote from everyday life, but three points relevant to later chapters arise. First, we take parental care for granted: it seems natural that parents should sacrifice themselves for their children. But to the sociobiologist this is just a special case of kin selection: it is advantageous to parents (in biological terms) to look after their children because they are thereby enhancing the chances of survival of their own genes. In Chapter 6, Goody focuses especially on prosocial behaviour within the family as determining and determined by social conventions concerning cooperation and helping others. And in general prosocial behaviour is often shown preferentially to kin, and it is more pronounced in cultures where extended families live together and where the production system favours mutual aid.

Second, this approach helps to integrate a number of findings about human behaviour. For example, in harmony with this view, Essock-Vitale and McGuire (1985), studying middle-class Los Angeles women, found that the greater the help a woman received the more likely it was to come from kin, that helping was an increasing function of the recipient's reproductive potential, and that helping of kin was less likely to be reciprocal than helping amongst friends.

Third, a precisely similar argument goes on in the social sciences: it is argued that people sacrifice themselves for others because, in one way or another, it makes them 'feel good'. Thus once again altruism is explained away as ultimately selfish.

Kin selection is not the only way in which sociobiologists explain apparently self-sacrificial behaviour in animals. Individuals may do well to help others if such help is later reciprocated. In humans this applies especially to help amongst friends as compared with relatives: amongst the latter, there is less expectation of reciprocation.

If individuals cooperated with strangers on first encounter, and subsequently behaved as the other had done, either cooperatively or selfishly, cooperation could become a generally accepted way of behaving. However, this could happen only in small groups, and humans cooperate in large groups of unrelated individuals. This is the problem addressed by Boyd and Richerson in Chapter 2: since individuals who behave prosocially to others in the expectation of reciprocation could be exploited by individuals who cheat, how can it come about that cooperation becomes part of a cultural tradition? Boyd and Richerson propose that groups in which individuals cooperate can compete successfully with groups whose members do not; that cooperation in such groups is maintained because individuals adopt the mode of behaviour which they encounter most frequently in others; and that such 'conformist tradition' will be favoured in heterogeneous environments.

Appreciation of a subtle terminological issue is necessary for resolving an apparent contradiction between these two contributions. Harcourt uses cooperation to refer to occasions on which two (or more) individuals mutually improve each other's chances of reaching a goal. Cooperation in this sense, he argues, is likely to spread, because individuals cannot compete with coalitions, and must therefore form coalitions themselves. This will be so, of course, only if the coalitions are inviolable or, in more general terms, if cooperators are more likely than non-cooperators to receive the benefits of the cooperative acts of others. For Boyd and Richerson the emphasis is slightly different: cooperators are individuals whose behaviour reduces their own welfare but benefits their group or increases the probability that their subpopulation will escape extinction. They emphasize that their model applies to differences in behaviour (or beliefs leading to behaviour) between groups, and that within group differences may conform to a selfish model of human behaviour.

Further reading

Essock-Vitale, S. M. & McGuire, M. T. (1985). Womens' lives viewed from an evolutionary perspective. 2. Patterns of helping. *Ethology & Sociobiology*, **6**, 155–73.

Hamilton, W. D. (1964). The genetical theory of social behaviour. *Journal of Theoretical Biology*, 7, 1–52.

Mussen P. & Eisenberg-Berg, N. (1977). *Roots of Caring, Sharing and Helping*. San Francisco: Freeman.

Trivers, R. (1985). *Social Evolution*. Menlo Park, California: Benjamin-Cummings.

1

Help, cooperation and trust in animals

A. H. HARCOURT

A crocodile opens its mouth to allow a plover to peck food fragments from between its teeth. A female rat crouches and remains still, so allowing her mate to mount her. A male pied flycatcher brings food to its young, instead of leaving its mate to do all the feeding, as its neighbour is doing. A year-old white-fronted bee-eater brings food to its parents' most recent brood, instead of breeding on its own. A dwarf mongoose stands exposed to sun and predators on a knoll, acting as a sentinel, while the other members of its social group forage in the shade. Four lionesses together stalk and kill a buffalo and then together defend the carcass from a rival pride. Two unrelated, middle-aged male baboons jointly threaten an otherwise dominant young male in his prime and drive him away from a female 'in heat'. A subordinate female macaque monkey joins in a fight between two other females on behalf of the dominant contestant, the one who was going to win anyway. A female vervet monkey grooms the unrelated but dominant female of the group having pushed away a more subordinate female who had been grooming the alpha animal.

In short, a wide variety of animals help one another, cooperate with one another, in a variety of ways. Common to all the examples is the concept that in some way one animal has improved the chances of another (help) or two animals have mutually improved each other's chances (cooperation) of reaching some sort of goal, whether it be winning a fight, raising offspring, or having clean fur. Much of the time I will not in fact distinguish between help and cooperation, for helping is so often merely one side of a cooperative interaction. The word 'goal' implies conscious intentions, but none should be inferred. They are

15

largely unknown, and indeed I hope that it will become clear that, if they exist, they might be different from what is apparent at first sight.

I first ask why animals cooperate, because only some species, and some individuals within species do so. Those individuals that do cooperate do not do so all the time: under what conditions do animals cooperate? When individuals do cooperate, they do not chose their partners randomly: what determines who cooperates with whom? Finally, I consider some of the consequences of cooperation for the nature of animal societies. Throughout, I confine the discussion to vertebrates, for they can more usefully be thought to have the option of deciding whether or not to cooperate, and to cooperation among members of the same species.

Why cooperate

The benefits of cooperation

Why do white-headed wood hoopoes feed nestlings that are not their own? One answer is that they are stimulated to do so by the begging cries of the young birds, a proximate explanation. Feeding others' offspring could increase their own chance of inheriting the parents' nesting hole and so being able to breed themselves, a functional explanation expressed in terms of beneficial consequences. Lastly, they might do so because they are wood hoopoes and all the wood hoopoe family feed others' youngsters, an evolutionary explanation. All the answers are equally valid, but it is the second that I will concentrate on here.

To biologists, an act is beneficial to an individual if it maintains or increases the representation in the population of copies of genes carried by that individual. The most obvious way to maintain or increase the number of copies of one's genes in the population is to breed, to produce offspring, because offspring necessarily carry copies of their parents' genes. Measuring the representation of genes in a population is not only rather an abstract concept, but also practically extremely difficult. Most biologists are therefore ready to consider an act to be beneficial if, with a reasonable degree of probability, it raises an individual's chances of surviving, mating, and rearing offspring to adulthood.

The interest in helpful and cooperative behaviours is that, on the face of it, they often involve increasing other individuals' chances of producing offspring, sometimes even apparently at the expense of one's own opportunities. First, then, how might helping others, cooperating with them, increase an individual's own chances of surviving, mating, and rearing offspring to adulthood.

Cooperation as selfish help: mutualism and reciprocation

The argument is that individuals cooperate because they increase their own chances of surviving, mating, and rearing offspring to adulthood above what they would be without cooperation. Three examples highlight the different ways in which this can happen.

The first demonstrates the concept of what might be called true mutualism, two partners benefiting in the same way at the same time from cooperating. Male lions fight each other for the opportunity to associate with females. David Bygott and coworkers found that four lions who cooperate in defence of a pride of lionesses retain tenure of the pride for over twice as long as do two cooperating lions. Mating rights are shared, and thus each lion produces more offspring through its cooperation than if it acted alone.

The second case indicates that the benefits to each partner from cooperation need neither be simultaneous nor of the same nature, the same 'currency'. It also emphasizes the point that cooperation is selfish. Male pied kingfishers without nesting holes visit unrelated breeding pairs' holes and attempt to feed their young. When there is plenty of food, the fathers aggressively drive away the potential helper. Uli Reyer and Klass Westerterp have shown that it is when parents are close to starvation themselves and cannot feed the nestlings on their own that they accept help. Why is the help initially refused, given that it can increase the young birds' chances of survival? The answer is that males with unrelated helpers are more likely to lose their nest-hole and mate – to the helper – than males without. The male parent benefits at the time help is given, but can suffer subsequent costs; the helper gains no benefit at the time, other than association with the female, but subsequently sometimes gains a very large benefit in the form of a mate and a nest-site.

In the last example, as in the second, one partner has to wait to acquire the benefits of its help. In addition, it also has to rely on active reciprocation of the help by the initial recipient. Adult vampire bats who have found an animal during their night's foray and fed well will, on returning to the communal roost, feed unsuccessful colony members. The act is costly to the feeder at the time, for it is giving up food, a serious consideration for an animal that will die if starved for more than three days. The benefit comes later, several days later perhaps, when the previous recipient has this time found food, but the donor failed to do so. Then the help will be reciprocated, for as Gerald Wilkinson has shown, bats are particularly likely to feed those who have previously fed them. The significance of the delay in reception of benefit is that the first helper opens itself to exploitation by its partner, especially when benefits come only via active reciprocation. The significance of active reciprocation is that some form of scorekeeping is occurring, even if the rule for the bats is as simple as 'feed the begging animal that last fed me.'

Cooperation for the benefit of others: kin selection

Organisms can preserve or increase their genetic representation in the population in two ways. One has been described, namely maintenance of the individual's own abilities to pass on genes. In addition, though, relatives carry copies of genes, hence family resemblances. Therefore, another way to maintain genetic representation is to improve a relative's chances of surviving, mating and raising offspring. Animals do just this.

When a silver-backed jackal helps its parents feed and care for their subsequent litter it more than doubles the number of surviving pups, from about 1.25/litter to 3.25 on average. Since the helper has 50% of its rare genes in common with its new siblings (the same as with its own pups), it is apparently doing equally as well by helping as by breeding on its own, given that it too would raise only 1.25 pups in the absence of help. The direct effect that helpers can have on survival of young has been demonstrated experimentally by Jerram Brown (1987) and coworkers. They removed helpers from babbler territories and showed that only one nestling fledged on average compared with 2.5 when helpers were present.

In this and many other cases, the increased survival of the immature animals is in part a direct result of the increased amount of food that the young receive. For example, immature white-fronted bee-eaters at Stephen Emlen's (1984) study site were fed 25 insects an hour when two helpers were present, compared with only 15/hr when just the parents fed the nestlings. It is not necessarily only the fed young that benefit from the extra help. Grey-crowned babbler nestlings are almost twice as likely to survive when helpers are present as when they are not, but in addition, the presence of helpers allows each parent to make about half the feeding trips it would otherwise undertake, and thus save on energy expenditure. Other studies have shown that such a saving can result in the parents being able to produce extra broods of young, and so increase the related helper's genetic representation in the next generation.

Individuals do not need to feed a related immature to ensure that it is well-fed. They can do this by supporting it in contests over food: I and Kelly Stewart found that the average immature gorilla was eight times as likely to gain access to a resource when it was helped in a contest as when it was not. If the support is frequent and effective enough, then in a stable social group, in which animals know one another, group members come to learn that a contest with the immature is equivalent to a contest with its relatives, particularly its mother. If they cannot defeat the mother, they defer to the immature. The immature animal then effectively has the competitive ability, the dominance rank, of its main supporters, even in contests with adults. This consequence of support in contests on dominance rank has been demonstrated experimentally by Bernard Chapais. He altered the composition of social groups of macaque monkeys in captivity and showed that monkeys with supporters present in the group (but not necessarily helping) could defeat others; the same individuals without their supporters were defeated by the others. Animals with supporters, and animals of high rank are, under certain conditions, more likely to produce surviving offspring than those without; hence the benefit to helpers from raising the competitive ability of relatives by supporting them in fights.

When to cooperate

The trade-off between costs, benefits and consanguinity

Although helping and cooperation can bring benefits, the behaviours, like almost any others, also probably carry risks. Animals do not, therefore, cooperate all the time. Costs have to be traded against benefits, and both have to be traded against consanguinity. Kin differ in the proportion of their genes that they hold in common, their degree of relatedness. Close relatives (e.g. siblings) are separated by fewer matings (chances for dilutions with other individuals' genes) and thus probably have more genes in common than do distant relatives (e.g. cousins). Individuals will help others if the increase in their relatives' ability to reproduce (B_k, benefit to kin) is more than the decrease caused by the helpful act in their own ability (C_s, cost to self) by a factor sufficient to offset the degree of relatedness (r) between them, i.e. $B_k r > C_s$. In other words, the more distant the relative, the greater does the benefit to it have to be to offset the costs that will probably be involved in any helpful act (the less the interest accrued from the investment, the greater must the probability be of the stock increasing in value). Alternatively, for a given cost and benefit, animals are more likely to help close relatives than distant ones (for a given risk, invest in stocks with the higher interest). Such trade-offs between risks and consanguinity occur among animals. My and Kelly Stewart's data show that a gorilla helps both close kin and distant kin against animals whom it can easily defeat in a contest (low risk investment), but supports only close kin in contests with animals whom it cannot easily defeat (high risk investment).

When partners are unrelated, and particularly when the benefit is received only after a delay, with the attendant large risk of not receiving the benefit, the contrast in costs and benefits needs to be especially great.

Availability of resources and intensity of competition

The interest for biologists in the phenomenon of cooperation is epitomized by Glen Woolfenden and John Fitzpatrick's finding that when a Florida scrub jay yearling breeds on its own, it adds four times as many 'gene equivalents' to the next generation than if it stays and helps its parents raise their subsequent brood of offspring. It appears that cooperation is disadvantageous, and yet it occurs. Survival, mating, and rearing offspring require access to resources, usually in competition with other members of the population. Sandra Vehrencamp and Stephen Emlen have argued that in many cases, animals cooperate because they cannot get sufficient access alone. One factor determining access is the quantity of resource present. Uli Reyer's pied kingfishers provide an example of shortage forcing animals to cooperate by accepting help. Vehrencamp and Emlen were more concerned with shortage forcing animals who could not breed on their own to make the best of a bad job by remaining with their parents and helping them raise the next brood, instead of not making any contribution to augmenting their genes in the next generation. Data from Stephen Emlen's studies of cooperatively breeding white-fronted bee-eaters provide substantiation. When rainfall is low in the month preceding breeding, and hence the number of insects is low, up to 40% of the young of the previous year will stay with their parents and help raise the next brood of young; when rainfall and the insect supply is good, as few as 20% of the previous year's young will stay.

Access to resources is determined by the number of other animals competing for them as well as by the amount of resource. If young animals face intense competition for space or food, so probably do their parents, in which case the parents would benefit by retaining the offspring as helpers instead of driving them out. Michael Taborsky experimentally increased competition for space in an aquarium and showed that young cichlid fish who had previously been driven out of the parental territory were allowed back in again. In this, as in some of the other cases of shortage or competition correlating with cooperation, it is not clear whether the parents' behaviour towards the potential helpers switches, or whether the parents simply lack the energy to drive them away.

The combined effects of food supply and intensity of competition on cooperation are very nicely demonstrated by Nick Davies' and Alistair Houston's (1984) study of territoriality in a pied wagtail population. Pied wagtails aggressively defend territories in which they feed, as long as the food is of high quality (worth defending) and clumped in distribution (defensible). Birds without territories of their own continually try to invade, and as the amount of food in a territory increases, so the number of intruders trying to get access to the food rises. Eventually there comes a point when the owner can no longer defend efficiently on his own, but at the same time his territory contains enough food for more than one bird. He then allows a second bird in, with whom food and also defence of the territory is shared.

Relative competitive ability of group members

The second bird is usually clearly subordinate to the territorial owner. If one animal can control the amount of help that the other gives, then the larger the difference in competitive ability between group members, the greater the likelihood that the subordinate will help the dominant, Sandra Vehrencamp argues. Coercion is exerted through threat of expulsion in the case of the wagtails and in the situations of help at the nest that Vehrencamp and Stephen Emlen discussed and modelled. Conversely, where coercion is not possible, and cooperation is used to gain access to resources, an egalitarian society might be associated with more frequent cooperation in contests than a despotic society, since the costs and benefits of the cooperation would be shared more equally between the partners (see later). In addition, if the dominance hierarchy is steep, two cooperating subordinate group members cannot overcome a dominant one. While baboon males are almost twice the size of baboon females, vervet and macaque sexes are closer in size, and coalitions of females against males are more common in vervet and macaque groups than in baboon groups.

Stability of association and intelligence

As far as is known the cooperating wagtails of Davies and Houston's study were strangers, because the flocks from which they came were unstable. However, familiarity and the stability of association on which it depends should promote help and cooperation. Computer simulations suggest that cooperation with a delayed benefit will evolve only if animals are together long enough to retaliate against partners who renege on reciprocating help, and to reward helpers with subsequent help, the tit-for-tat principle of Robert Axelrod and William D. Hamilton (1984). In effect, animals are more likely to cooperate with those whom they have learnt to trust. As Robert Trivers (1985) originally pointed out in his seminal discussion of reciprocal helping, these concepts of trust, retaliation and reward imply the ability to recognize and remember individual group members and their acts of helping and cheating, i.e. to scorekeep. The more stable the group, the greater the chance to learn the identity and nature of group members. As yet, however, the influence of stability of association and its corollaries on propensity to cooperate has not been tested quantitatively in animals.

The more information that an animal can process, the more refined can its scorekeeping be. Information processing ability, or intelligence, might thus be a constraint on the evolution of reciprocation of help as a competitive strategy. A difficulty with testing this idea of Robert Trivers' lies in defining and quantifying intelligence. Nevertheless, primates are generally considered to be more intelligent than non-primates, and certainly have larger brains for their body size. A detailed review of the literature has led me to suggest that the cooperative relationships of primates might indeed take more account of individuals' identities than do those of non-primates (see below). Furthermore, Frans de Waal's (1989) work indicates that chimpanzees, who might be the most intelligent primate, might be alone in distinguishing not only those that supported them in fights and subsequently reciprocating, as a number of primate species do, but also in remembering those that joined others against them in fights, and subsequently punishing them.

Who to cooperate with

Kinship

An obvious preference throughout the animal kingdom is for kin, not only for genetic reasons, but presumably also for reasons of familiarity and trust. As an example, the black-tailed prairie dogs studied by John Hoogland were twice as likely to give a call warning colony members that they had seen a predator if kin were present in the colony than if they were not, whether the relatives were descendant kin (offspring) or not (siblings or parents). Consanguinity is not the only characteristic of importance in an ally, however.

Quality of partner

Not all individuals are equal, and biologists have long suggested that courtship, for example, is in part a process of assessment of the potential mate's quality. The choice appears to be passive, with animals accepting or rejecting suitors. However, individuals who actively solicit, establish and work to maintain partnerships with more able group members, for instance dominant ones, should be at an advantage over those who make no distinction. One way to maintain a partnership is to help the partner. Among primates, help is indeed sometimes directed more often at dominant than subordinate group members, other things being equal. Joan Silk found that adult female macaques groomed dominant group members three times as frequently on average as they groomed subordinates; 80% of incidents of grooming of adult females by immature females in a wild baboon group observed by Dorothy Cheney were directed at the top ranking 10% of females; and in all 28 cases of support in contests among 15 unrelated female macaques observed by Bernard Chapais, the more dominant contestant was supported. Among primates such instances are common, but the literature on non-primates does not report such active choice of partners of high quality.

Equitability of competitive ability among partners

While it might be advantageous for low-ranking animals to associate with high-ranking ones, it is likely that high-ranking animals would benefit more from associating with another high-ranking partner: in partnered wrestling matches, a heavyweight needs another heavyweight on his side, not a featherweight. Furthermore, the more similar the ability of the partners, the less likely is it that one will monopolize any resource gained as a result of the cooperation. Male baboons cooperate with one another in acquiring females 'in heat' from other males, and the partners in a very high proportion of such coalitions are males who are adjacent in rank, 58% of 13 successful coalitions in a group of 12 males in a study by Barbara Smuts (1985), and 83% of 24 coalitions in a group of 12 males in another study by Fred Bercovitch. Although Ronald Noe found that the dominant partner obtained the female in all the coalitions between males that he saw, Smuts, Bercovitch and others have seen a near 50 : 50 split.

Constraints on choice

Very able allies might provide more useful help, but allies of similar ability (and especially kin) might provide more reliable help. Thus John Colvin found that of eight immature male rhesus macaques that he studied, seven associated most with their subordinate kin, rather than with dominant non-kin. A choice of a closely ranked partner might be enforced for another reason. If dominant group members are valuable as allies, they become a resource over which subordinate animals compete. The subordinates can offer the dominant animal more services, for example by grooming it more frequently than their rivals, or they can prevent their rivals from interacting with it. Primates do prevent group members from being friendly with one another, and Robert Seyfarth found that among wild vervets the individuals that females most frequently stopped others from grooming were the most dominant females. Low ranking animals therefore find it difficult to get access to high-ranking ones, and are forced to associate with other low ranking group members.

Consequences for the structure of society

Once two animals cooperate and as a result improve their chances of survival, mating or raising offspring, all other animals in the group, perhaps even the population, are going to have to start to cooperate also. A cooperative act can therefore have ramifications and implications for the nature of society that extend way beyond the act itself.

In the face of coalitions, individuals cannot compete well on their own, and hence the society is organized into groups of cooperating individuals. Tim Caro's data show that 15 of 17 territories defended by male cheetahs were held by coalitions of males; and Richard Wrangham has argued that individuals of most primate species live in groups, not alone, because of the necessity to join forces and cooperate in competition over food. If individuals need to form alliances, then it is especially advantageous to do so with kin, and thus many primate species are organized as groups of cooperating relatives. The help in contests that results in offspring acquiring the ability to defeat all those whom their main supporters, principally their mother, can defeat means that not only individuals, but whole families, have dominance ranks with respect to one another in taxa as diverse as swans, hyaenas and Old World monkeys. Once group members compete to form rival alliances, subordinate animals might have to ally with dominants simply to maintain their own position in the hierarchy, with the secondary result that the dominant's position is reinforced and the hierarchy is rigidified, probably to the detriment of the subordinates. Dorothy Cheney has suggested that the tendency to cooperate with dominant group members might result in high ranking families being more stable than low ranking ones. Members of dominant families are friendly to one another both because they are kin, and because they are powerful; low ranking animals have only the first reason to be friendly to group members, and indeed abandon relatives in favour of partnerships with dominant non-kin. The hypothesis has not been proved, and indeed one can think of reasons why competition to cooperate might result in subordinate families being more stable, for example they have to cooperate closely if they are to have any chance of competing with dominant animals. Whichever, the point is that cooperation between two group members sets in train a process that can influence the structure of the whole society.

In conclusion, hundreds of species of animal help and cooperate in a wide variety of ways. Primates especially make what seem to be complex calculated decisions about when to cooperate, who to cooperate with, and who to cooperate against. Since the effects of such help and cooperation on society extend far beyond the initial cooperative act, understanding of the nature of animal societies, or indeed of human society, requires deep appreciation of the helpful, cooperative and trustful behaviours that are such integral aspects of animal society.

Further reading

Alexander, R. D. (1984). *The Biology of Moral Systems.* Hawthorne: Aldine de Gruyter.
Axelrod, R. & Hamilton, W. D. (1984). *The Evolution of Cooperation.* New York: Basic Books.

Brown, J. L. (1987). *Helping and Communal Breeding in Birds*. Princeton: Princeton University Press.

Davies, N. B. & Houston, A. I. (1984). Territory economics. In J. R. Krebs & N. B. Davies (eds), *Behavioural Ecology*, 2nd edn. Oxford: Blackwell Scientific Publications.

Emlen, S. T. (1984) Cooperative breeding in birds and mammals. In J. R. Krebs & N. B. Davies (eds), *Behavioural Ecology* 2nd edn. Oxford: Blackwell Scientific Publications.

Harcourt, A. H. & de Waal, F. (in press). *Us against Them: coalitions and alliances in animals and humans*. Oxford: Oxford University Press.

Smuts, B. B. (1985). *Sex and Friendship in Baboons*. New York: Aldine.

Trivers, R. l. (1985). *Social Evolution*. Menlo Park, California: Benjamin–Cummings.

de Waal, F. B. M. (1989). *Peacekeeping in Primates*. Cambridge, Massachusetts: Harvard University Press.

2

Culture and cooperation[1]

ROBERT BOYD AND PETER J. RICHERSON

> I am dying not just to attempt to end the barbarity of the H-block, or
> gain the rightful recognition of a political prisoner, but primarily what is
> lost here is lost for the Republic ...
>
> Bobby Sands, from his strike diary

On May 5, 1981, Bobby Sands died after 66 days on hunger strike in Long
Kesh prison, Northern Ireland. He was the first man to die in a strike
aiming to force the British to treat IRA inmates as political prisoners
rather than common criminals, and to galvanize public opinion against
British policies in Ulster. Nine more men died before the strike ended
without achieving its stated goal on August 10. The strikers had all
volunteered, from a group of 60 IRA prisoners. All but three of the fasters
either died or were still on fast when the strike ended. Of these three, two
were fed on the instructions of relatives after they had lapsed into a coma,
and the third was removed by the strike leadership because a bleeding
ulcer would have caused him to die too rapidly.

The hunger strikes at Long Kesh provide evidence that people sometimes make choices that are personally costly in order to benefit the group
or society to which they belong. The strikers chose to participate, while
many other IRA prisoners did not. They had to reaffirm their choice three
times a day when food was brought to their hospital rooms, and later by
resisting the pleas of relatives that they accept medical treatment. The
choice was clearly costly – it resulted in weeks of terrible suffering,
followed often by death. These men were not serving life sentences; if they
had not volunteered, they would eventually have been released. It is also
clear that the strikers believed that their deaths would benefit the other
IRA inmates and the Republican cause in general. Borrowing terminology from game theory, we say that such behavior, intended to benefit a
group at a cost to the individual, represents cooperation.

[1] This chapter was originally published in *Beyond Self-interest*, edited by Jane
Mansbridge. Chicago: University of Chicago Press (1990).

The hunger strikes at Long Kesh are not an isolated instance of human cooperation. Self-sacrifice in the interest of the group is a common feature of human behavior. Amongst hunter-gatherers, game is typically shared among all members of the group regardless of who makes the kill. In stateless horticultural societies, men risk their lives in warfare with other groups. In contemporary societies, people contribute to charity, give blood, and vote – even though the effect of their own contributions on the welfare of the group is negligible. Indeed some authors argue that the existence of complex urban society is evidence for at least a degree of unselfish cooperation. Psychologists and sociologists have also shown that people cooperate unselfishly under carefully controlled laboratory conditions, albeit for small stakes.

Human cooperative behavior is unique in the organic world because it takes place in societies composed of large numbers of unrelated individuals. There are other animals that cooperate in large groups, including social insects like bees, ants, and termites, and the naked mole rat, a subterranean African rodent. Multicellular plants and many forms of multicellular invertebrates can also be thought of as eusocial societies made up of individual cells. In each of these cases, the cooperating individuals are closely related. The cells in a multicellular organism are typically members of a genetically identical clone, and the individuals in insect and naked mole rat colonies are siblings. In other animal species cooperation is either limited to very small groups or is absent altogether.

The observed pattern of cooperation among non-humans is consistent with contemporary evolutionary theory. Evolutionary biologists reason that cooperative behavior can evolve only when cooperators are more likely than non-cooperators to receive the benefits of the cooperative acts of other cooperators. Suppose that this condition is not satisfied, and non-cooperators receive the same benefits as cooperators from the cooperative acts of others. Then, because non-cooperators receive the same benefits but do not suffer the costs of cooperative behavior, they will, on the average, leave more offspring than cooperators – over the long run cooperation will disappear.

Kinship is one important source of non-random social interaction. Relatives have a heritable resemblance to one another. Thus, cooperators whose behavior disproportionately benefits genetically related individuals have a greater than random chance of benefiting an individual who is similarly predisposed.

Reciprocity is the other important source of non-random social interaction. Here cooperators discriminate based on the previous behavior of others – I will cooperate with you only as long as you cooperated in previous encounters. If individuals commonly employ some such rule, cooperative interactions among reciprocators will persist, while interactions of reciprocators and non-cooperators will quickly cease. Thus reciprocators will be more likely than non-cooperators to receive the benefits of the cooperative acts of others. Theoretical work suggests that reciprocity can readily lead to the evolution of cooperation, but only in quite small groups.

Thus the fact that humans cooperate in large groups of unrelated individuals is an evolutionary puzzle. Our Miocene primate ancestors presumably resembled non-human primates in cooperating only in small groups, mainly made up of relatives. Such social behavior was consistent with our understanding of how natural selection shapes behavior. Over the next 5 to 10 million years something happened that caused humans to cooperate in large groups. The puzzle is: What caused this radical divergence from the behavior of other social mammals?

Several authors have suggested that selection among cultures is the solution to the puzzle. People's behavior is shaped by the beliefs, attitudes, and values they acquire growing up in a particular culture. Cultures differ in the extent to which they motivate people to behave cooperatively. If cooperators discriminate in favor of group members, costly acts of cooperation are likely to be directed toward helping fellow group members who are inclined to cooperate with members of the group. Because economic and military success requires a high degree of interdependence, cooperative cultures will tend to persist longer than less cooperative ones as a result of a selective process acting at the level of cultural groups. Human behavior thus represents a compromise between genetically inherited selfish impulses, and more cooperative culturally acquired values.

This cultural argument, however, does not account for either the mechanisms that maintain cultural variation among groups or for the ways these mechanisms may have arisen in the course of human evolution.

We need an account of the mechanisms that maintain cultural variation because group selection works only when cultural differences persist among groups. Yet many processes, like intermarriage, cultural borrowing, and people's self interested decisions erode variation among groups. When non-cooperative beliefs arise in largely cooperative groups they will increase in number because they are beneficial to individuals. Intermarriage, selfish decisions, and cultural diffusion will cause selfish beliefs to spread to infect other groups of cooperators. Models of group selection in population biology suggest that the two processes of growing non-cooperation and eroding variation will work faster than the processes promoting cooperation, which depend on group extinction. In short, selection among groups, which produces cooperation, will have little influence without social or psychological mechanisms that maintain cultural differences among groups.

We also need an account of the ways difference-maintaining mechanisms may have arisen, because cooperation is maladaptive from a genetic point of view. We need to explain how whatever social or psychological mechanisms led to cultural group selection could have evolved in the face of countervailing evolutionary tendencies at the level of genes.

Several recent models of cultural change assume that the transmission of culture in humans constitutes a system of inheritance. Humans acquire attitudes, beliefs, and other kinds of information from others by social learning, and these items of cultural information affect individual behavior. Cultural transmission leads to patterns of heritable variation within and among human societies. While individual decisions are important in determining behavior, these decisions depend on individuals' beliefs, often learned from others, about what is important and valuable, and how the world works. Human decision-makers are enmeshed in a web of tradition; individuals acquire ideas from their culture, and in turn make modifications of what they learn, which modifications become part of the cumulative change of the tradition.

To understand cultural evolution, we must account for the processes that increase the frequency of some variants and reduce that of others. Why do some individuals change traditions or invent new behaviors? Why are some variants transmitted and others not? As in the analogous case of organic evolution, individual level processes have to be scaled up to the whole population and marched forward in time to understand how a behavior like cooperation might become part of a cultural tradition despite the costs that non-cooperators can impose.

In the analysis that follows, we illustrate how this sort of model can be used to evaluate the cogency of the cultural group selection arguments. The model suggests that certain specific factors must be present in the cultural evolutionary process for group selection to be any more important than it seems to be in the genetic case.

Cultural transmission and group selection

We begin by imagining that individuals live in subpopulations embedded in a larger population. The subpopulations represent the 'societies' whose extinction drives cultural group selection. Suppose, for simplicity, that each individual in the population is characterized by one of two cultural variants. One variant causes individuals to place a high value on group goals compared with personal gain. Under the right circumstances, these 'cooperators' will act in the group interest even if it is personally costly. The other cultural variant causes individuals to place a low value on group goals compared with personal gain. In the same circumstances, these 'defectors' will not cooperate. This assumption does not entail the view that culture can be broken down into independent, atomistic traits. Each cultural variant could easily represent a large, tightly coupled complex of beliefs. Moreover, change in any one trait might easily depend on other beliefs that people hold. Evolutionary analysis requires constructing a framework for evaluating each process that affects whether the variants will increase or decrease relative to each other. In a 'life-cycle' of cultural transmission, we need to specify when and from whom an individual first acquires a belief, what later events might cause an individual to change beliefs, and finally, how holding different beliefs affects the chance that an individual will serve as a model for others. Let us assume the following life cycle:

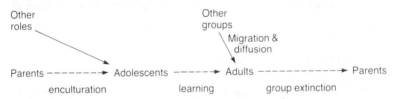

In this model, we want to focus on four processes: enculturation, individual learning, flow of ideas among societies, and the extinction of societies. Real human life-cycles are much more complex than this. However, more realism would complicate matters greatly, and distract from our attempt to see the essence of how the group selection process might work in human societies. The model is like a caricature, but, like a good caricature, it is intended to highlight some real attributes of our subject.

Next we describe what happens to individuals in a subpopulation during each step of the life-cycle. We begin with the enculturation of children. In this initial model, we assume that enculturation involves faithful copying. That is, children initially acquire a disposition to be either cooperators or defectors by unbiased imitation of the adults among whom they grow up. Many different models can have this property. For simplicity, suppose that children imitate only one of their biological parents. Then, if both cooperators and defectors have the same number of children, cultural transmission will not change the proportion of cooperators in the population on the average. However, if subpopulations are small there may be random changes due to sampling variation. (The Appendix contains a brief sketch of how the ideas in this section are formalized.)

Next, children mature to become adults. During this period, they learn about the consequences of holding different beliefs about the importance of group goals. Part of their information will come from their own experience. In accordance with the assumption that cooperation is costly to the individual but beneficial to the group, we postulate that they will learn that their own cooperative acts do not yield personal benefits. Part of their information will come from observing the consequences of the behavior of others. They will see that defectors do not suffer as a result of their non-cooperation. As a result, some people will change their beliefs. Some who began adolescence cooperating by copying their parents will learn that they can get by without cooperating, and will switch to defection. If the observations people make in the course of individual learning are prone to error, some people may decide (incorrectly) that cooperation is individually beneficial. Thus, if cooperation is costly to the individual, as we assume, cooperators will be more likely to switch to defection than vice versa. This imbalance will usually cause the proportion of cooperators in the population to decrease.

To say more than this one must specify the details of the individual learning process. Let us suppose that in the course of adult individual learning each person encounters a single other 'role model,' and, if the role model's beliefs are different from the beliefs that an individual has copied as a child, people then evaluate the costs and benefits of the alternative beliefs and adopt the one that seems better. The magnitude of the change will depend on how easily individuals can discern the consequences of the alternatives available to them. If it is easy to see that adopting cooperative beliefs is costly, then many cooperators will switch to defection and few defectors will switch to cooperation. If it is hard for individuals to discern the costs and benefits of alternative beliefs, there will be only a small increase in the fraction of defectors.

No society is completely isolated from other societies. People leave and are replaced by immigrants who bring with them different culturally acquired beliefs and values. People also borrow ideas from people in other societies. Both kinds of inputs lead to changes in the cultural composition of subpopulations. The exact nature of change depends on the details of the mixing process. For simplicity, let us assume that cooperators and defectors are equally likely to emigrate, and emigrants are drawn randomly from the population as a whole. Then if a fraction of each subpopulation emigrates or shares their beliefs, mixing will cause the proportion of cooperators in each subpopulation to approach the average proportion in the population as a whole. In other words, migration will cause subpopulations to become more alike.

The final step in the life-cycle is the extinction of subpopulations. The group selection model assumes that cooperative societies are more likely to persist than non-cooperative societies. Let us assume, then, that the probability that a subpopulation becomes extinct is negatively related to the proportion of cooperators – the more cooperators, the lower the probability of extinction. Empty habitats are recolonized by immigrants drawn from other subpopulations.

By combining all of these steps together, we obtain a model of events during one generation that changes the cultural composition of the population. This set of equations (one for each subpopulation) captures the nature of the evolutionary process on a microscopic time-scale. To predict the longer-run course of evolution, the equation is iterated recursively, stepped forward from one time period to the next, to determine how the frequency of different variants changes through time, and perhaps reaches some equilibrium, a value at which the proportions of the two variants become constant. Thus we can ask: Will the cooperators or the defectors increase and come to dominate the population? How will changing the size of subpopulations, or the ease with which individuals can discern the costs and benefits of alternative cultural variants, affect the outcome?

The model we have just sketched is closely analogous to 'interdemic group selection' models studied by population geneticists, differing only in that learning rather than natural selection leads to the spread of selfish behavior within subpopulations. Extensive analysis of such models by population geneticists indicates that populations will eventually become composed almost completely of defectors unless subpopulations are extremely small, there is little mixing among groups, and it is very hard for individuals to discern the costs and benefits of alternative beliefs. The existence of culture is, by itself, no guarantee that cooperation will evolve.

To understand this result it is useful to distinguish among processes that act within and processes that act among societies. Processes within societies, according to our assumptions, produce the result that, if there were no extinction of subpopulations, cooperation would disappear. However, the differential extinction of societies creates a process of selection among groups that increases the frequency of cooperators.

Selection among groups in this model fails to increase cooperation because in order for group selection to work, groups must differ persistently in degree of cooperation. Yet the tendency of migration to make all groups the same by mixing is powerful, while processes that generate differences among groups – random fluctuations due to small population size and small numbers of individuals colonizing empty habitats – are weak unless groups are very small. Given the cultural processes that we have assumed to operate, this logic will cause selection among culturally different groups to be ineffective at generating the scale of cooperation observed in humans.

At this point, the reader may point out that in the real world differences sufficient to provide the raw material for group selection do seem to persist. We agree, but we also believe that the preliminary model just presented is useful nevertheless. It shows that one must specifically account for the mechanisms that maintain cultural differences among groups. Our model incorporates two basic features of the enculturation process – children faithfully copy the beliefs and values of adults in their society, and people through individual learning tend to modify their beliefs in the direction of their own self-interest. Surprisingly, even with small amounts of mixing, these processes do not maintain enough variation among groups for group selection to be important. This cultural model has the same behavior as genetic models of interdemic group selection, despite the assumed differences between cultural and genetic transmission incorporated into the model. Extensive analysis of genetic models indicates that many other elements can be added without changing the qualitative conclusion: group selection for cooperation is a weak effect.

Most biologists, and social scientists influenced by evolutionary biologists, are properly impressed with the argument against group selection. If culture plays some role in human cooperation, it must be due to some specific attributes, not simply to the existence of a non-genetic means of acquiring behavior by tradition.

Conformist cultural transmission

Let us now consider an additional cultural process that could act to maintain variation among groups. In the model of group selection described above, we assumed that social learning was unbiased, meaning that it involved faithful copying plus random error. Social learning, however, could be biased in many ways, as individuals exercise conscious and unconscious choices about which variants to adopt. One form of biased transmission, which we call 'conformist cultural transmission' is interesting and potentially important because it can maintain enough variation for group selection to be effective.

Conformist transmission occurs whenever children are *disproportionately* likely to acquire the variant that is more common among the adult role models. A naive individual uses the frequency of a variant among his role models to evaluate the merit of the variant. For example, in the most extreme form of conformist transmission, children would always adopt the variant exhibited by a majority of their adult role models. Conformist transmission causes frequency of the more common variant in the population to increase, all other things being equal. For example, if more than half of the parents are cooperators, then there will be more cooperators among children than among their parents. If less than half of the parents are cooperators, then cooperation will decrease in frequency.

Group selection again

Adding conformist transmission to the model of group selection completely changes its results. When conformist transmission is strong, it can create and maintain differences among groups, and, as a result, group selection can cause cooperation to predominate even if groups are very large and extinction infrequent.

To see why, consider a subpopulation in which cooperation is common. When conformist transmission is weak compared with the effect of learning and mixing, the result is the same as in the previous model – these within-group forces act rapidly to reduce the frequency of cooperators in each subpopulation below the average frequency in the population as a whole. As in the previous model, the only stable long-run state is total defection. (Once again, see the Appendix for more detail.) When the effect of conformist transmission is strong enough, the outcome is quite different – the effect of conformist transmission will exactly balance combined effects of learning and mixing so that subpopulations with a large fraction of cooperators persist indefinitely.

If conformist transmission is sufficiently strong, and if the mode of colonization is by single or relatively unmixed subpopulations, the process of group selection will increase the frequency of cooperators even if extinction rates are low, mixing is substantial, and subpopulations are very large. Once cooperation becomes sufficiently common in a subpopulation, conformist transmission acts to counter the processes of learning and mixing, leading to a stable equilibrium at which cooperation is common. The existence of such equilibria means that the long-run outcome in the population depends only on the among-group processes: extinction and colonization. Subpopulations in which cooperators are more common are less likely to go extinct than those in which cooperators are less common. Thus, differential extinction acts to increase the frequency of cooperation in the population as a whole. If the colonization of empty habitats is by a single subpopulation, then variation among groups is heritable at the group level. Groups in which cooperation is common can produce similar 'offspring' groups. When these conditions hold, the results are strikingly different from those in the previous model, because conformist transmission acts to preserve variation between groups.

The evolution of conformist transmission

Conformist transmission, therefore, might allow the evolution of beliefs and values leading to self-sacrificial cooperation. To an evolutionist there is an obvious objection. The human cognitive capacities that allow the acquisition of culture must have been shaped by natural selection. If conformist transmission leads to excessive cooperation from the genetic point of view, it could evolve only if it has some compensating benefit. The following model of the evolution of conformist transmission suggests such a benefit.

We begin with exactly the same life-cycle and model of conformist transmission used above. However, we now assume that the environment is heterogeneous, meaning that one cultural variant is better in some habitats, while the other cultural variant is better in others. To model the evolution of biased transmission we assume that individuals vary in the extent to which they are prone to adopt the more common type, and that this variation has a heritable genetic basis. We then combine the effects of biased social learning, individual learning, and natural selection to estimate the net effect of these processes on the joint distribution of cultural and genetic variants in the population. To project the long-run consequences, we repeat this process over and over. We then ask, what amount of conformist transmission will be favored by natural selection?

In heterogeneous environments the answer is: Always as strong a bias as possible. In the simplest model in which there are two cultural variants, the favored cultural transmission rule is: Adopt with certainty the more common variant among your role models. Natural selection will favor such a transmission rule because in a heterogeneous environment both individual learning and selection will tend to make the most fit in a subpopulation the most common variant in that subpopulation. Commonness among role models provides naive individuals with an estimate of the commonness of the variant in the local environment. The rule 'imitate the common type' (or 'When in Rome, do as the Romans do') provides a good way to increase the chance of acquiring the best behavior without costly individual learning.

As an example of how conformist inclinations might have evolved, consider a population of early humans in the process of expanding their range. Environments at the margin of their range differ from those at the center of their range. People at the center live in a tropical savanna, those at the margins in temperate woodland. Different habitats favor different behaviors. This is obviously true of subsistence behaviors – the foods that have the highest pay-off, the habits of prey, how to construct shelter, and so on. However, different habitats will also favor different beliefs and values affecting social organization: What is the best group size? When is it best for a woman to be a second wife? What foods should be shared? It will be difficult and costly for individuals to learn what is best regarding many of these questions, particularly those affecting social behavior. Pioneering groups on the margin of the range will thus converge only slowly on the most adaptive behavior. Their convergence will be opposed by the flow of beliefs and values from the savanna, with the result that a minority will often hold beliefs more appropriate to life in the savanna than life in the woodland. Any system of enculturation that involves unbiased reproduction will simply reproduce these locally inappropriate beliefs and values of the individual's value structure. If individuals do not simply copy the beliefs and values of their elders faithfully, but instead adopt the most common cultural variants they see, they will adapt to the new environment more quickly. Since individual learning will eventually make the best variants the most common, those who in their social learning copy the most common variant will have no chance of acquiring beliefs that are more appropriate elsewhere. If this 'conformist' tendency is genetically heritable, it will be favored by natural selection.

To sum up, group selection can lead to the evolution of cooperation if humans inherit at least some portion of the values, goals, and beliefs that determine their choices by way of conformist cultural transmission. To explain the evolution of human cooperation, we therefore need to account for the evolution of a human capacity for conformist culture transmission. We have seen that conformist transmission may be favored in heterogeneous environments because it provides a simple, general rule that increases the probability of acquiring behaviors favored in the local habitat. Averaged over many traits and many societies, this effect could plausibly compensate for what is, from the genes' 'point of view,' the excessive cooperation that may also result from conformist transmission. There are other peculiarities of the cultural inheritance system that may lead to a similar point. We believe that such explanations of human altruism are attractive because they give us a plausible account for why we may be an exception to the powerful argument against group selection.

Cultural endogamy

People belong to many different groups based on age, size, skin color, language, and so on. Often the interests of different groups conflict. Our model can be interpreted to yield predictions about which groups should cooperate. When individuals (1) acquire behaviors culturally from other members of the group via frequency-dependent cultural transmission, and (2) the flow of cultural traits from outside the group is restricted (the group is 'culturally endogamous') group selection acting on cultural variation will favor cooperators – individuals whose behavior reduces their own welfare but increases the probability that their subpopulation will escape extinction.

Different social groupings may be culturally endogamous for some traits but not for others. For traits acquired by young children from members of their family, the culturally endogamous group might resemble the genetic deme. That is, an individual's cultural parents would be drawn from the same social grouping as its genetic parents. The culturally endogamous group for a trait acquired disproportionately from parents of one sex may differ from the culturally endogamous group for a trait acquired from parents of both sexes. For example, suppose that beliefs about what constitutes acceptable behavior during warfare are acquired exclusively from males. In patrilocal societies, where wives move to the kinship group of the males, the culturally endogamous group for these beliefs could be very small (sometimes so small as to require substantial amendments to the model). In the same societies, the culturally endogamous group for behaviors acquired from both sexes – for example, language or religious belief – could be very large. In contrast, in matrilocal societies, where husbands move to the kinship group of the wives, the culturally endogamous group might be the same for warfare, language, and religion. For traits acquired as an adult, the culturally endogamous group may be different again. For example, in modern corporations people acquire through cultural transmission from individuals who precede them many aspects of individual behavior, including professional goals, work norms, and beliefs about the nature of the product and the marketplace. For these behaviors, the culturally endogamous group is the firm. For other traits, the culturally endogamous group may be a fraternal organization, craft guild, or academic discipline.

Cultural extinction does not require morality

When conformist cultural transmission prevails, the extinction of a group need not entail the physical death of individuals; the breakup of the group as a coherent social unit and the dispersal of its members to other groups will suffice. Imagine that the members of a subpopulation are dispersed randomly to all the other subpopulations. Because the members of any subpopulation are either mostly cooperators or mostly defectors, their dispersal will change the frequency of cooperators in the other subpopulations during the generation in which the dispersal takes place. The small change will perturb each of the subpopulations from its equilibrium value, but if the number of migrants is small compared with the host population, and if they do not enter at a moment in which a crisis in individual learning is moving the host population toward a new equilibrium point, the perturbation will be overcome by the forces tending toward the original equilibrium. Each subpopulation will eventually reach the same equilibrium that it would have reached if the members of the dispersal group had never entered it. The dispersal of a group is thus equivalent to extinction because conformist transmission favors the more common variant. Cooperators persist in cooperative groups because they are most common in those groups. If they are dispersed, their numbers will usually be insufficient to cause the frequency of cooperators in defector groups to exceed the threshold necessary to cause a change to a cooperative equilibrium, and vice versa.

As we have seen, the model suggests that group selection will be more effective when vacant habitats are recolonized by individuals drawn from a single subpopulation. This model of colonization seems plausible in the human case for several reasons: first, in a social species in which division of labor and cooperative activities are important for subsistence, individuals may often immigrate as a social unit to colonize empty habitats. Second, in warfare individuals from a single victorious group may disperse defeated groups and replace them. Finally, even if a vacant habitat is colonized by groups that originated in more than one subpopulation, behavioral isolating mechanisms may prevent them from mixing to form one larger group.

Cooperation is most likely for hard-to-learn traits

As we have seen, the model suggests that group selection will be important only if conformist transmission is strong compared to learning. This condition is most likely to be satisfied regarding cultural traits about which it is difficult and costly to make adaptive choices in real environments. The natural world is complex, hard to understand, and varies from place to place and time to time. People can make some intelligent guesses about complex questions like the reality of witchcraft, the causes of malaria, whether natural events are affected by human pleas to their governing spirits, or whether an afterlife exists in which virtue is rewarded. But compared with the variation we observe in others' behavior, the number of alternatives we can investigate in detail is quite limited.

When determining which beliefs are best is costly or difficult, individual learning will tend to be weak, and thus group selection may be important. Consider, for example, belief in an afterlife in which virtue is rewarded. If external signals are weak and individual learning difficult, some atheists will, through individual learning, interpret some natural phenomena as confirming the existence of the afterlife and become believers. Some believers, through the same process, will despair for lack of convincing proofs and apostasize. The net effect of such decisions on the fraction of the population that believes will be relatively small, and therefore neither belief will spread rapidly at the expense of the other. If, however, such beliefs have strong effects on group persistence – if, for example, a society of believers is more able to mobilize individuals to provide education, participate in warfare, and so on – then group selection may cause believing to increase.

So far we have concentrated on behavioral differences between people who hold different culturally acquired beliefs. If the cultural environment is taken as given, however, much behavioral variation may still be explained in terms of self-interest. People who hold the same beliefs in different circumstances will make different choices. Because such choices may often be self-interested, much observed behavioral variation may conform to a selfish model of human behavior. If, for example, first sons have more to lose than second sons by joining the Crusades, then second sons will be more likely to join even if eldest sons' beliefs in Christianity are equally fervent. Given Christianity, such choices are self-interested. However, it may be very difficult to account for militant Christianity in terms of self-interest.

Evolution of ethnic cooperation

One human grouping that seems generally to satisfy the requirements of the model is the ethnic group. The flow of cultural traits within the ethnic group is often much greater than the flow between ethnic groups. The model predicts that group selection acting on culturally transmitted traits will favor cooperative behavior within ethnic groups and non-cooperative behavior toward members of other groups. Table 2.1 lists the traits of ethnocentrism that, according to anthropologist Robert LeVine and psychologist Donald Campbell, characterize human ethnic groups. These traits seem consistent with the predictions of the model. Sanctions against theft and murder within the group provide civil order, a public good that benefits group members. Few sanctions protect outgroup members. Cooperative behavior typifies interactions between group members, while lack of cooperation typifies interactions between members of different groups. Individuals are willing to fight and die for their own group in warfare against other groups. In recent times, actions on behalf of the ethnic group, like movements for ethnic autonomy, have even been taken in direct opposition to the authority and power of the modern state. These ethnic groups are often very large, too large to suggest that reciprocal arrangements are responsible for the observed behavior.

Variation in behavior among ethnic groups also provides support for the hypothesis of group selection acting on culturally transmitted behavior. LeVine and Campbell point out that in 'socially divisive' ethnic groups, patrilocality or local group endogamy helps develop a parochial loyalty structure and generates warfare among segments of the ethnic community. In 'socially integrative' groups, the dispersion of males

Table 2.1. *Traits identified with the syndrome of ethnocentrism by LeVine and Campbell (1973)*

Attitudes and behaviors toward ingroup	Attitudes and behaviors toward outgroup
See selves as virtuous and superior	See outgroup as contemptible, immoral, and inferior
See own standards of value as universal, intrinsically true. See own customs as original, centrally human	
See selves as strong	See outgroup as weak
	Social distance
	Outgroup hate
Sanctions against ingroup theft	Sanctions for outgroup theft or absence of sanctions against
Sanctions against ingroup murder	Sanctions for outgroup murder or absence of sanctions against outgroup member
Cooperative relations with ingroup members	Absence of cooperation with outgroup members
Obedience to ingroup authorities	Absence of obedience to outgroup authorities
Willingness to remain an ingroup member	Absence of conversion to outgroup membership
Willingness to fight and die for ingroup	Absence of willingness to fight and die for outgroups
	Virtue in killing outgroup members in warfare
	Use of outgroups as bad examples in the training of children
	Blaming of outgroup for ingroup troubles
	Distrust and fear of the outgroup

fosters loyalties to wider groupings and prevents such warfare. They go on to argue that while socially divisive societies are characterized by extensive feuding and violence, they are rarely involved in large-scale warfare, and when they are involved in warfare, alliances are formed on the basis of immediate military contingencies. In contrast, while socially integrative societies have much less violence within their own society, they readily cooperate in large-scale conflict. Again, it appears that the unit upon which group selection works is the culturally endogamous group. If this unit is small, as in the case of socially divisive societies, then so is the unit within which social cooperation takes place. In socially integrated societies, the culturally endogamous unit is larger (at least with regard to traits transmitted by males) and the scale of conflict larger as well.

Conclusion

The simple model of cultural group selection outlined here is not, strictly speaking, verified by data concerning ethnic cooperation. Nor does it claim to account for all cooperative behavior in humans. The model illustrates what we believe is a crucial property of the evolution of cultural species: if the rules of cultural transmission are different from the rules of genetic transmission, similar selective regimes will result in different evolutionary behavior. The model also suggests one transmission rule, a rule of conformist transmission, that might explain human cooperative behavior. Such a rule allows group selection to be a strong force in determining human behavior in different societies. While it may still be possible to defend an egoistic theory of human society, this and similar models of cultural evolution undermine any purely deductive argument against group selection. Conformist transmission provides at least one theoretically cogent and empirically plausible explanation for why humans differ from all other animals in cooperating, against their own self-interest, with other human beings to whom they are not closely related.

Appendix

Group selection with unbiased transmission

Cultural transmission. Suppose that a subpopulation is made up of N individuals. At the end of generation $t - 1$, c_{t-1} of these individuals are cooperators, and n_{t-1} are non-cooperators. These individuals mate and have f offspring per mating. Then the number of cooperators among offspring at the beginning of generation t, c_t, is:

$$c_t = \frac{1}{2}fc_{t-1}$$

The right side of this equation is multiplied by one-half because each child has two parents, and so under our assumed rule of transmission, each parent will on average transmit the trait to half of them. The equation describing the socialization process for defectors is similar:

$$n_t = \tfrac{1}{2} f n_{t-1}$$

We can simplify the model considerably, without sacrificing anything for most purposes, by keeping track of frequencies or proportions of the variants instead of the numbers. Let q_t be the frequency of cooperators among offspring. Then

$$q_t = c_t/N$$

Since at this or any other state of the life-cycle the proportion of defectors is just one minus the proportion of cooperators $(1 - q_t)$, we need an equation for only one of the two types to specify the model completely. We assume that cooperation is costly to the individual but beneficial to the group. Thus it is plausible that cooperators may, on the average, have fewer children than defectors. If so, the proportion of cooperators among children will be smaller than among parents, a change due to natural selection acting on cultural variation. For simplicity, however, let us suppose that both cooperators and defectors have the same number of children. A little algebra then shows that the cultural transmission does not change the proportion of cooperators in the population, or:

$$q_t = q_{t-1}$$

Thus cultural transmission does not change the frequency of cooperators in a subpopulation. This result can be obtained for a much wider range of models.

Learning. Assume that each individual encounters a single other role model. If the role model has the same behavior as the focal individual, the focal individual does not change his/her behavior. However, if the role model has a different behavior, the focal individual evaluates the two behaviors and adopts the one that seems best. Then the frequency of cooperation in the subpopulation after learning, q_t, is

$$q_t' = q_t - B q_t(1 - q_t)$$

The magnitude of the parameter B depends on the accuracy of the evaluation process. If it is easy to evaluate the merit of the two variants then B will near one, and learning will substantially reduce the frequency of cooperation within the subpopulation. If identifying the best variant is difficult, then B will be near zero, and learning will cause only small reductions of cooperation during a single generation.

Migration. If m is the fraction of each population that emigrates, then the frequency after migration, q_t'', is:

$$q_t'' = (1 - m)q_t' + m\bar{q}_t$$

where $\bar{q}_t$ is the average frequency of cooperators in the population as a whole.

Group selection and conformist transmission

Conformist transmission. The following model is a very simple example of conformist transmission. Assume two cultural variants, cooperate (c) and non-

Model			Probability of acquiring	
1	2	3	*c*	*d*
c	*c*	*c*	1	0
c	*c*	*d*		
c	*d*	*c*	$(2 + D)/3$	$(1 - D)/3$
d	*c*	*c*		
d	*d*	*c*		
d	*c*	*d*	$(1 - D)/3$	$(2 + D)/3$
c	*d*	*d*		
d	*d*	*d*	0	1

cooperate (*d*), but now suppose that children acquire their initial beliefs from three adults, say their parents and a teacher. The probability that an individual acquires variant *c*, given that he/she is exposed to a particular set of adults, is shown in the above table.

This model of cultural transmission is particularly simple because each of the adults is assumed to have the same role. It does not matter which adults are cooperators, only how many are cooperators. The parameter *D* measures the extent to which cultural transmission is biased. If *D* is equal to zero, the probability that a child acquires either cultural variant is proportionate to the frequency of that variant among the child's adult role models. Assuming that each set of adults is a random sample of the population, then it can be shown that the frequency of *c* among naive individuals after transmission is given by:

$$q_t = q_{t-1} + Dq_{t-1}(1 - q_{t-1})(2q_{t-1} - 1)$$

When $D > 0$, cultural transmission creates a force increasing the frequency of the more common variant in the population.

Interaction of conformist transmission, learning, and migration

To see how conformist cultural transmission can preserve variation among groups, we compare the net effect transmission, learning, and migration when transmission is conformist with the case in which transmission is unbiased.

First consider the case in which conformist transmission is unbiased. Fig. 2.1 plots the change in the frequency of a subpopulation over one generation as a function of the frequency of cooperators in the subpopulation in a subpopulation that does not suffer extinction. If the frequency of cooperators in a particular subpopulation is higher than the average frequency in the population as a whole, they will either decrease in frequency – both because of immigration and, more importantly, because people will individually learn the deleterious consequences of holding cooperative beliefs and will switch to defecting. If cooperators make up a lower than average proportion of the subpopulation, cooperators will either increase in that subpopulation due to immigration or will decrease if the consequences of individual learning are greater than those of immigration. Thus, within each subpopulation the proportion of cooperators moves toward the value

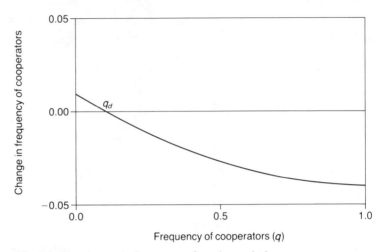

Fig. 2.1. The change in frequency of a subpopulation over one generation as a function of the frequency cooperators in the subpopulation. It is assumed that the subpopulation does not suffer extinction and enculturation involves faithful copying.

labeled q_d. Notice that q_d is less than the average frequency of cooperators in the population as a whole. Within-group processes tend to lower the average frequency of cooperators in the whole population because the same processes act in every subpopulation.

Now suppose that there is strong conformist transmission. Fig. 2.2 plots the change in the frequency of a subpopulation over one generation as a function of the frequency cooperators in a subpopulation that does not suffer extinction. Notice that now there are three values of q_t for which the change in the proportion of cooperators over one generation is zero. If the focal subpopulation were to reach exactly one of these values it would remain there, in equilibrium, until some external event occurred. When cooperators are rare, conformist transmission and learning both act to reduce their numbers. As long as there are some subpopulations composed primarily of cooperators, mixing will prevent the processes from completely eliminating cooperators, and instead the subpopulation will stabilize at a low frequency of cooperators, here labeled q_d. This equilibrium is 'stable,' meaning that the subpopulation will return to this value after small departures from it. There are two other equilibria, labeled q_u and q_c, at which the effects of conformist transmission, learning, and migration also balance. The equilibrium labeled q_c is also stable. In contrast, the equilibrium labeled q_u is unstable – if the frequency of cooperators is increased a small amount, the subpopulation will evolve toward q_c; if the frequency of cooperators is reduced a little, eventually the subpopulation will evolve toward q_d. Thus, if for some reason the frequency of cooperators in a subpopulation increases above q_u, the subpopulation will reach a stable equilibrium at which cooperators are common. If conformist transmission is strong enough, this may occur in a subpopulation even if every individual in the larger population is a defector.

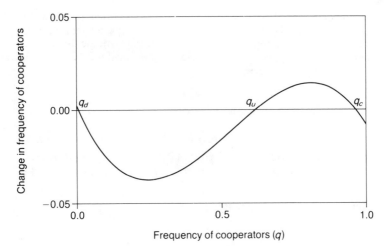

Fig. 2.2. The change in frequency of a subpopulation over one generation as a function of the frequency cooperators in the subpopulation. It is assumed that the subpopulation does not suffer extinction, enculturation involves conformist transmission that is strong compared with mixing.

The effect of extinction and recolonization.

Consider a population made up entirely of defectors. Even in these circumstances, if conformist transmission is strong, cooperation can increase whenever colonists of a vacant habitat are drawn from a single other population and that population has a majority of cooperators.
Algebraically,

$$
\begin{array}{c}
\text{probability of} \\
\text{extinction of} \\
\text{subpopulations at } q_c
\end{array}
<
\dfrac{
\begin{array}{c}
\text{number of} \\
\text{extinctions in} \\
\text{population}
\end{array}
\times
\begin{array}{c}
\text{probability frequency of} \\
\text{cooperators among colonizers} \\
\text{is greater than } q_u
\end{array}
}{
\text{number of subpopulations at } q_c
}
$$

Any empty habitat colonized by more than q_u cooperators will eventually end up at the equilibrium frequency q_c. Thus the numerator on the right-hand side gives the number of q_c subpopulations that are 'born' during a particular generation. Dividing this by the number of subpopulations at q_c gives the 'birthrate' of these subpopulations. This birthrate depends critically on the way that empty habitats are colonized. If they are typically colonized by people drawn from a single subpopulation, the probability that the frequency of cooperators among colonizers will be greater than q_u is just the proportion of subpopulations that are at q_c (since $q_d < q_u < q_c$). This means that the right-hand side of this inequality is just the probability that a randomly chosen subpopulation will suffer extinction. Since the extinction rate decreases with the frequency of cooperators, it follows that cooperation always increases over time via extinction when colonists are drawn from a single other randomly chosen subpopulation, either mainly cooperators or mainly defectors. If colonists are drawn from more than one subpopulation, the probability that the frequency of cooperators is greater than q_u will be smaller. The more mixing among colonists, the lower the birthrate of cooperative subpopulations, although cooperation could still increase if the effect of cooperation on extinction rates is sufficiently great.

Acknowledgments

We thank Jane Mansbridge for inspiring this paper, and for her extensive, and very helpful editorial comments. This research was partially funded by R. Boyd's John Simon Guggenheim Fellowship.

References and further reading

On the genetical evolution of cooperation

Axelrod, R. & Hamilton, W. D. (1981). The evolution of cooperation. *Science*, **211**, 1390–6.
Boyd, R. & Richerson, P. J. (1988). The evolution of reciprocity in sizable groups. *Journal of Theoretical Biology*, **132**, 337–56.
Trivers, R. (1985). *Social Ecology*. Menlo Park, CA: Benjamin Cummings.
Wade, M. J. (1978). A critical review of group selection models. *Quarterly Review of Biology*, **53**, 101–14.

On models of cultural evolution

Boyd, R. & Richerson, P. J. (1985). *Culture and the Evolutionary Process*. Chicago: University of Chicago Press.
Campbell, D. T. (1975). On the conflicts between biological and social evolution and between psychology and moral tradition. *American Psychologist*, **30**, 1103–26.
Cavalli-Sforza, L. L. & M. W. Feldman. (1981). *Cultural Transmission and Evolution*. Princeton: Princeton University Press.
Lumsden, C. J. & E. O. Wilson. (1981). *Genes, Mind, and Culture*. Cambridge, MA: Harvard University Press.
Richerson, P. & R. Boyd. (1989). The role of evolved predispositions in cultural evolution: Or human sociobiology meets Pascal's Wager. *Ethology and Sociobiology*, **10**, 195–215.

On cooperation among humans

Dawes, R. (1980). Social dilemmas. *Annual Review of Psychology*, **31**, 169–91.
Kaplan, H. & Hill, K. (1985). Food sharing among Ache foragers: Tests of explanatory hypotheses. *Current Anthropology*, **26**, 223–45.
LeVine, R. A. & D. T. Campbell. (1973). *Ethnocentrism: Theories of Conflict, Ethnic Attitudes, and Group Behavior*. New York: Wiley.
Peoples, J. E. (1982). Individual or group advantage? A reinterpretation of the Maring ritual cycle. *Current Anthropology*, **23**, 291–310.

B.

THE DEVELOPMENT OF PROSOCIAL PROPENSITIES

Editorial

Prosocial behaviour is shown more by some individuals than by others. This section is concerned primarily with the developmental issue of how individuals come to be predisposed to show prosocial behaviour. However, prosocial behaviour is itself complicated, depending on more than one capacity in the individual. Miller, Bernzweig, Eisenberg and Fabes (Chapter 3) mention two ways in which cognitive abilities are important – in enabling one individual to understand the point of view of the other, and to engage in moral reasoning as to the right course of action. In addition they emphasize the capacity to respond vicariously to the emotions of others, and the interrelations between this and cognition. These abilities are influenced by the methods used by parents to bring up their children, and the relative effectiveness of these are discussed in some detail.

The child's relationships, and especially parent–child relationships, are of critical importance in the development of prosocial behaviour in two ways. First, certain specific techniques, such as reinforcement, reasoning, moral exhortation and modelling, may be used by parents to various degrees and in various combinations. Second, such procedures are more likely to be effective if there is an harmonious relationship between parent and child. Considerable evidence suggests that 'harmonious' involves a proper balance between parental warmth and parental control, neither alone being sufficient.

The first contribution in this section is thus concerned with the nature of prosocial behaviour and with the methods which may be used by rearing agents to inculcate a propensity to be prosocial. However what parents and others do will depend in part on what they perceive to be

51

desirable behaviour in the child – that is, on their cultural values. Triandis (Chapter 4) distinguishes between 'collectivist' and 'individualistic' societies. In the former, people are socialized to conform to the values, norms and goals of the group to which they belong, with little emphasis on personal goals or self-realization. In individualistic societies, by contrast, conformity is downgraded and people are encouraged to develop as individuals. Of course no culture is a pure example of either extreme, but the conceptual distinction is important. Collectivists, though readily behaving prosocially to members of their own group, tend to be negative and hostile to outsiders, while individualists make much less distinction between ingroup and outgroup members. Thus prosocial behaviour becomes progressively less likely in the order: collectivists with ingroup members; individualists with ingroup members, individualists with outgroup members; and collectivists with outgroup members.

The distinction between collectivists and individualists is illustrated by Stevenson (Chapter 5), who describes child-rearing practices in China, Japan and Taiwan. In these countries much greater emphasis is placed on prosocial behaviour than in the West, and their socialization practices differ accordingly. This contribution reminds us of the important issue that ingroup cooperation may be linked to outgroup hostility. Stevenson's chapter should also be associated with his recent findings of a marked difference in achievement between schoolchildren in schools in certain Eastern cities and comparable schools in the USA (Stevenson & Lee, 1990) – a finding which may be ascribed to greater motivation and parental involvement in the former, or possibly to the role in cognitive development of peer cooperation over problem-solving (e.g. Doise, 1985).

In contrasting the West with China, Japan and Taiwan we are dealing with large-scale societies. Goody (Chapter 6) focuses on much smaller, egalitarian societies, and demonstrates an even greater diversity in socialization practices and their outcome. In these societies the means used to inculcate prosocial behaviour are diverse, and some of the generalizations apparently valid for large-scale societies require adjustment. For instance, antipathy towards a few outgroup members may lead not to hostility but to intense prosocial behaviour towards them.

Implicit in these accounts of different societies is the manner in which societal values both influence, and in the long run are influenced by, the child-rearing practices in the society. 'Culture', Goody points out, 'sets premises for the meanings of events', and thereby influences the behaviour of individuals. But in learning the meaning of events, individuals are perpetuating the culture. Goody indicates that the importance placed on prosocial behaviour, and the methods by which it is inculcated, are intimately linked to other aspects of the sociocultural structure (the presence of political and kin structures that induce ingroup/outgroup differentiation), to environmental conditions, and to the presence or absence of hostile or competitive neighbours. As indicated in the Introduction, each culture is a structure whose separate parts cannot be considered in isolation.

Further reading

Doise, W. (1985). Social regulations in cognitive development. In R. A. Hinde, A.-N. Perret-Clermont & J. Stevenson-Hinde, (eds.). *Social Relationships and Cognitive Development*. Oxford: Clarendon Press.

Stevenson, H. W. & Lee, S.-Y. (1990). *Contexts of Achievement*. Monographs of the Society for Research in Child Development, No. 211, 55, 1–2.

3

The development and socialization of prosocial behavior

PAUL A. MILLER, JANE BERNZWEIG, NANCY EISENBERG, AND RICHARD A. FABES

Although developmental psychologists have studied negative behaviors such as aggression and dishonesty for many years, it was not until about 1970 that sizable numbers of developmental psychologists began to examine the development and maintenance of positive behaviors such as helping, sharing, donating, and comforting. In recent years, these types of behaviors have been frequently examined, and even given a new name: *prosocial* behaviors, that is, voluntary behaviors intended to benefit another.

Of particular interest to many psychologists, including those studying development, is one particular subtype of prosocial behavior, *altruistic* behavior. Although prosocial behaviors can be performed for a variety of reasons, including self-gain, social approval, and concern for others, altruistic behaviors are defined as those prosocial behaviors that are motivated by other-oriented concern (e.g., sympathy) and internalized values. Additionally, altruistic behaviors are defined as *not* being motivated primarily by the desires to obtain concrete or social rewards or to reduce aversive internal states (e.g., anxiety, guilt) due to being exposed to another's distress or knowing that one has not behaved appropriately. Unfortunately, because motives and emotional reactions such as sympathy are internal phenomena, it usually is difficult to ascertain whether people's prosocial behaviors are altruistic or not. Thus, behavioral scientists generally study prosocial behaviors in general, even though they frequently are most interested in the development of altruism (although the mere performance of prosocial actions, regardless of motive, is of interest to some investigators).

In the past two decades, we have learned a great deal about the

54

development of positive behavior: however, there are still many inconsistencies in the empirical literature and there is much to be learned. In this chapter, we attempt briefly to review some of the major findings with regard to the development of prosocial behavior, with an emphasis on the cognitive, emotional, and social variables that appear to affect its development. Due to space limitations, we often are unable to delineate some of the interesting complexities in the existing research; rather, we try to provide an overview of the available research and thinking.

Our chapter begins with a discussion of cognitive capabilities that appear to influence prosocial functioning; then we review emotional factors and age and gender differences in prosocial behaviors. These sections are followed by a review of the ways in which parents, teachers, peers, and other people can shape and socialize prosocial behaviors in children.

Cognitive and emotional factors in prosocial development

The role of cognition

Cognitive capabilities seem to play a variety of roles in the development of prosocial, especially altruistic, behavior. Clearly, children's understanding of others' perspectives and emotional states, and the ways in which they conceptualize and resolve moral dilemmas, would be expected to affect the development of prosocial behavior. In addition, more intelligent children might be expected to be more prosocial simply because they should be better able than less intelligent children to understand others' perspectives and emotions, to reason in morally mature ways, and to have competencies related to enacting helping behaviors. As we will see, the empirical data provide some support for these assumptions.

Perspective taking and prosocial behavior   Among psychologists, there is a commonsense expectation that as one's capacity to understand the perspectives of others develops, so does one's ability to react sensitively and appropriately to their needs and problems. In the developmental literature, perspective taking generally is viewed as taking the cognitive, perceptual, or affective perspective of others. Cognitive or perceptual perspective-taking skills have been assessed via tasks that tap the ability to infer the intentions or motives of a story character, to tell a story to an uninformed person, or to make decisions or inferences based upon the other person's cognitive or perceptual (e.g., visual) view of a situation. Indexes of affective perspective taking generally require the ability to identify or infer the emotional state or situation of another person.

Researchers who have reviewed the empirical data pertaining to the relation between perspective taking and prosocial behavior have found that the data are not highly consistent. Positive relations have been found in a number of studies, but no relations, or occasionally even a negative relation, have been found in others. However, when the results of various studies have been combined statistically using meta-analytic procedures, an overall positive, but modest, relation between perspective-taking abilities and frequency of prosocial responding has been found.

A number of conceptual and methodological problems have been identified that may account for the mixed pattern of findings between perspective taking and prosocial behavior. For example, researchers have sometimes assumed that all prosocial behaviors are equivalent when, in fact, some prosocial behaviors reflect an understanding of others' perspectives whereas others do not. Acting prosocially does not always require the ability to take the perspective of another; other motivations (e.g., values, principles/norms, the desire to reduce aversive stimuli emanating from a needy other) may be sufficient for acting prosocially, and different types of perspective taking may be relevant for certain prosocial behaviors but not others. In addition, measures of prosocial behavior sometimes may tap compliance or the desire for social approval rather than the desire to assist another, and measures of perspective taking may not be appropriate for the age group studied or relevant to prosocial responding. Moreover, one would expect the relation between perspective taking and prosocial behavior to be higher if both were assessed in the same setting (so that the index of perspective taking is relevant to the decision of whether to assist another person). Usually, however, perspective taking is assessed with a measure pertaining to hypothetical people in types of settings unrelated to that in which prosocial behavior is assessed.

Although perspective-taking skills obviously can be used to manipulate and take advantage of other people, numerous theorists have argued that such skills generally are associated with morally mature ways of resolving moral conflicts. This is because individuals with higher level perspective-taking skills would be expected to be capable of considering the views of other people and of the larger society, as well as abstract principles based on the higher level perspective taking, when making moral decisions. Thus, it frequently has been hypothesized that the effect of perspective taking on moral behavior is by means of its effect on the process of moral reasoning.

Moral reasoning and prosocial behavior   The development of moral reasoning – reasoning about how one should resolve moral dilemmas – has been a topic of interest to developmental psychologists for at least 25 years. Interest in this topic was stimulated by the seminal writings of Jean Piaget and by Lawrence Kohlberg's research in which he expanded upon Piaget's conceptualization of developmental stages in moral reasoning.

Moral reasoning is typically assessed by presenting people with hypothetical moral dilemmas (or, occasionally, real life moral dilemmas) and asking individuals to explain their reasoning concerning how the moral dilemmas ought to be resolved. For example, in the well-known 'Heinz' dilemma used by Kohlberg, Heinz was forced to choose between allowing his wife to die of cancer and stealing a new medicine from a druggist who refused to sell the drug for a price that Heinz could afford. In Kohlberg's system for coding individuals' moral reasoning, blind, unthinking compliance with authorities' dictates and acting to avoid punishment or to obtain material gains are considered lower level reasoning. The desire to behave in ways that would result in social approval and the desire to conform with global stereotypic notions of good and bad behavior are viewed as reflecting intermediate level reasoning. Higher level reasoning is that based on the principle of the greatest good for the most people and on internalized, universal moral principles pertaining to justice.

In most of Kohlberg's moral dilemmas, issues related to obeying authorities and their dictates, compliance with laws, rules, and formal obligations, and potential punishment are salient. Even when prosocial behavior is an option in a dilemma (e.g., Heinz could assist his wife by stealing the drug), the moral reasoner usually is forced to choose between violating a law, rule, authority, or formal obligation and assisting someone in need. Only in the last decade has there been much research concerning moral dilemmas in which individuals reason about a conflict between their own needs and the needs of others (henceforth labeled prosocial moral reasoning).

This relative lack of attention to prosocial moral reasoning is somewhat surprising. Prosocial actions necessarily involve a certain degree of cognitive understanding of others' needs, and also usually are the outcome of decision-making processes in which various motives, values, principles, and personal needs and desires are weighed. Thus, one would expect individuals who use higher levels of moral reasoning in resolving such dilemmas frequently (albeit not always) to exhibit more, and higher quality (i.e., more altruistic), prosocial behavior.

In general, there does seem to be a modest relation between individuals' prosocial moral reasoning or reported motives for actual prosocial actions and their prosocial behavior. In a typical study, children are asked to reason about hypothetical prosocial moral dilemmas (e.g., a situation in which a child has to choose between helping an injured child and going to a party) and prosocial behavior (e.g., helping or sharing) is assessed in another context. Alternatively, children's self-reported motives for their own actual prosocial behavior (assessed either in an experimental or natural setting) are examined in relation to the quantity or quality of their helping or sharing behavior. Most researchers consider moral reasoning or self-reported motives that reflect a clear other-orientation or internalized values to be more mature than self-oriented reasoning and stereotypic reasoning. Consistent with this perspective, children who exhibit higher level moral reasoning tend to engage in more sharing and helping than do other children, although sometimes no relations between behavior and reasoning or reported motives are found. Also, positive associations sometimes have been found between levels of moral reasoning in prohibition-oriented conflicts (e.g., Kohlbergian dilemmas) and the frequency and amount of prosocial behavior. In addition, low level moral reasoning has been linked with delinquent and dishonest behaviors, as well as with low levels of prosocial behavior.

The relation between moral reasoning and prosocial action generally is more consistent when individuals are asked to reason about conflicts that involve a prosocial dilemma (e.g., a dilemma concerning helping someone in need or distress versus pursuing one's own goals) rather than other types of moral dilemmas. Indeed, the relation between moral reasoning and morally relevant behavior generally appears to be higher if individuals are asked to reason about moral situations that are similar to those in which moral behavior is assessed. This may be because moral reasoning about certain types of moral conflicts frequently may be of little relevance to morally relevant behaviors performed in a different context.

Additionally, the relation of prosocial moral reasoning to behavior appears to vary as a function of the cost of the prosocial behavior that is assessed. Higher level prosocial moral reasoning is more highly associated with costly than low cost prosocial actions. For example, preschoolers' prosocial reasoning is more highly related to their giving possessions to another child than simple acts of helping such as handing objects (e.g., blocks) to another child using the objects. This is not really surprising; there is no reason to expect individuals' moral reasoning to be relevant to prosocial action in situations in which there is little cost to the self and, consequently, no real moral conflict. Indeed, people seem to perform many low cost prosocial behaviors without much cognitive processing (e.g., helping someone pick up dropped papers).

In summary, the empirical literature is consistent with the conclusion that the relation between level of moral reasoning and prosocial behavior tends to be modest, but is significant and positive. The fact that these relations are only modest may reflect the fact that prosocial behavior is, no doubt, often determined by multiple motives, some of which may not involve moral-cognitive components (e.g., emotional reactions may motivate prosocial actions). We return to this point shortly.

Intelligence and prosocial behavior   Thus far, we have seen that prosocial behavior is associated with higher level perspective taking and moral reasoning. Given the relation between these cognitively-based skills and prosocial behavior, one also might expect general intelligence (IQ) and level of logical reasoning to be associated with prosocial behavior, although they probably affect prosocial behavior indirectly (e.g., through their effects on perspective taking and moral reasoning).

In fact, according to recent research, children who score higher on tests of IQ and logical thinking sometimes are more prosocial, but the overall effect is weak and relations are found only occasionally. It seems likely that the relation is stronger in contexts in which cues related to the other's need are subtle; in such situations, more intelligent children may be more likely to detect the other's need, perspective take, and figure out a way to assist. In addition, more intelligent children may assist more frequently in settings in which specific competencies are needed to help (e.g., academic skills). However, intelligence alone does not appear to account for individual differences in prosocial responding in many types of settings.

If cognitive capabilities alone are insufficient for explaining prosocial behavior, one must look to other factors. As noted previously, some theorists have suggested that cognitive awareness of the other person's need or problem may lead to vicarious emotional responding, or empathy, and to feelings of concern about the distressed or needy other (i.e., sympathy), which may act as motivational bases for prosocial actions. We now turn to the role of empathy and sympathy in prosocial behavior.

Vicarious emotional responding and prosocial behavior

Unlike perspective or role taking (which is sometimes viewed as a form of cognitive empathy), emotional empathy refers to a vicarious emotional reaction to another's emotional state or situation that is congruent with the other's emotional state or situation. For example, if a child feels sad when observing a sad person, the child is empathizing. Sympathy refers to feelings of sorrow or concern for another person's plight or situation (rather than merely reflecting the same emotion as the other person) and often stems from perspective taking and empathy. Although there is some disagreement on the exact definitions of empathy and sympathy, the view that empathy or sympathy, or both, serve as a motivational basis for altruistic action has been longstanding.

In a recent review of the empirical data pertaining to the relation between empathy (defined as including sympathy) and prosocial behavior, there was a moderate relation between the empathy and prosocial behavior for older children and adults, but an inconsistent relation for younger children. The lack of consistent findings for young children appears to have been due to the fact that children's self-reported empathic (and sympathetic) reactions generally are not associated with their prosocial behavior. In contrast, children's prosocial behavior has been associated with empathy when empathy has been assessed with measures that do not require children to report their own emotional reactions (i.e., facial/gestural reactions to films/lifelike stimuli of distressed people, parents' or teachers' report). For adults, the relation of empathy to prosocial responding appears to hold across a variety of methods for assessing sympathy/empathy (e.g., questionnaires, physiological indices, experimental simulations in which adults think they have viewed real people in need or distress and then are asked how they felt). Moreover, investigators recently have shown that physiological reactions (i.e., heart rate responses) in children also can be predictive of their prosocial behavior in certain contexts.

The notion that empathy or sympathy frequently motivate prosocial behavior is also supported by other data. Empathy and sympathy appear to be associated with a variety of indices of socially competent behavior such as sociability and cooperativeness – behaviors that one would expect to enhance the likelihood of individuals engaging in a prosocial manner. Moreover, people who score high on indices of empathy tend to exhibit relatively low levels of aggression and other antisocial behaviors. Finally, in a study of the role of vicarious emotion in real life acts of heroic altruism, nearly half of a group of people who attempted to rescue Jews from the Nazis in Europe mentioned feelings of sympathy or empathy as the basis for their actions. Thus, data from a variety of sources are consistent with theoretical assertions that feelings of empathy or sympathy for others frequently motivate helping, sharing, and comforting behaviors.

In fact, much of the existing research may underestimate the relation of empathy and sympathy to prosocial behavior. Empathy frequently has been assessed in one context (e.g., using a questionnaire index) whereas prosocial behavior has been assessed in another – a procedure which is likely to underestimate the relation between sympathy and prosocial behavior directed toward the target of that sympathy. Furthermore, indices of empathy and sympathy frequently have been examined in relation to prosocial behaviors that were likely to be due to other sorts of motives (e.g., the desire to act in a manner consistent with one's principles or the egoistic desire to gain social rewards or approval). The failure of many investigators to try to differentiate between other-oriented concern and self-focused, aversive emotional reactions (e.g., discomfort, anxiety) to others' emotional states or situations probably has resulted in some of the non-significant findings in the empirical literature. Sympathy, but not a self-focused, personal distress reaction, has been found to motivate prosocial behavior when it is easy for those experiencing personal distress simply to leave the distressing situation. One can expect the strongest relation between vicarious emotional responding and prosocial behavior when the vicarious response is sympathy and the prosocial behavior is altruistic in motive.

Integration of cognition and affect in relation to prosocial behavior

The importance of integrating cognitive and emotional components in the study of the development of prosocial responding is reflected in research employing causal attributional approaches to examining empathic and prosocial responding. It appears that the attribution or interpretation one makes regarding another person's situation is critical in shaping one's subsequent emotional and behavioral responses to them. For example, if an observer concludes that a needy or distressed person is not responsible for his or her condition (e.g., has fallen down due to illness), the observer is likely to feel sympathy for the person and to help. In contrast, if the observer believes that a needy other is responsible for his or her plight (e.g., is sick because he or she drank too much), the observer is relatively unlikely to experience sympathy; rather, he or she often will experience disgust or other negative reactions, and is unlikely to assist. Clearly, vicarious emotional reactions to others are affected by the individual's cognitive evaluations of the specific situation, as well as by individuals' perspective taking and moral reasoning capabilities (see prior sections of the chapter).

Indeed, the development of relevant cognitive skills and vicarious emotional responding appear to be intertwined in childhood. Martin Hoffman, a prominent developmental psychologist, has argued that children cannot empathize until they can comprehend the affective states of others and cannot sympathize until they realize that others often have feelings different from their own. In addition, as children become better able to take the perspectives of others, they also are better able to assist in an appropriate and sensitive way. Hoffman further suggests that the cognitive realization in the late elementary school years that one's own and others' existence is continuing, and that others' conditions can be chronic and ongoing (even if their condition is not immediately visible), has important implications for empathy. This realization enables the child to empathize and sympathize with others' general conditions (e.g., those of the economically deprived, the sick, the politically oppressed). Although some aspects of Hoffman's theory have not been tested, they provide a conceptual basis for examining the interrelated roles of cognition and emotion in early prosocial responding.

Given the age-related changes in many cognitive capabilities such as perspective taking and moral reasoning and, perhaps, in the capacity for empathy and sympathy, one would expect age-related changes in prosocial behavior. In addition, based on the stereotypic conception that females are more emotional, sympathetic, and other-oriented than are males, one might also predict gender differences in prosocial behavior. We now examine the data related to these issues.

Age-related changes in prosocial behavior

Until fairly recently, the common notion was that young children are very self-centred and unable to act altruistically, and become more prosocial with socialization and cognitive development. Children were not expected to be capable of altruism until five to seven years of age, when they were viewed (wrongly) as first developing perspective-taking skills.

Contrary to this view, one of the most interesting recent developments in the research on prosocial behavior is the consistent finding that very young children (i.e., from 14 months of age) are capable of performing many different types of prosocial behaviors. These behaviors are directed toward both peers and adults. Young children's prosocial actions are not, in the main, haphazard or based upon the young child's own egocentric perception of the situation, but frequently are at least somewhat appropriate to the needs of the other, follow a logical sequence of execution, and often resemble actions modeled by adult caregivers. This suggests that children have the abilities to recognize and interpret others' perspectives, at least to some degree, long before what has been traditionally recognized.

The presence of prosocial behaviors in very young children introduces new and provocative questions about age-related changes in prosocial action. If prosocial behavior does increase with age, what is the nature of this change? Are there changes in the frequency or the quality of prosocial behavior (or both) with age, or are the changes manifested in different types of prosocial behavior? It turns out that the picture of age-related changes in prosocial behavior is an unclear one, and appears to vary as a function of (a) the type of prosocial behavior, (b) the definition and index of prosocial behavior, and (c) whether assessments are made in laboratory or naturalistic settings.

When prosocial behavior is defined in terms of frequency of occurrence, the following patterns are found. First, prosocial responding involving caregiving and/or comforting shows a mixed pattern. Positive relations are found with age in some research but, in other studies, more of these behaviors have been performed by younger children. The context probably plays an important role in the pattern of age-related changes in comforting, but the nature of contextual effects has not yet been adequately delineated.

When the prosocial index is helping, there appears to be a slight increase with age (when differences are found), especially when helping is defined as incorporating a variety of different positive social behaviors rather than being narrowly defined. However, this age trend is very weak. In contrast, sharing behavior more clearly increases with age, although the strength of the pattern appears to be affected by the method of measurement. The age-related increase is strongest when sharing is assessed with donating tasks in laboratory settings (usually involving a hypothetical recipient), but is less consistent when assessments involve real peers as recipients or are in naturalistic, real-life situations.

Although the vast majority of researchers have used frequency of occurrence as their index of prosocial behavior, some have attempted to assess qualitative dimensions such as intensity, duration, intention, or the effectiveness of the prosocial behavior. When this is done, investigators tend to find increases in prosocial responding with age. For example, older children are more likely than younger children to help or share when they are not pushed to do so by an adult or are not promised a prize for doing so.

In sum, the weight of the studies suggests that there is an increase in prosocial behavior with age; however, conclusions must be drawn cautiously due to the fact that the pattern of findings is dependent upon the type of prosocial behavior and the method of measurement. The relation between age and prosocial behavior would likely be more consistent if researchers paid more attention to the qualitative dimensions of prosocial behavior (e.g., tried to isolate altruistic from less moral modes of prosocial behavior). The procedure of relying merely on counts of prosocial behaviors would be expected to produce a measure of altruism only if prosocial behavior is assessed in a context in which concrete or social rewards for assisting are unlikely, and if it is easy to escape from dealing with the unpleasant cues emanating from the other's need or distress. Moreover, age-related changes may be apparent only for those prosocial acts involving perspective taking, sympathy, or other capabilities that increase with age.

Gender differences in prosocial behavior

There is considerable evidence that boys and girls are socialized to think and act in gender-stereotypic ways, and that they sometimes tend to react in ways consistent with sex-role stereotypes (e.g., boys tend to exhibit more aggression than do girls). Thus, it seems reasonable to expect gender differences in the frequency and the type of prosocial behaviors performed. Few researchers, however, have attempted to assess directly whether gender differences exist in prosocial behavior, although relevant data are reported in many studies.

Overall, in both reviews of the literature and in recent studies, no consistent differences have been found between the sexes in children's sharing, cooperation, comforting, or helping behavior. When differences have been found, they tend to favor girls. For example, there is some evidence that girls may be somewhat more responsive and nurturant toward younger children and are more socially oriented and sensitive to others. However, these few differences appear to be due, at least in part, to the type of prosocial index used and the methodology used to assess prosocial responding.

Specifically, consistent with the gender-role socialization literature, prosocial responses that involve expressiveness and caring actions tend to favor females, whereas those requiring more active or instrumental response may be enacted more often by males (especially for adults). Also, females may score more highly than males on indices of prosocial behavior when it seems to be socially desirable for one to assist others or when gender stereotypes are salient. For example, females tend to score higher than males on self-report indices of responsiveness to others, and in situations in which it is clear that they are expected to behave in a prosocial manner. In these types of situations, it is obvious what is being assessed or there are social pressures to behave prosocially. Thus, gender-related stereotypes are easily accessed from memory. In contrast, no differences or mixed patterns of findings (i.e., favoring either males or females) have been found in situations in which people are unaware of what is being assessed and physiological or behavioral indices of responsiveness to others are used. Thus, it appears that people often try to present themselves in ways that are consistent with gender-stereotypes concerning sympathetic and nurturant behavior.

In addition, measures of boys' and girls' prosocial reputations (e.g., teachers' or peers' ratings) often indicate that girls are considered more altruistic. However, these ratings often are not good predictors of actual altruistic behavior. Moreover, the gender difference in rated prosocial behavior appears to be due, in part, to the rating scales including more items that are stereotypically feminine (i.e., involve nurturance or comforting or other prosocial acts deemed more appropriate for girls) than items that are stereotypically masculine (e.g., involve instrumental helping).

Overall, then, there is a need for careful consideration of factors associated with gender differences in prosocial responding. Gender differences in prosocial behavior may be found to the extent that children are rewarded or encouraged for certain modes of prosocial action, and to the extent that the index of prosocial behavior is tapping individuals' stereotypes about gender differences in prosocial behaviors. Future investigations would benefit from greater attention to the situational and psychological factors that are responsible for gender differences in prosocial actions, when they are indeed found.

Socialization of prosocial behaviors

Thus, far, we have considered the relation among various cognitive, affective, and person variables (i.e., age and sex) relevant to individuals' performance of prosocial behaviors. In this section, we focus on socialization processes that influence the acquisition of cognitive, affective, and behavioral responses relevant to prosocial responding.

Socialization generally is defined as the process by which parents (or other agents) attempt to influence children's prosocial values and behaviors. Children may or may not conform to socializers' expectations, and conformity may take one of two forms.

The first form of conformity is labeled 'compliance' and refers to situations in which a child conforms with parents' requests or pressures merely to gain rewards or avoid punishment. The child who merely complies is likely to revert back to the original behavior (or lack of it) when rewards and/or punishments are no longer forthcoming. The second form of conformity is labeled 'internalization'; it involves change that is more lasting and is independent of the immediate instrumental value of the conforming prosocial attitude or behavior. Thus, children may conform prosocially for at least two different reasons: because of the pragmatic value of the behavior (i.e., compliance) or based on the internalization of prosocial values that guide behavior in the absence of salient external pressures (i.e., rewards or punishments). It is the latter kind of conformity that is the long-term goal of most socialization agents.

Dimensions of parenting

Many researchers have attempted to delineate the parenting behaviors associated with children's prosocial compliance and internalization. Two critical components that account for parental influence have been identified: parental support and parental control. Supportive parenting is characterized as behavior that makes the child feel comfortable in the presence of the parent and confirms in the child's mind that he or she is, at a basic level, accepted and approved of as a person by the parent.

The control dimension of parenting refers to the amount of autonomy that parents allow their children. Restrictive parents limit their children's expression by imposing many demands and by monitoring their children's behavior to ensure that parental rules are followed. Permissive parents are much less controlling; they make relatively few demands of their children and allow them considerable freedom to explore the environment and express their opinions and emotions.

Control and internalization   Focusing primarily on the control dimension, researchers have suggested that different parental control practices have different influences on children's internalization of prosocial values. For example, Mark Lepper has argued that permissive parenting involved parental control and pressure that is insufficient to promote children's compliance to parental desires. In contrast, parents who use overly sufficient (i.e., obvious) external control when inducing compliance undermine the internalization process because the salient nature of the external control leads children to believe that their behavior was a function of the external inducement rather than intrinsically (i.e., self-) motivated. Lepper concluded that internalization is most likely to occur when parental control not only is successful in producing compliance, but also is sufficiently subtle to prevent the child from inferring that his or her conformity with socializers' expectations was solely a function of external rewards or punishments.

With regard to prosocial behavior, excessive use of rewards to induce conformity appears to result in children's helping or sharing merely for rewards rather than because they have internalized prosocial values. Children who believe that they engaged in prosocial behavior because of rewards are relatively unlikely to perform another prosocial act when left alone. However, the effects of rewards seem to vary as a function of the child's prior socialization experiences. For example, researchers have found that children's unsupervised helping was undermined by rewards for prior helping; but this was true primarily for children whose mothers used rewards consistently to induce conformity with their desires. Thus, rewards had a detrimental effect on children's prosocial motivation, but only for those children who had frequent experience with rewards as inducements within the family.

Configurations of parental dimensions   Parental control and warmth do not occur in isolation; parents (and other socializers) use varying levels of both in interactions with children. Thus, it is important to consider configurations of parenting behaviors, and not just isolated, single dimensions of parenting, when studying socialization.

Using a classification scheme that includes both parental control (i.e., demandingness) and parental support (i.e., responsiveness), Diana Baumrind has delineated three styles of parent–child interaction that may be related to the development of prosocial behavior. The first pattern is the authoritarian parent who is very restrictive and who places strict limits on allowable expression of the child's needs. Authoritarian parents expect strict obedience and rarely, if ever, explain their actions; rather, they often rely heavily on power assertive tactics (i.e., punishments, threats, deprivation of privileges) to gain compliance.

The second pattern is the authoritative parent. Authoritative parents expect mature behavior and set clear standards for these behaviors. They also firmly enforce rules and standards, using commands and sanctions when necessary. In addition, however, authoritative parents encourage the child's independence and individuality, and value open communication between parent and child – recognizing the rights of both parents and children.

The third style is the permissive parent. Baumrind characterizes permissive parents as those who make relatively few demands on their children and rarely exert firm control over them. These parents are also relatively warm and encourage their children to express their feelings and impulses.

One important distinguishing characteristic of permissive parents relative to authoritative and authoritarian parents is the degree of control exerted over children's behavior. Both authoritative and authoritarian parents exert firm control, whereas permissive parents do not. Although both authoritative and authoritarian parents are similar on the control dimension, authoritarian parents emphasize the power assertive aspects of parental control whereas authoritative parents emphasize independence of action and expectations for mature behavior.

In terms of the effects that these parenting styles have on the development of prosocial behavior, parental demandingness has been associated with higher social responsibility in boys whereas parental responsiveness has been related to higher social responsibility in children of both sexes. These findings suggest that if parents are high in demandingness but low in responsiveness (the authoritarian pattern), boys but not girls should be relatively high in social responsibility. However, high demandingness and high responsiveness (authoritative pattern) appears to be associated with high levels of social responsibility in both boys and girls.

There also is evidence that both boys and girls of authoritarian parents show less evidence of 'conscience' and are more likely to have an external rather than an internal moral orientation when discussing situations involving moral conflicts. Moreover, permissive parenting appears to have negative effects in that it is associated with low levels of social responsibility in children.

Thus, general styles of parental behavior have been found to be related to their children's prosocial tendencies (broadly defined). In the next section, we examine some of the specific child-rearing techniques that are facets of parental control and support.

Parental disciplinary techniques

Inductions   Inductions are defined as socializers' use of reasoning in disciplinary encounters. For example, parents may react to undesirable behavior in their offspring by pointing out the consequences of the child's behavior for others, encouraging the child to imagine the victim's perspective, or discussing aspects of the situation that may have influenced the child's or others' behaviors.

Parental inductions have been found to relate positively, albeit not entirely consistently, with children's prosocial responding. Moreover, they can be effective with children as young as 15 to 20 months old. In very young children, however, inductions may be effective only if verbalized with emotional force (i.e., with anger, concern, or some emotional intensity), perhaps because the emotional charge attracts the child's attention and indicates to the child that the issue at hand is important. Moreover, inductions with school-aged children appear to be more effective if used in a generally supportive environment, and if children have had a history of exposure to inductive discipline.

Inductions appear to promote prosocial behavior because they frequently highlight the consequences of the child's behavior for others, thereby enhancing perspective taking, empathy, and sympathy. If inductions are administered in a supportive context and accompanied by strong expression of parental emotion, the emotion is more likely to be interpreted in terms of the value that the parent assigns to the situation rather than in terms of physical punishment. The child is therefore able and motivated to attend to the information provided by the parent about the situation and is not overly concerned about their own welfare. In addition, because the child is not induced to focus on external pressure such as punishment, they are relatively likely to internalize the values communicated by the socializer. Finally, when inductions are accompanied by statements regarding parents' expectations and instructions for reparation, children not only learn that they are responsible for the consequences of their behavior, but are also provided with a means to act responsibly in the future.

Power assertion   Power assertion is defined as the use of actual or threatened force, punishment, or withdrawal of privileges to induce compliance. These techniques involve considerable external pressure on the child to behave according to the parents' desires.

Somewhat inconsistent associations have been found between power assertive techniques and prosocial behaviors. Overall, however, frequent use of power assertive techniques have been negatively associated or unrelated to children's sharing, understanding of kindness, cooperation, or helping. However, moderate use of power assertive techniques in a generally supportive context does not appear to have detrimental effects on children's prosocial development. Diana Baumrind found that parents who provided a nurturant, responsive child-rearing environment, yet maintained high demands for compliance and occasionally used power assertion (i.e., were authoritative parents), tended to rear socially responsible boys. However, when parental demands were administered in a punitive, authoritarian context, boys exhibited less socially responsible behavior.

Frequent use of power assertive discipline appears to inhibit internalization of prosocial values. The use of such techniques focuses children's attention on the consequences of their behavior for themselves rather than on the consequences of their behaviors for others. Thus, power assertive techniques are unlikely to capitalize on the child's capacity for empathy. In addition, power assertive discipline may arouse children to such an extent (because of their fear) that they have difficulty attending to parents' inductions. Finally, parents who model hostile, punitive behavior communicate that it is an acceptable interpersonal response.

Love withdrawal   Love withdrawal is expressed in those disciplinary encounters in which socializers give direct but nonphysical expression to their anger or disapproval of the child for engaging in some undesirable behavior (or for not engaging in desired behaviors). Like power assertion, love withdrawal has a highly punitive quality; in addition, it generally involves the implicit or explicit message that love will not be restored until the child changes her behavior. Love withdrawal includes such parental behaviors as ignoring or isolating the child, as well as explicit indications of rejection and disappointment in response to something the child has done.

Generally, the use of love withdrawal has been found to have weak and inconsistent relations with indices of children's moral (including pro-social) behavior. However, in one study, mothers reported that love withdrawal was the single most effective technique they had for securing their children's compliance, but they used it infrequently and always in combination with some other technique. In this study, withdrawal of love was also associated with avoidant behavior in children, suggesting that it is an especially aversive event. Withdrawal of love, therefore, may motivate immediate conformity with socializers' desires, but it does not seem to be effective in promoting autonomous, internalized standards of behavior.

Parental warmth and nurturance

A warm and nurturant parent is someone who is deeply commit-ted to the child's welfare, responsive and sensitive to the child's needs, and shows enthusiasm over the child's performance of desired behaviors (including acts of altruism). Although parental nurturance and support are generally positively related to children's prosocial behavior in empiri-cal studies, when considered apart from other socialization techniques, they often are not predictive of children's prosocial behavior.

The effects of parental warmth appear to be moderated by other socialization practices. Recall that although Baumrind found no corre-lation between parents' nurturance and preschoolers' social responsibility scores, authoritative parenting (involving warmth and control) was asso-ciated with children's social responsibility. Parental warmth and nurtu-rance seem to have an indirect influence on children's prosocial behavior by influencing the effectiveness of a particular disciplinary strategy. For example, in an early study conducted by Sears, Maccoby, and Levin, power assertion (spankings) was most effective when administered by a warm, affectionate mother. Cold, hostile mothers were more likely to report that spanking was ineffective. In this same study, mothers' use of love withdrawal was more likely to be positively associated with moral development when administered by a warm, affectionate rather than a cool, aloof mother. Thus, parental warmth and nurturance seems to be a background variable that influences the effectiveness of parental discipline.

The indirect effect of parental warmth on morally relevant behaviors probably is due to several factors. A warm, nurturant parent is likely to be an empathic parent – one who attends to the child's point of view and feelings. Thus, the warm, empathic parent not only fosters a close parent – child relationship, but also models nurturant, prosocial behavior and is likely to reinforce children's empathic responses.

Parental warmth also appears to enhance the development of prosocial behavior because it results in the child being positively oriented toward the parent and fosters the child's receptiveness to parental influence. It is likely that parental nurturance is especially critical when the child is distressed. How a parent deals with a child's distress may have consequences for the child's capacities for empathy. When parents respond nurturantly to the child's feelings of helplessness and distress, the child learns to express distress without shame and responds sympathetically to distress in others. If parents respond to their child's distress with anger or contempt, the child will learn to suppress their own feelings and avoid dealing with others' distress.

Other socialization practices associated with prosocial behavior

A number of other practices used by socializers have been associated with enhanced levels of prosocial behavior. Some of these are now reviewed briefly.

Modelling   Modelling is related to the development of prosocial behavior in that the moral character of the child, depends, in part, on the moral character and conduct of parents. There is both experimental and nonexperimental evidence for the role of modelling in children's prosocial behavior. Exposure to generous models has been associated with enhanced prosocial responding, whereas exposure to stingy models appears to disinhibit selfish behavior. Moreover, the effects of modelling persist over time and generalize in some situations. For example, people who have been involved in real life altruistic activities frequently were raised in families in which the parents engaged in humanitarian activities.

The degree of children's imitation of prosocial behavior is influenced by the nurturance of the model. In studies in which children are exposed to noncontingent nurturance for a brief period of time, children are unlikely to imitate generosity. Noncontingent nurturance appears to be interpreted by children as indicating permissiveness. Consequently children do as they please after contact with noncontingent, warm models. In contrast, when nurturance is part of an ongoing relationship and is not entirely unconditional, it has been found to increase the effectiveness of a prosocial model.

The effect of a nurturant prosocial model has been illustrated in studies of real life altruists. In a retrospective study of the effects of modelling, activists who made great sacrifices to participate in the civil rights movement in the 1960s maintained warm, cordial relationships with at least one parent. In contrast, those activists whose commitment involved little personal sacrifice described their relationships with their parents as hostile during childhood and as cool during the adult years. Moreover, parents of the fully committed activists had themselves been activists at an earlier time and had passed altruistic values along to their children. Parents of the partially committed activists had verbally endorsed altruistic values but rarely practiced altruism. Thus, parents who are warm and nurturant and practice the prosocial values they preach appear to foster the development of altruism in children.

Prosocial attributions Socializers' verbalizations can affect children's prosocial behavior by modifying children's self-attributions regarding their own previous actions. When adults attribute children's positive behaviors to internal factors (e.g., kindness), children are more likely to assist on subsequent occasions. For example, children that are told that they helped because they are helpful people and like to help others are more prosocial on subsequent occasions than children who are told that they assisted someone merely to comply with an adult's expectations. The effectiveness of providing children with internal attributions increases with the age of the child; it is likely that this is due to the fact that young children find it difficult to think in terms of enduring personality traits that produce consistency in behavior.

Moral exhortations, direct instructions, and assignment of responsibility Sometimes parents attempt to influence their child's future prosocial behavior in nondisciplinary situations by asserting how they themselves plan to act or by discussing the merits of a given course of action. Such verbalizations have been called moral exhortations or preachings, and their effect on children's subsequent prosocial behavior varies as a function of the content of the message. Children's sharing seems to be enhanced by preachings that either provide symbolic modelling (i.e., that involve the message that the parent himself or herself plans to act, or would act, in a prosocial manner) or are likely to elicit perspective taking and an empathic response (e.g., 'The poor children are sad because they don't have any toys'). In contrast, appeals to general social norms (e.g., 'It's good to give') and exhortations that involve verbal threats of disapproval or refer to self-oriented motives for assisting (e.g., indicate that the child will get something out of helping) appear to be less effective.

Whereas exhortations provide information about what children ought to do, instructions are used to communicate directly what the child is expected to do. The effectiveness of direct instructions varies according to the type of instruction used. For example, constraining instructions ('What I'd like you to do ...') appear to enhance children's donating in the immediate situation more than permissive instructions ('You may give her some pennies if you like but you don't have to'). However, the effectiveness of constraining instructions may decrease with the age of the child.

Rehearsal or practice in performing prosocial behaviors also has been found to promote prosocial tendencies. Children assigned the responsibility to teach other children prosocial behaviors display more prosocial behaviors in other situations, and children who are subtly induced to share (so that children won't attribute their sharing to external pressure) are more helpful on subsequent occasions. Similarly, children assigned to perform household chores, particularly chores involving responsibility for others, tend to be more prosocial than are other children.

Characteristics of the socializer, the child, and the context

What conclusions can be made about socializer's practices and their effects on the development of prosocial behavior? The empirical findings indicate that the more parents make use of power assertion, the less children will internalize enduring standards pertaining to prosocial behavior. However, the deleterious effects of power assertion arise mainly from the extreme use of it. A lesser degree of power assertion is part of most parental disciplinary techniques and does not seem to have a negative impact.

Parental warmth or nurturance generally is positively related to the development of prosocial behavior, but not in isolation from other parental practices. Socializers who combine warmth with other positive child rearing practices such as the use of inductions, assignment of responsibility, and the modeling of prosocial behavior seem to be particularly effective at fostering prosocial development. Thus, it is important to consider the influence of constellations of socialization practices rather than only a single practice.

In addition, it is important to recognize that parental practices can vary as a function of the socializer's psychological state. For example, transient negative affective states may make social relationships difficult for the parent, resulting in reduced levels of parental warmth and modelling of prosocial behavior. Moreover, chronic states of anger or depression may affect parents' functioning as disciplinarians. Manic-depressive mothers sometimes manifest an intense need for nurturance and turn to their child to provide it. Children from these families tend to show intense preoccupation with adults' suffering and frequently exhibit disturbances in their social and emotional responses to others. It appears that intense involvements with the emotional needs of other family members make the development of social relationships outside the family difficult for the child.

Of course, transient and enduring characteristics of the child also affect parents' use of various child-rearing techniques. For example, adults are more likely to use reasoning techniques to induce prosocial behavior for children who attend to them and answer promptly whereas they more frequently use material rewards to elicit helping behavior from children who do not attend to them. Cause and effect in regard to socialization practices and children's social behavior, including prosocial behavior, clearly are not unidirectional.

Finally, it is important to note that parents are not the only people who socialize children with regard to prosocial proclivities. Teachers, other extrafamilial adults, peers, and even the media act as socializers. Many of the same parental practices that foster prosocial development seem to be useful when used by other adults. In addition, prosocial modelling by peers and the media (especially television) have been found to influence children's prosocial responding. However, there is a need for additional research on the role of socializers other than parents in prosocial development.

Summary

Social behaviors are determined by multiple factors, and no single determinant of social behavior (including prosocial behavior) has an overriding influence in all situations. Thus, our statements regarding the correlates and determinants of prosocial behavior apply in general, with the qualifier of 'all other things being equal.' But because other things are almost never equal, we cannot predict with certainty the factors that will influence prosocial responding in any particular child. Nonetheless, based on an ever-growing body of literature, we are able to make predictions that are substantially better than chance regarding children as a group.

What kinds of children are more prosocial than others? In part, the answer is those who are advanced in sociocognitive skills (e.g., perspective taking and moral reasoning) and who sympathize with others' needs and distresses. In addition, children who are exposed to socializers who use practices that foster perspective taking, concern for others, and responsibility, and who provide models of prosocial behavior, are likely to be relatively altruistic in their orientation and behavior. Although not discussed in this chapter, cultural values, norms, and practices also have a socializing influence; prosocial and cooperative behaviors are expected, valued, and performed more in some cultures than others (see Chapters 4–6).

Although the issues of personality and situational influences on prosocial behavior are not discussed in this chapter, it also is clear that children assist more in some situations than others and that personality variables (some of which may be partially biological in their bases) affect the frequency and mode of children's (and adults') prosocial behaviors. For example, sociable children are more likely than less sociable children spontaneously to approach other people and offer to help or share. What is not very clear is the ways in which situational cues and contingencies, personality factors, sociocognitive capabilities, emotional tendencies, and socialization history interact in their influence on the development and maintenance of prosocial behavior, particularly altruism. Specifying these linkages is a challenge for the future.

Acknowledgement

This research was supported by grants to Nancy Eisenberg from the National Science Foundation (BNS–8509223 and BNS–8807784) and the National Institute of Child Health and Development (K04 HD00717).

Further reading

Baumrind, D. (1971). Current patterns of parental authority. *Developmental Psychology Monographs*, **1**, 1–103.
Eisenberg, N. (1986). *Altruistic Emotion, Cognition and Behavior*. Hillsdale, New Jersey: Erlbaum and Associates.
Eisenberg, N. & Mussen, P. (1989). *The Roots of Prosocial Behavior*. Cambridge: Cambridge University Press.
Eisenberg, N. & Strayer, J. (1987). Critical issues in the study of empathy. In N. Eisenberg & J. Strayer (eds.), *Empathy and its Development* (pp. 3–13). Cambridge: Cambridge University Press.
Hoffman, M. L. (1984). Interaction of affect and cognition in empathy. In C. E. Izard, J. Kagan, & R. B. Zajonc (eds.), *Emotions, Cognitions, and Behavior* (pp. 103–31). New York: Cambridge University Press.

Moore, B. S. & Eisenberg, N. (1984). The development of altruism. In G. Whitehurst (ed.), *Annals of Child Development* (pp. 107–74). Greenwich: Jai Press.

Oliner, S. P. & Oliner, P. M. (1988). *The Altruistic Personality: Rescuers of Jews in Nazi Europe*. New York: Free Press.

Radke-Yarrow, M., Zahn-Waxler, C. & Chapman, M. (1983). Prosocial dispositions and behavior. In P. Mussen (ed.), *Manual of Child Psychology. Vol. 4, Socialization, Personality, and Social Development*, (E. M. Hetherington, ed.) (pp. 469–545). New York: John Wiley.

Staub, E. (1979). *Positive Social Behavior and Morality. Vol 2, Socialization and Development*. New York: Academic Press.

Zahn-Waxler, C., Cummings, E. M. & Iannotti, R. (eds.) (1986). *Altruism and Aggression: Biological and Social Origins*. Cambridge: Cambridge University Press.

4

Cross-cultural differences in assertiveness/competition vs. group loyalty/cooperation

HARRY C. TRIANDIS

Most humans experience common fate with one or more groups, to be called 'ingroups' in this essay. Ingroups include the family, the tribe, or even the nation. When these groups require of their members behaviors that are incompatible with the personal goals of the members, the conflict is resolved sometimes in favor of the group (in collectivist cultures) and sometimes in favor of the individual (in individualist cultures).

An essential attribute of culture is that it includes shared meanings that are not debated. The correctness of an action is so obvious that people do not need to discuss it. Thus in 'purely collectivist' cultures people are socialized to obey and conform to ingroup norms and goals, and there is little debate about the desirability of such action. Similarly in 'purely individualistic' cultures individual goals are given priority and normally no objections to self-serving actions are voiced. People are assumed to have the right to 'do their own thing' and if that does not harm others they are allowed to do so, even when this means not doing much to promote the goals of ingroups.

Most cultures are not 'pure' forms of either of these syndromes. Rather, they are a unique mixture of each. However, since the time of the ancient Greeks various forms of individualism have been noted in Europe, and became more pronounced in England after the 12th century. The Renaissance provided a substantial move toward individualism. With the industrial revolution, and the complexity of post modern societies, additional shifts toward individualism can be noted, with an extreme form, such as narcissism, found in many post modern societies.

78

Attributes of collectivists

Collectivists pay more attention to the views, needs and goals of their ingroup, than to their own personal views, needs and goals. Their behavior is more responsive to ingroup norms than to personal pleasure. They pay much attention to the beliefs, attitudes and values they share with some ingroup rather than to their personal beliefs, attitudes and values. They trust and are ready to cooperate with ingroup members in more situations than are individualists; they distrust, and are ready to behave competitively with outgroup members in more situations than are individualists.

The socialization of collectivists emphasizes the importance of loyalty to the ingroup, trust and cooperation with members of that group. *Hierarchy* and *harmony* within the ingroup are highly valued. Ingroup authorities are supposed to know what is good for all ingroup members and to act for the benefit of all. But the influence of the ingroup on behaviour varies with the type of collectivism. In extreme forms of collectivism, such as in theocracies or Mao's China, the ingroup is allowed to control aesthetic judgements, economic, political, and religious decisions, and most social life. 'Truth' is valid only if it conforms to ingroup ideology. Decisions such as where to live, what job to do, what education to get, how many children to have, what medical treatment to obtain, are legitimately seen within the province of the ingroup. Duty is highly valued; achievement is conceived as ingroup rather than personal achievement; competition is conceived in intergroup rather than interpersonal terms.

In collectivist cultures ingroups are conceived as homogeneous, and ingroup authorities assume that individuals have very similar needs, so that the distribution of resources should be in equal shares.

In such societies the self is defined in ingroup terms (e.g., I am a son; I am Roman Catholic; I am Chinese; I am a member of the Communist Party). Social behavior is best predicted by knowing the *norms* of the ingroup, and the *role* definitions that the ingroup uses in socializing its members. Social behavior takes the *context* of the interaction into account much more than the *content* of the interaction. How something is said is more important than what is said. Saving face is very important. Most behavior can be explained as an attempt to promote the goals of the ingroup, to maximize the integrity, welfare and well-being of the ingroup. People are extremely concerned about the fate of members of their ingroup.

Since people are so strongly committed to their ingroups, and the ingroup consumes so much of the energies of its members in collectivist societies, the behavior of these members toward outgroups is distant, suspicious, distrustful, and even hostile. Outgroups consist of people with whom one does not perceive common fate, or at least the potential of common fate developing in the future. This set often includes very many people. For example, if the family is the ingroup, everyone who is not family may be the outgroup. Relationships with the outgroup are often characterized by the perception of 'limited good'. Thus, if something desirable happens to an outgroup member, this is perceived as a threat to the ingroup. In such societies, since 'good' is assumed to be limited, if the outgroup gets the good (e.g., wins a lottery) this is seen as an undesirable event for the ingroup.

If conflict develops with the outgroup, cruel treatment is very likely. Mass executions (e.g., note the history of Spain during the civil war or the Nazi treatment of Jews), the sacking of cities with mass rape and murder (e.g., the 1937 rape of Nanking by the Japanese), the events of 4 June 1989 in Tiananmen Square, etc. can be seen as examples of collectivist ingroup–outgroup conflict.

Attributes of individualists

Definitions of individualism emphasize that the individual is an end in him or her self, and ought to realize the 'self' and cultivate his or her own judgement, notwithstanding the weight of pervasive social pressure in the direction of conformity.

Personal rights, the pursuit of pleasure, achievement defined in personal terms, competition defined in interpersonal rather than intergroup terms, the definition of the self in terms of traits (e.g. I am distinguished, I am hardworking), moderate attachment to many ingroups, behavior under the control of attitudes, and the perceived value of the consequences of that behavior characterize this cultural syndrome. Emphasis on both freedom and equality means that ingroup authorities do not have the right to impose their views. Resources are to be distributed according to each individual's contributions, rather than equally, or by taking into account the needs of individuals.

While ingroups and outgroups do exist, the difference between them is not seen as nearly as great as it is seen by collectivists. One may have to fight outgroups, but one does this 'rationally', i.e., aiming to produce minimum damage to innocent bystanders, or those not responsible for the conflict.

Individualists are more likely than collectivists to trust people they do not know, strangers, and outsiders. They are more likely than collectivists to show prosocial behaviour toward unknown others. They are more likely to cooperate with unknown others with whom they share attitudes and values, or ideological principles. They are more likely than collectivists to become committed to abstract entities, particularly if the entities are personal, such as 'my ideas', 'my career', 'my future'.

Some determinants of collectivism and individualism

When resources are limited, the presence of a food ingroup acts like insurance that most ingroup members will have something to eat. Thus, in many collectivist cultures one is not supposed to eat the produce of one's own garden, or what one has been able to find in food gathering, or caught during a hunt. Rather, one has to distribute this food among relatives and friends. In return, since relatives do the same thing, one obtains food from others. In such schemes the probability that all will eat is high. If there is no food all starve together, so common fate is very clear. During wars, when a city is under seige, or under enemy attack, collectivism increases further. In any ingroup–outgroup confrontation the effect is to increase collectivism.

As societies become more affluent, and more complex, individuals are able to detach themselves from their groups. The ingroup is no longer needed as an 'insurance instrument'. Since there are many ingroups one has to decide as an individual which ones to join and which ones to ignore, and how much energy to devote to the promotion of the goals of each ingroup. One usually promotes the goals of those ingroups that have goals similar to one's own.

Ingroups are no longer seen as homogeneous. On the contrary, since people emphasize that they are 'distinguished' and they know more members of their ingroups than their outgroups, they see their ingroups as more heterogeneous than their outgroups.

The upper classes in most societies are more individualistic than the other social classes. Thus, it may well be the case that even in the most collectivist cultures the top leadership is highly individualistic. In addition, the emergence of cities, with the numerous potential ingroups that they offer, results in more individualism. Small families can bring up each child somewhat differently, thus increasing individualism. Social and geographic mobility means that people become detached from their ingroups. On the frontier individuals decide for themselves what is the best course of action. Children exposed to diverse ideologies become individualistic, since they have to decide for themselves which ideology is valid.

Certain occupations lead to collectivism (e.g., where group action in hierarchically organized groups is optimal, e.g. building a canal) and other occupations lead to individualism (e.g., writing a book). Usually, the upper classes and professionals are socialized to be independent and self-reliant and hence individualists, while the lower social classes are socialized to obey, and become collectivists. The older segments of the population are most likely to become acculturated to the dominant cultural pattern of their society, so that socialization and aging in a collectivist culture results in higher levels of collectivism, and socialization and aging in an individualistic culture results in higher levels of individualism. Thus, within culture, there is much variance in the qualities reflected in collectivism and individualism.

I find it convenient to use a different terminology to describe within-culture and between-cultures variations in these tendencies. Corresponding to collectivism I use the term *allocentrism*; corresponding to individualism I use the term *idiocentrism*. This permits easy discussion of the behaviour of allocentrics and idiocentrics in individualistic cultures (e.g., allocentrics report receiving more social support than idiocentrics; idiocentrics report feeling lonely) and the behavior of allocentrics and idiocentrics in collectivist cultures (e.g., idiocentrics try to escape collectivist cultures, such as has happened in East Germany before the recent political changes there; note the profile of those who left East Germany: they were young, relatively affluent, from cities, many with occupations requiring solitary work).

The contrast between collectivism and individualism is not mysterious. We see these patterns within culture, and know something about their determinants. First, differences in socialization are important. The collectivist emphasis on obedience and being a 'good child' contrasts with the individualist emphasis on self-reliance, exploration, creativity, and 'finding yourself'. Second, certain relationships are more 'communal' in all cultures. For example, parent–child, or close-friend relationships are communal, while client–customer or government official–citizen are exchange relationships. Collectivists spend more time in communal relationships and are good at them, so they try to convert all relationships into communal, personalized relationships. Individualists spend more time in exchange relationships, and try to convert all relationships into them, e.g., giving money to a child in situations where giving time or love are more appropriate. Third, we know from work with open groups (whose membership constantly changes) and closed groups (whose membership is stable) that behavior in open groups is similar to behaviors that occur frequently in individualistic cultures, and behavior in closed groups is similar to behaviors that occur frequently in collectivist societies.

Behavioral consequences of the cultural syndromes

Collectivists behave so that they will receive maximum approval from members of their ingroups. While all humans crave for ingroup approval, individualists are perfectly capable of acting independently of their ingroups, particularly if a 'principle' guides their behavior. They are much more concerned with 'what works', 'what is most profitable', and 'what feels good' than with 'what others say'. Thus the behavior of individualists is under the control of the 'profit and loss' occurring in interpersonal relationships. The behavior of collectivists appears much less 'calculating' since loyalty to the ingroup can demand self-sacrifices that can not always be traced to logic.

Collectivists are generally more intimate in their relationships within the ingroup, and spend much time building relationships within their ingroup. Individualists are especially good in getting in and out of ingroups, and so thrive on relatively superficial relationships. The cocktail party is an individualist invention! Note also that in cocktail parties people talk to one or two others, rather than in groups of six to eight, which is more common in collectivist gatherings. When spending their leisure time, individualists are more likely to engage in solitary pursuits, do things in pairs or in very small groups, and even if a large group is involved they break it up into small groups. Collectivists spend such times in groups of six to eight or larger, if that is at all possible (it is not always). The tête-à-tête dinner party is very individualistic; even if the table has 50 guests the conversation will be mostly between two persons. The collectivists are more likely to organize banquets for a dozen, and to interact from person to group.

Collectivists are more committed to their ingroups than individualists; commitment by its very nature guides behavior in the direction of altruism and self-sacrifice. But collectivist socialization also emphasizes reciprocity more than does individualist socialization. Thus, while collectivist parents sacrifice themselves for their children rather extremely, they also depend on their children during their old age. As a result children are perceived in 'instrumental terms' (guaranteeing a good old age) in the case of collectivists, while they are seen much more in 'expressive terms' (fun to have) by individualists. Marriage is also more instrumental for the collectivists and expressive for the individualists, since the former emphasize 'chastity' and the latter 'an exciting personality' as important traits in mate selection.

Collectivists usually interact with people they have known for a long time. As a result they know precisely how much they can or cannot trust them. They also share much with the people with whom they interact, and hence they are able to communicate 'associatively', or use the situational context as part of the communication. For example, they may say 'This is like the relationship of Joe and Bill' and that communicates a great deal, but only to people who know a lot about Joe and Bill. By contrast, the individualists' communication depends very little on context. Individualists must define their terms (e.g., what are the critical attributes of Joe, and what are the essential attributes of Bill, and thus what is it about this relationship that can be used as a prototype for understanding other relationships). The communication is much more abstract in the case of individualists.

Morality in collectivist cultures is framed in ingroup terms – if it is good for the group it is good. Morality in individualist cultures is more likely to be framed according to certain principles and one's own conscience.

When a collectivist society is also very homogeneous, it tends to have very clear norms for action, and deviation from these norms is severely punished. The result is that people are very careful to conform precisely to proper action. Gossip and other mechanisms for shaming those who deviate have the effect of keeping people in line.

Collectivists emphasize proper behavior, and care very little about how one feels about it. As long as one acts as one is expected, one is a good member of the ingroup. If one dislikes what one is doing this is of little importance. The individualist's idea is for the action to be isomorphic with the person's attitudes and values; if a person acts one way and thinks differently that person is a hypocrite. The collectivist does not care about action–feeling discrepancies. What matters is the action. This is so, in part, because the collectivist needs more predictability and consistency in social behavior than the individualist. Such predictability allows one to trust others. If a person is unpredictable, that person cannot be trusted. Also, since collectivists are more likely to interact with members of the same ingroup over time, they have to think of interpersonal relationships 'in the long run'. Thus, harmony, saving face, and being predictable are especially important. One must not acquire a bad reputation by making waves, speaking up one's mind, or acting unpredictably. The individualist does not have such concerns to the same degree, since if relationships within the ingroup become spoiled, one can always find another ingroup. If people are not trustworthy, one drops them from one's circle.

Individualism in its extreme forms is difficult to distinguish from selfishness, and often involves much assertiveness, speaking one's mind, even if the feelings of others are hurt, and interpersonal competition. As cultures become affluent and shift from collectivism to individualism (e.g., South Korea, Taiwan, Japan), many observers (e.g., the press in the People's Republic of China) report 'selfishness' and other undesirable symptoms. In fact, in the People's Republic of China even symptoms of schizophrenia are sometimes misdiagnosed as 'individualism'!

What are the bases for understanding and cooperation between collectivist and individualist societies?

Understanding the attributes of the two kinds of societies can be achieved through 'training' and can result in members of each society getting along better with members of the other type of society. For example, if one understands that in collectivist cultures the self is defined in ingroup terms and in individualist cultures it is defined in personal attribute terms, one can change one's behavior to take that insight into account. It means that individualists will take the time to find out more about the ingroups of collectivists. They need to learn more about the attitudes, norms, roles and values of these groups, and the position of ingroup authorities on the questions under discussion. On the other hand, collectivists have to *un*learn paying so much attention to the ingroups of individualists and learn to pay more attention to individual beliefs, attitudes, and predispositions.

Individualists are proud of being 'distinguished'; collectivists are proud of their groups, but do not want to stand out from them. This can lead to misunderstandings, when the collectivist behaves too modestly, and the individualist appears too boastful. Training the collectivist to stand out and the individualist to be modest may be helpful.

Since people in the two types of cultures are differentially attached to their ingroups, that is relevant information for behavior. Individualists will be emotionally detached from their ingroups, and the collectivist must learn not to see this as 'neurotic behaviour'. On the other hand, collectivists will be more attached to their ingroups than is normal for individualists so individualists must learn not to see them as 'excessively dependent'. If saving face is important in one type of society, and speaking up one's mind is in the other, that must be taken into account. If hierarchy really matters in one and is de-emphasized in the other, one must note this. If cooperation is more common in collectivist ingroups and competition is more common in all kinds of interpersonal relationships in individualist cultures, training the collectivists to tolerate more competition, and making sure that competition is not perceived as hostility, can be beneficial.

If the unit of analysis is the individual in the one case and the group in the other case, interaction has different meanings. If collectivists behave very differently toward their ingroup than toward their outgroup members, an individualist must learn how to become a member of the ingroup. Intimate behaviors, and a communal orientation are likely to help an individualist to become a member of the ingroup. But people need to be trained in intimate behavior. Individualists tend to be formal in more situations of intermediate social distance than collectivists (e.g., with coworkers), so to avoid misunderstandings they have to learn to become more intimate (e.g., reveal more about themselves).

Status is defined as a function of accomplishments by individualists, and as a function of membership and formal position in groups by collectivists. Unless this is understood, people can come to different conclusions about the status of those in interaction. Expectations concerning what one has to do to receive respect will also need to be learned.

The way one constructs persuasive arguments is different in the two kinds of cultures. In individualistic cultures one stresses logic, principle, and attempts to give information that will change opinions. In collectivist cultures one stresses the importance of cooperation, harmony, and the avoidance of confrontation.

Behaviour that takes the needs of others into account and equal distribution of resources, self-sacrifice and pleasant, smooth relationships will go a long way toward making an individualist an ingroup member. Conversely, if the emphasis is on pleasure, personal achievement, and personal competition the collectivist has to learn how to deal with individualists along such lines. Learning to use more exchange relationships, and to avoid communal relationships in situations that call for exchange relationships (e.g., citizen to authorities), can be helpful.

Collectivists must learn to expect more superficial, short-term but good-natured interactions. Individualists must learn to be more intimate in their relationships, and not to be upset when the other is more intimate than they see as appropriate for the situation.

Collectivists must learn to do business without getting to know a lot about the other, and individualists must learn to take time to get to know the other before doing business. Collectivists must learn that relationships will only last a limited time, and to depend on favorable reward–cost balances; individualists must learn that relationships are expected to be long-lasting and people are expected to be loyal over long periods of time in collectivist cultures. Problems with hierarchy are also important. Collectivists have to learn to de-emphasize it and individualists to emphasize it.

Conclusions

In all cultures there are two tendencies present in some mixture. One tendency is toward organizing information around ingroups, in which case social behavior is highly dependent on what happens in these ingroups. The other tendency is toward organizing information around individuals, in which case social behavior is highly dependent on individual attitudes, and cost/benefit computations. These two patterns shift in importance, depending on the situation, but are also influenced by social class, occupation, gender, rural vs. urban residence, the heterogeneity of the society and other variables. It is useful to learn to recognize these two cultural syndromes. Those taking into account this information can improve their relationships with members of societies using the cultural syndrome that is less common in their own culture. They can also develop different expectations concerning when people will cooperate or compete, when they will trust, and when they will be committed to their groups.

Further reading

Chinese Cultural Connection (1987). Chinese values and the search for a culture-free dimension of culture. *Journal of Cross-Cultural Psychology,* **18**, 143–64.

Doumanis, M. (1983). *Mothering in Greece: From Collectivism to Individualism.* New York: Academic Press.

Hofstede, G. (1980). *Culture's Consequences.* Beverly Hills: Sage.

Hsu, F. L. K. (1983). *Rugged Individualism Reconsidered.* Knoxville: University of Tennessee Press.

Naroll, R. (1983). *The Moral Order.* Beverly Hills: Sage.

Triandis, H. C. (1988) Collectivism v. individualism: A reconceptualization of a basic concept of cross-cultural psychology. In G. K. Verma and C. Bagley (eds.). *Cross-Cultural Studies of Personality, Attitudes and Cognition.* London: Macmillan, 60–65.

Triandis, H. C. (1990) Cross-Cultural Studies of Individualism and Collectivism. *Nebraska Symposium on Motivation, 1989.* Lincoln, Nebr: University of Nebraska Press.

Triandis, H. C., Bontempo, R., Villareal, M., Asai, M. and Lucca, N. (1988) Individualism–collectivism: cross-cultural perspectives on self-ingroup relationships. *Journal of Personality and Social Psychology,* **54**, 323–38.

Triandis, H. C., Leung, K., Villareal, M. and Clack, F. L. (1985). Allocentric vs. idiocentric tendencies: Convergent and discriminant validation. *Journal of Research in Personality,* **19**, 395–415.

Ziller, R. C. (1965). Toward a theory of open and closed groups. *Psychological Bulletin,* **64**, 164–82.

5

The development of prosocial behavior in large-scale collective societies: China & Japan

HAROLD W. STEVENSON

Anthropologists have described the prosocial behavior of adults in various cultures, but there has been little systematic exploration of the ways different cultures promote the development of this kind of behavior in children. For example, in a recent volume on the roots of prosocial behavior by Eisenberg and Mussen (1989), only a slim chapter of 13 pages is devoted to the discussion of culture and prosocial behavior. The chapter is brief, not because the authors lack interest in the topic, but because so little cross-cultural research is available. In reporting the information about children that does exist, the authors relied on ethnographic accounts, such as those of Mead (1935) and the Whitings (1975), Bronfen-brenner's (1970) well-known discussion of the social behavior of Soviet children, and a few experimental studies of cooperation among children of different cultural groups. The remaining chapters in the book describe a great deal of research on prosocial behavior, but the content is derived primarily from studies conducted by psychologists in North America.

Our knowledge about the origins of prosocial behavior in different cultures is obviously very limited, and if we are to escape from the parochial views resulting from the study of American families we must extend our interest to other cultures of the world.

The purpose of this chapter is to describe the ways in which three societies – those of China, Japan, and Taiwan – have fostered the develop-ment of children's prosocial behavior. These societies are very different from that of the United States, and are of special interest because they place extraordinary emphasis on children's socialization. They make explicit and well articulated efforts, beginning with the child's earliest years, to inculcate prosocial behaviors.

89

The chapter begins with a discussion of the philosophical background that has guided Asian efforts in regard to prosocial behavior. To illustrate one of the ways in which this philosophy has been put into practice, we will describe what was perhaps the most concerted effort ever undertaken by a nation to instill prosocial behavior in its children – that of China during the Cultural Revolution. We then turn to a very different setting – Japanese classrooms – and to the techniques used by teachers to foster the development of prosocial behavior. The data used to illustrate these examples come from analyses of textbooks and from classroom observations. A final section describes how the behaviors developed in children are reflected in the success of Asian cultures in education and commerce.

Asian concepts of prosocial behavior

Asian considerations of what constitutes prosocial behavior are very similar to those of the West: altruism, kindness, considerateness, sympathy, aiding or benefitting another person or group, and promoting general welfare in society by reducing inequalities. Such forms of behavior are considered to be prosocial because there is no expectation of external rewards on the part of the person performing the acts. However, Asian considerations differ from those in the West because of their conception of the role of the individual in relation to family and society. We in the West place great emphasis on the importance of the individual and on the development of an independent, self-directed child. Raising an independent child is not a major concern for Chinese and Japanese parents; much more importance is given to establishing interdependent relations between the child and other members of the family and society. In Chinese societies emphasis is placed primarily on kinship; in Japan it is on the social group. All three forces – those arising from the individual, from the family, and from society – play important roles in all social behavior, but the American, Chinese, and Japanese societies appear to place different degrees of emphasis on each.

In most Western societies an important agent for children's moral education, including training for prosocial behavior, is organized religion. This is not true in Chinese or Japanese societies. Formal religion seldom impinges on the lives of children, not only in China, where the Communist government has sought to de-emphasize religion, but also in Taiwan, where government leaders have been more sympathetic to religion. In Japan, Buddhism and Shintoism have had some influence, but religion tends to have a ceremonial function rather than being a guide for children's daily lives. In our study of elementary school children in Taiwan and Japan, for example, only 3% of the Chinese children and 4% of the Japanese children in our samples attended any formal religious classes. In strong contrast, 69% of the American elementary school children attended religious classes each week. Moral education of Chinese and Japanese children thus devolves upon parents and teachers, rather than being the province of religious leaders.

Moral education of Chinese children is also based upon a different conception of childhood from that held in the West. The Chinese distinguish a stage in which there is great freedom on the part of the child and great tolerance and indulgence on the part of adults. This stage is distinguished from a later stage in the Chinese terms *budongshi* and *dongshi*. The first term describes a period when the child does not 'understand', and the second, a period of understanding. These periods are also referred to as the stage of innocence and the stage of reason. The stage of innocence, usually considered to encompass the child's first six years, is a time when children are believed to lack the cognitive competence necessary to understand moral concepts and the functions of prosocial behavior. It is believed that there is little utility in attempting to get the child to learn certain types of behavior or social concepts before the age of reason begins. During the period of innocence, direct imitation of prosocial models and the development of affection for the model are considered to be the major ways for acquiring prosocial behavior.

The age of reason begins quite abruptly when the child enters school. The previously nurturant, permissive parents quickly become transformed into authorities who expect obedience, respect, and adherence to their rules and goals. Once the age of reason begins, the child is faced with well-defined expectations in terms of moral and intellectual accomplishments. Now children are expected to go beyond imitation and to be able to comprehend and internalize the precepts that guided the model's behavior.

Similar stages are posited in Japan, but the transition from a period of innocence to one of greater reason appears to occur less abruptly in Japanese families. Whereas Chinese parents forthrightly express new demands on their school-age children from those expressed earlier, the change in parental attitudes appears to be more subtle and prolonged on the part of Japanese parents.

Asian goals for children and society

As recently noted by Ho (1981), Chinese parents focus primarily upon their child's moral development and academic achievement. There is little concern for personality development, or for other aspects of parent–child relations. Parents are considered to be successful if their child has a high moral character as reflected in proper conduct, good manners, humility, and respectfulness – and receives good grades at school.

Supreme among the typical Chinese and Japanese citizen's goals for society is the preservation of order and harmony. The interdependence of individuals means that there is constant striving for smooth interpersonal relations. Thus, harmony and balance among social forces are sought-after goals. This is well illustrated in the Japanese term, *nemawashi*. Visitors to Japanese business firms are often surprised at the readiness of the group to make decisions, even important ones involving large amounts of money and long-term obligations, with little discussion and little disagreement among group members. Public confrontation among individuals within the firm does not appear; no one loses face by supporting a losing position. What the foreign visitor may not be aware of is that all discussion and necessary compromises have occurred prior to the public meeting. The meeting is held to announce the decision, not to make it. That is, like the preparations that are necessary for transplanting a tree or bush – to which the term *nemawashi* refers – the group has prepared itself by considering all important options and constraints before the meeting takes place. Disagreements may still exist, but group harmony, at least for the present, has been preserved.

Another example of the importance of the preservation of harmony is found in notes published by a Japanese business firm for foreigners about how they should comport themselves in Japan in order to be socially successful. The notes, titled 'Advice to Foreigners in Japan,' begin with the suggestion that 'In order to get along well in Japan, you will have to try continually to create harmonious relationships, avoid disputes, and make an effort to understand the way people express their emotions.'

The Asian orientation, then, is consistently directed toward the group. The individual is typically defined through participation in family, school, community, company, and the nation. In fact, polite address in Japanese conversation is typically stated in terms of the recipient's role, such as mother or teacher, and the use of the pronoun 'you' is considered to be brusque and insensitive. Primary obligations are to the group, and actions that preserve group harmony, whether in the family or the nation, have moral virtue. Competition is between groups, not among individuals within groups. This is true in Japan, and it is also true in socialist China and in capitalist Taiwan. The message in all three cultures is that societies are composed of interdependent individuals whose current and future health and security depend upon concern for each other.

This group orientation is evident in interviews with mothers of nursery school children, whose major purpose in enrolling their children, they explain, is so that the child will be initiated to *shudan seikatsu* (group living), and it appears in the motto often found in the playrooms of Chinese preschools: 'Cooperation (friendship) first, competition second.' In both places, children are expected to learn that their personal desires must be secondary to the activities and harmony of the group.

Identification with the group results in a strong common bond among its members. This means that group- or ethnocentrism is almost inevitable. Many instances of this can be found. It appears in benign contexts and in more serious ones. For example, children of Japanese employees of automobile firms who are temporarily residing in the United States discriminate among each other at times on the basis of their father's companies. Thus, the Mazda children have a group identity and may find it inappropriate at times to associate at school with the Honda children. (In Japan, they would seldom be in the same schools because they would reside in company housing located in different areas.) It can also appear in more serious forms, as we will later see in the disregard for those who are not part of the national majority, such as has been the case with the *burakumin*, the 'outcasts' of Japan. It is evident, too, in a guide to Chinese school teachers that suggests, 'China should be perceived by children as superior to any other country.'

The Confucian roots

Central to Eastern views about prosocial behavior is the moral philosophy of Confucianism. The fundamental Confucian assumption is that the human being is defined within a social context, and the identity of the individual is determined to a great degree by reference to his or her status within the group. The family is considered to be the basic group, but the individual always belongs to other social groups as well. All of these groups guide behavior and absorb or reflect the individual's accomplishments or failures, or as Bond (1986) has described them, the individual's glory or shame. In Confucian thought, relations within the group are structured hierarchically and each person is expected to understand the requirements of his or her relationships to this hierarchy.

Within the Confucian tradition it is assumed that human beings learn primarily through the imitation of models. Munro (1977) has described how this learning occurs:

> The learning can occur unintentionally, through the unconscious imitation of those around one, or it can occur intentionally, through the purposive attempt to duplicate the attitude and conduct of a virtuous model. Although the behavior of a negative model may be absorbed, most people are positively attracted to and consciously seek to imitate virtuous models.   (pp. 135–36)

It is important for individuals not only to follow the behavior of other models, but to become models themselves. The respect that is obtained from the imitation of one's behavior is regarded as far more desirable than money or other material rewards. In contrast to efforts made to inculcate the development of moral behavior by means of abstract principles, the model in Confucian thought is always represented through concrete actions that can be readily imitated.

Confucian ideas about the acquisition of behavior are often similar to those of Western social learning theorists. Both approaches stress the importance of learning through observation and imitation, and emphasize the critical role played by the characteristics of the model. Western researchers have proposed, and have demonstrated in experimental settings, that imitation occurs more readily if the model controls resources, is nurturant, and is competent (Radke-Yarrow, Zahn-Waxler & Chapman, 1983), characteristics that are in close conformity with those promoted in Confucian thought.

Analyses of textbooks

Mao Zedong's vision was that China should become 'a totally selfless society in which all individual, group, and factional interests are absent.' Singular efforts were made in pursuit of this goal to develop identification of the individual with the masses and abdication of self-interest. Although these were presented in the context of Communism, they were goals that, in line with older Confucian doctrines, relied heavily upon the use of models.

One way in which Mao's goals were taught to children was through stories written for their elementary school readers. Many of the stories contained prosocial messages transmitted through the behavior of models.

A child finds manure on the road, gathers it up and puts it in the manure pit; a child discovers a heavy parcel that has dropped from the load of an old man and carries it to him; children forego playing in order to help their teacher with her tasks; a boy repairs a torn page of another child's textbook; children learn to stop killing insects for fun and to kill flies to promote hygiene; children acknowledge that two, three, or even four children can play joyously with a jump rope that previously was one child's possession. There are stories of children repairing trees, washing the handkerchief of another child, taking food to a sick neighbor, giving up a seat on the bus to an old worker, and of many more ways in which the models were able to serve others.

Formal analyses of the content of the children's elementary school readers from the People's Republic of China provide some quantitative means of evaluating this emphasis on prosocial behavior. Ridley, Godwin, and Doolin (1971) categorized the content of the stories according to the behavioral themes that were depicted. These themes included such topics as social and personal responsibility, achievement, altruistic behavior, and so forth. The frequency with which each of these themes appeared in the children's textbooks was then tallied.

There were 135 instances in which themes or subthemes of the stories were concerned with social and personal responsibility (e.g., devotion to duty, performance of social obligations, honesty), 97 instances of altruism, 75 of collective behavior (e.g., cooperation in a common endeavor), and 43 dealing with prosocial aggression. None of the remaining themes, other than 118 instances of themes related to achievement, had frequencies anywhere near those describing these social and prosocial forms of behavior.

The overall pattern of behavioural themes in these textbooks was summarized by the authors as follows:

> [The elementary school child] is taught that he has obligations to society at large, and he should strive to achieve not for himself, but for the common good. Further, at this young age, the individual is taught to value labor not personal achievement, and labor is seen as primarily physical labor as a worker, peasant, or soldier. The model citizen, seen through the behavioral themes, is a person who works well with others, accepts as his goals in life the modest ones of being a worker, soldier, or farmer, and approaches the world and its problems in a spirit of objective inquiry.

(Ridley, Godwin & Doolin, 1971, p. 155)

Lest the content of the textbooks be attributed solely to the influence of Communism, we look next at a similar analysis of textbooks used in the elementary schools of Taiwan (Wilson, 1970). In Taiwan, as in China, textbooks represent the official position of the government and, in fact, are published only by the government.

The stories and examples described by Wilson are much like those found in the Chinese textbooks. A typical prosocial story described by Wilson is the following: A child asks an old worker why he is planting walnut trees when he will not be able to eat the walnuts. The old man replies, 'Child, I eat walnuts now, and aren't these from trees planted by people who went before? If we eat the walnuts of those who went before, we ought to plant some for those who come later to enjoy. If people thought only of themselves, we would not be able to eat walnuts now.'

There are examples in the textbooks from both China and Taiwan of how control by the group is more powerful and potentially more severe than that provided by any individual. In both sets of textbooks there also are examples of how criticism and self-criticism may be effective in molding behavior. Through criticism by the group and subsequent self-criticism, individuality is conquered and conformity to the group is produced.

Models appear in the Taiwan textbooks, and their functions are similar to those in the Chinese textbooks. Young children are told they should imitate the model, and older children are expected to be able to discuss the principles that were demonstrated in the model's behavior. A fifth-grader, for example, in describing the content of one of the stories, said: 'Dr. Sun Yat-sen said we should do things for others. Therefore, we should be like that.' This theme has been popular for many years in China. In fact, in a study in 1932, Webster found that in China the two most common answers to a question about the best thing a boy or girl could do were to help people and to study.

Yamaguchi (1989) conducted a similar analysis of textbooks used in moral education classes in Japanese elementary and secondary schools. The 201 stories that were written for these books consistently fostered collectivism in Japanese society. The ideal Japanese depicted in the stories is one who considers the welfare of the society, who makes contributions that require great effort or are against their own personal interests, and who does not place personal success before social contribution. The examples of contributions made to society are not dissimilar to those in the Chinese stories: individuals clean public places voluntarily, they use private property for the community, and they help others in need. Yamaguchi points out, however, that despite the current efforts at the internationalization of Japan, there are no stories involving tolerance for racial diversity, there are no stories depicting the importance of freedom, and there are no stories about the discrimination that exists against Koreans and the *burakumin*, the 2 million citizens of Korean ancestry who are consigned to a low social status and to jobs that others in Japan do not wish to perform.

The story of Lei Feng

No model is more famous in China than Lei Feng – a model youth whose diary and life story have been read by Chinese children since the early 1960s. Children are exhorted to learn from Lei Feng, but the degree to which this is emphasized depends upon the political climate in China. During the ten years of the Cultural Revolution, countless books, posters, pencil boxes and other objects carried the picture and the message of Lei Feng; by the 1980s, when the political climate changed, it was difficult to find anything referring to Lei Feng in Chinese bookstores or schools. In late 1989, however, the *People's Daily*, the official government newspaper, again pointed to the importance of looking to the life of Lei Feng as a guide to behavior, and his name has appeared in many official publications since then.

During his youth, Lei Feng, the son of a poor family, modelled himself after the local Party secretary, and throughout his life demonstrated a multitude of prosocial acts. As a youth, he voluntarily worked to help construct a dam that would help his county; he gave all of his savings so that the county could buy a tractor; he gave his shirt, jacket, and blanket to a worker to cover bags of cement that were getting wet. When he was older and a member of the People's Liberation Army, he continued his long series of prosocial acts. He gave his lunch to a comrade who forgot his lunch; he gave another soldier some workbooks and pencils to help him study; he washed and mended his PLA comrade's clothes; he patched burned trousers; he sent money to the sick father of a comrade in the comrade's name. Each new section of the books (e.g., Chen's (1968) *Lei Feng, Chairman Mao's Good Fighter*) describes another series of laudable deeds.

In addition to the stories, reference is often made to Lei Feng's diaries. These contain such entries as : 'I feel that a real revolutionary is never selfish. Whatever he does is always for the benefit of the people.' 'Man's life is finite, but the cause of serving the people is infinite. It is my wish to devote my life to the infinite cause of serving the people.' 'A Communist party member is a servant of the people. He should regard other people's troubles as his own and other people's happinesses as his own.' Children are encouraged to follow the thoughts of this hero, and, above all, to follow the dictum, 'Serve the People,' which was demonstrated so frequently in the life of Lei Feng.

There are many other models for Chinese children. During the years of the Cultural Revolution, the children heard about the two heroic sisters of the grasslands, who saved their commune's flock of sheep in the face of a bitter snowstorm. In 1990, they were reading about Lai Ning, the primary school boy who died putting out a potentially devastating fire, and the two young women named Lan who became national heroines for thwarting the robbery of the bank in which they worked. There are not only models for the boys and girls, but also for men and women. Each serves the function of demonstrating behavior and attitudes that merit emulation.

Prosocial aggression

Nothing has been said thus far about the use of aggression with a prosocial goal. We usually think of positive forms of prosocial behavior, but aggression can also be considered to be prosocial when it occurs in an effort to promote the goals of the group. There are many examples of prosocial aggression in the textbooks from both China and Taiwan. To foster hatred of Americans by Chinese children during the Korean war, the following scene is described in a Chinese textbook: 'The pedicab driver, holding the American soldier's head down, clamped him firmly by the neck so that the children could punch and kick him. They gave the rotten thing a savage beating.'

Chinese textbooks make frequent reference to the enemy in such sentences as, 'A debt of blood must be returned in blood' and 'When the children learned of the destruction, killing and invasion . . . they wanted to gnash their teeth in hatred for the enemy.' Similar themes appear in the Taiwan textbooks: 'The thought of opposing the Communists and Russians is everywhere. Everything should reflect our dedicated effort to love the country and help the people.'

It is evident from the analyses described earlier that reference to prosocial aggression is not uncommon, but what is unusual are the vivid descriptions of the aggressive acts performed by children. It is often a fine line that defines when aggression is considered to be directed toward prosocial ends and when it offers little more than examples of children engaging in bitter, destructive forms of behavior.

The Social outcomes

We have no idea whether reading these stories, or seeing posters, cartoons, films, and programs depicting prosocial behavior had a significant effect upon children and youth of China. We do know that many of those who were exposed to these materials became members of the Red Guards during the Cultural Revolution. The Red Guards, under the charge to destroy revisionistic elements in society, and to weaken the position of the 'rotten egg' Confucius, roamed the cities and countryside, engaging in the destruction of monuments, museums, books, and other vestiges of the 'Old Society,' and behaving aggressively toward many of their peers, elders, and sometimes even their parents.

The times of the Cultural Revolution have passed, and until very recently efforts to develop prosocial behavior have been less fervent. Nevertheless, in discussions of the goals of education, moral development still usually occupies first place among the Chinese trio of goals. What was always the second most important goal several years ago – intellectual development – tended in the 1980s to slip into first place. There seem to be indications in the early 1990s, however, that moral development will resume its firm status as the primary goal and that efforts will be made to show that prosocial behavior means primarily adherence to social goals defined by the government. The third of the goals, physical development, has always been last on the list.

Japanese classrooms

Japanese classrooms are organized to foster prosocial behavior. There is no streaming in Japanese schools, and no separation of children either within or between classrooms according to their levels of achievement. Nor are there special schools for other than those with profound sensory defects, such as blindness or deafness, or severe forms of mental retardation or emotional disturbance. Even so, the visitor to the Japanese classroom does notice that the class splits up into small groups for various activities. These groups, called *han*, contain from four to eight children. In contrast to the small groups formed in American classrooms, children are selected so that all levels of achievement and other characteristics are represented in each of the *han*.

Because the members of the *han* operate as a group and because rate of progress, access to privileges, and social acknowledgements are dependent upon the group's activities rather than upon those of any individual, prosocial forms of behavior are often required for the group's success. All children participate on an equal footing, but it is acknowledged that each child will not always contribute equally to the *han*'s progress.

It is assumed that if the children apply themselves diligently to their tasks all will eventually be successful. Thus, the slow learner, the over-talkative student, the clumsy athlete – all are likely to elicit helpful responses from other children in the *han* as the group attempts to accomplish the tasks with which they have been faced. Because the children operate as a group, they realize how important it is for those who are more capable in certain activities to help those who are not, and they generally regard their responsibility to the group to be more important than their personal pleasure and advancement.

The likelihood of cooperation and other positive social responses in Japanese classrooms is also increased through each child's service as *toban*. Each *toban* has experienced how hard it is to manage a group and how important it is for a group to work together toward a common goal.

Teachers in the early grades make every effort to bond the members of each *han* tightly together. They strive to make students identify with the school, the classroom, and with other members of their *han*. As this process proceeds, the child becomes ever more strongly controlled by members of the *han*. As Peak (1989) has pointed out, it becomes nearly impossible to rebel or escape, for this means that the child must sever contact with the companionship and warmth of social life.

Increasingly, reliance in Japanese classrooms is placed on the group for the arbitration and resolution of social conflicts. Japanese teachers tolerate a wide range of behavior, and the extent of this tolerance often astounds the Western observer. The teacher's purpose is to get the group to learn how to take on this function. Lewis (1989) describes the efforts of a teacher to get the group to stop two boys from fighting. When initial efforts by the members of the group were not successful, she left the scene, which required the members of the group to devise a means of ending the fight. After they had done this, the teacher got the children to describe how they accomplished this goal and thanked them for solving the class's problem.

Rather than expecting prosocial behavior to emerge spontaneously, Japanese teachers create different types of settings in which children can learn prosocial behavior. At times they may purposefully set up situations where antisocial behavior is likely to occur and its resolution by prosocial means is necessary. For example, the teacher may provide toys in such a fashion that the number of toys for a nursery school group may be less than the number of children. This discrepancy often leads to arguments and fights, but the teacher intervenes in the resolution of these conflicts only under the most extreme circumstances. The children, left alone to solve this problem and to return order to this situation, must learn to cooperate with each other and to take responsibility for making arrangements that are equitable for all members of the group. When this occurs, the teacher helps the children understand the situation by reviewing what happened, who did what, and the kinds of solutions that did and did not work. Catherine Lewis (personal communication) describes the process this way: '[Teachers] model and stimulate a great deal of self-reflection on a daily basis. The assumption seems to be not that children will learn from their experience, but that they will learn from thinking about their experience.'

A general hypothesis in the literature on prosocial behavior is that children who are given responsibility early are then capable of channeling their behavior into the service of others. If this is so, there should be a high incidence of prosocial behavior among Chinese and Japanese children – a level, perhaps, that is higher than that among American children, where much less effort is directed toward the development of these forms of behavior. Many efforts are made in Japanese schools to give children responsibility. One seldom sees janitors or cafeteria workers in Japanese schools. The children are usually responsible for cleaning the classrooms and halls, for serving lunch, and for assisting their teacher in other ways. Japanese parents also expect their school-age children to assume greater responsibility than do American parents for such things as their choice of playmates and television programs, spending money and getting their homework done.

A study of kindergartens

One of the few comparative studies of social interaction among Chinese, Japanese, and American children that we are aware of is one that we conducted in kindergartens of Sendai (Japan), Taipei (Taiwan), and the Twin Cities of Minneapolis and St. Paul, Minnesota. Although the primary purpose of the study was other than to investigate prosocial behavior, some of the information is related to this topic.

The children were from 24 kindergarten classrooms in each city, and were chosen to constitute a representative sample of each city's kindergartens. We interviewed the mothers of 288 children (6 boys and 6 girls in each class) in each city and we observed each of the classrooms for four hours with a time-sampling method.

Mothers were asked why they sent their children to kindergarten. There were many reasons, but the predominant one expressed by Japanese mothers was for their child's social experience; 91% of the Japanese mothers, but only 55% of the American mothers and 58% of the Taiwan mothers mentioned this goal for their kindergarten children. A second set of reasons included education, cognitive development and preparation for school. Japanese mothers were far less interested in these areas than were the Chinese and American mothers. Only 4% of the Japanese mothers, compared to 34% of the Chinese mothers and 39% of the American mothers sought these experiences for their children. When asked about the particular types of social experiences they wanted their children to have, nearly all mothers – between 85% and 89% in all three cities – mentioned the development of social skills with other children. However, few Americans and many more Chinese and Japanese mothers (13%, 26% and 29%, respectively) thought that attendance at kindergarten should improve children's adaptation to society in terms of such factors as the child's learning to follow social norms.

We found some indication that the five-year-olds from the three cultures were beginning to diverge in their tendency to display positive forms of group interaction. The incidence of prosocial behavior was high in all three locations. Prosocial behavior was observed in 91% of the observational periods in the Taiwan kindergartens, in 88% in Japan, and in 84% in the United States – high, but the differences were statistically significant. Prosocial behavior was considered to occur when a child demonstrated a positive response to another child by sharing, comforting, or helping the other child.

The incidence of other forms of positive response to each other also differed. Smiling and/or making a friendly gesture toward another child occurred in 72% of the Japanese, 73% of the Chinese, and 42% of the American observations. Similarly, there was a significant difference in how frequently children talked to each other. Chinese and Japanese children talked to each other more frequently than did the American children. This occurred in 77% of the observations in Japan, in 73% in Taiwan, and in 52% in the United States. Aggressive behavior was low – between 2% and 3% of the observations – in all three locations, but there was a large difference in the incidence of negative behavior (e.g., being off-task, out of their assigned place, or behaving aggressively). Perhaps reflecting agreement with a lax approach to the management of young Chinese children, negative behavior occurred in 53% of the observational periods in Taiwan, in 15% in Japan, and in 14% in the United States.

These observations did not explore the behavior of the children or of the teachers in a fine-grained manner, but the general findings offer evidence of differences in the incidence and forms of prosocial behavior among young children of these three cultures.

Conclusions

Children's attitudes and behavior reflect what is valued by the societies in which they live. The emphasis in Asian cultures on group harmony, with the accompanying concern for prosocial behavior, stands in sharp contrast to the individualism of the West. From the examples given in this chapter, it is evident that strong, explicit efforts are made in Chinese and Japanese societies to transmit positive attitudes about group loyalty and participation and about the critical role of prosocial behavior for the advancement of members of these groups.

The stereotype that Asians are 'naturally' group oriented or cooperative fails to take into consideration children's early experiences in being taught these forms of behavior. Children are not expected to learn prosocial behavior incidentally, as is typically the case in the West, but there are purposeful and definite efforts to train children from their early years to demonstrate such behavior. These attitudes acquired in childhood have broad ramifications for the social interactions of children and for their later participation in adult society.

The high level of academic achievement of Chinese and Japanese children stems in part from their cooperative attitudes. The children are attentive and responsive to their teachers and thereby are better able to benefit from the teachers' instruction. The children also are helpful to each other. In this atmosphere, even slow learners come to believe that with their own effort and the assistance of their peers, they, too, will be able to accomplish the goals of the curriculum.

Westerners also have been astounded during the past several decades at the ascendance of Japan and Taiwan as economic forces in the world and by the strides China has made in industrialization. Here, too, the positive effects of group effort are evident. The high levels of productivity are often traced to the workers' cooperative attitudes and to their dedication to the company by which they are employed.

Efforts to understand these 'we' cultures of the East may help the 'I' cultures of the West to gain insight into edsome of the problems of individualism: the sense of isolation, disengagement from the group, and the lack of common effort and ideals. At the same time, there are many contradictions in a society's efforts to foster prosocial behavior. These efforts are often accompanied by problems: the ethnocentrism that often accompanies strong identification with a group, the lack of individual initiative that occurs when advancement of the group has greater priority than individual achievement, the frustration that exists at viewing the conflict between the ideal behaviors of the models and the realities of everyday life, the conflicts that arise between ingenuity and obedience to rules and authority, and the tension produced by pressure for individual achievement and the simultaneous need to work for the advancement of the group. However, given the problems that exist in the absence of ideals and the positive features of prosocial behavior, compromises between the two approaches and efforts to find ways to reduce the problems must be examined.

References and further reading.

Bond, M. H. (ed.) (1986). *The Psychology of the Chinese People.* New York: Oxford University Press.
Bronfenbrenner, U. (1970). *Two Worlds of Childhood: U.S. and U.S.S.R.* New York: Russell Sage Foundation.
Chen, K.-S. (1968). *Lei Feng, Chairman Mao's Good Fighter.* Peking: Foreign Languages Press.
Eisenberg, N. & Mussen, P. (1989). *The Roots of Prosocial Behavior in Children.* New York: Cambridge University Press.
Ho, D. Y. F. (1981). Traditional patterns of socialization in Chinese society. *Acta Psychologica Taiwanica,* **23**, 81–95.
Lewis, C. C. (1989). From indulgence to internalization: Social control in the early school years. *The Journal of Japanese Studies,* **15**, 139–57.
Mead, M. (1935). *Sex and Temperament in Three Primitive Societies.* New York: Morrow.
Munro, D. J. (1977). *Concept of Man in Contemporary China.* Ann Arbor, Michigan: University of Michigan Press.
Peak, L. (1989). Learning to become part of the group: The Japanese child's transition to preschool life. *Journal of Japanese Studies,* **15**, 93–124.
Radke-Yarrow, M., Zahn-Waxler, C. & Chapman, M. (1983). Prosocial dispositions and behavior. In P. Mussen (ed.), *Manual of Child Psychology, Vol. 4: Socialization, Personality, and Social Development,* pp. 469–545, (E. M. Hetherington, ed.). New York: Wiley.
Ridley, C. P., Godwin, P. H. B. & Doolin, D. J. (1971). *The Making of a Model Citizen in Communist China.* Stanford, California: The Hoover Institution Press.
Webster, J. B. (1932). *Interests of Chinese Students.* Shanghai: Shanghai University Bureau of Publications.

Whiting, B. B. W. & J. W. M. (1975). *Children of Six Cultures: A Psycho-Cultural Analysis.* Cambridge, Mass: Harvard University Press.
Yamaguchi, Y. (1989). *What is the Ideal Japanese? Moral Education in Japan.* Unpublished paper. University of Michigan.

6

The learning of prosocial behaviour in small-scale egalitarian societies: an anthropological view

E. GOODY

Introduction

Definitions of actors and analytical definitions

Because anthropologists seek to compare and analyse as well as to describe, it is important to distinguish between definitions within a society – definitions which its members use in daily life – and our own analytic definitions. In studying human behaviour there is always this duality between the meanings which people have created in structuring their lives, and the concepts which anthropologists, outsiders, use to understand this very process of social living. As an analytical definition we can say that prosocial behaviour is 'behaviour carried out for the benefit of others' or more generally, 'a principle of action opposed to egoism and selfishness'. But cultural definitions of 'egoism' and 'the benefit of others' are certain to vary. The Utku Inuit (who lives in the Arctic) learns to respond with protective concern to any person s/he meets who is alone, cold, hungry or ill, but does so partly from fear of anger. A Gonja youth (from northern Ghana) must respectfully carry out the work required by elders of the household, kin group, and village, and by chiefs; this respectful obedience is clearly 'unselfish' and 'for the benefit of others'. It is inseparable from membership of household, kin group, village and state, and at the same time integral to the Gonja conception of what it is to be a fine person, to have a good character. (Whether or not it is perceived as exploitation by those in powerful roles depends on whether the youth is a commoner or a member of the ruling elite.) The Japanese child is carefully taught to want to participate energetically in groups

106

based in the neighbourhood, school and work, and to seek the good opinion of group members. In order to be sure of acceptance s/he learns to be highly sensitive and responsive to the feelings and wants of the other members of these groups.

However as long as we work within definitions of particular cultures it is not possible to make very interesting observations about learning, since what is learned differs so widely. We tend to find ourselves considering the learning of norms *per se*. Is it possible to go beyond this relativity? Are there any general patterns which can be recognized as 'prosocial behaviour' and which appear in all kinds of society?

Nurturance, sharing and protection

There are certainly behaviours which would be generally accepted as 'for the benefit of others' and unlikely to be selfishly motivated: the nurturance and care of infants; the sharing of scarce and valued things; and protection of others from harm. In fact these three areas of behaviour are quite different from one another, and might even be represented as concentric rings encircling the individual. At the centre, each of us depends for survival on having been the object of nurturance and care by *someone*. Every society arranges for at least a minimal level of infant nurturance, and this requires some roles which include nurturant prosocial behaviour. In certain societies, like the Utku Inuit, the Ifaluk of the southwestern Pacific and the Arapesh of New Guinea, nurturance is made the focus of socio-cultural elaboration. In others like the New Guinea Fore and the Gonja it may not receive much explicit attention, but nevertheless be the focus of institutional arrangements which have significant *de facto* consequences. If nurturance lies at the core, sharing creates social fabric, links between the individual and others; sharing defines who has rights of participation, of membership – and who is excluded. Sharing is both social cement and social dynamite, as many ethnographic accounts testify. These metaphors might seem so contradictory as to be useless, but since social structures are emergent, continually being built and challenged, they are in fact apposite. Finally, patterns of protection define domains of safety and erect barriers against external threat. In some societies this is done sharply and consistently; *we* are permanently and clearly distinguished from *them*. In other societies individuals or groups who are allies on one occasion are later enemies, but may again be friends. Protection is a strategic necessity, but carries complex implications. This is perhaps most obvious where there are marked differences in power within a society, and is evident in a different way where groups are implicated by the actions of their individual members, as where the feud is institutionalized.

Case studies and comparison as a technique for exploratory analysis

A general view of a pattern such as prosocial behaviour is drawn from two very different but complementary forms of analysis. The anthropologist's key resource is the close study of particular societies, based on personal involvement with ordinary daily life and the normal contexts for that behaviour. But a single such study is only the first step. For a general picture it is necessary to consider several in-depth cases, to make a comparison across societies, in order to establish both patterns of consistency and of variation. Ideally, then, it is possible to move back and forth between the detailed intensive study, from which come insights about the dynamics of the behaviour, and the comparison, which reveals patterns, and may suggest possible conditions for different forms.

A major problem, of course, is the choice of which societies to compare. We have accounts of only a tiny proportion of the world's present societies, and none for the past millennia of human time. Of the accounts we have, only a few are detailed enough to permit real understanding of interpersonal behaviours. Inevitably then, in this consideration of prosocial behaviour the cases used are those known to me which are rich enough to provide the information needed. Several of these are truly excellent, but we have no way of assessing how representative they are, and this does raise problems when one turns to a comparison for evidence of general patterns. At best we can look for patterns to explore further.

Modelling, shaping and training in the learning of prosocial behaviour

It would probably be wrong to imply that there is an accepted view among anthropologists of the nature of prosocial learning. The major body of research that touches specifically on this issue is that of the Whitings and their colleagues. This focuses on the systematic recording of interaction in selected settings which are critical for the lives of children of different ages. There are extremely interesting findings from this work regarding the influence of settings and participants on patterns of sharing and delegating nurturance, particularly in relation to male and female siblings of different ages. However because the richest descriptive in-depth case studies are for very small-scale, egalitarian societies it is these which are the focus in this paper. Because this is an exploratory exercise in comparative analysis rather than an account of established findings, it has seemed important to present enough of the ethnography itself to convey the flavour of the case material. In such an analysis, the move from description to analytic characterization is critical. To simply assert the results would almost certainly be to lose the meaning of the processes they reflect.

One of the most significant differences in the learning of prosocial behaviours among the various societies considered here is at the level of formality. Much of this learning occurs because, as the Whiting and Edwards analysis shows, settings construct contexts which shape learning. *Shaping*, then, is the learning which occurs in the course of ordinary interaction in a socio-culturally specific setting. Having to attend school at the same time every weekday shapes the Western child's behaviour in relation to time and routines. *Scaffolding* is the process in which a child participates jointly with an 'expert' in an activity which is gradually mastered through practical action. Scaffolding can often occur as a result of the way a setting is structured, without being intentional. *Modelling* occurs when a child observes the behaviour of others s/he respects, and internalizes this as a standard against which his or her own behaviour is evaluated. *Training* is a special sort of shaping that involves repeated intentional coaching on the part of a teacher. Most of the prosocial learning described in the following accounts occurs through shaping, practical scaffolding or modelling. Where training occurs this suggests an important extra element of intentionality on the part of the members of the society.

To anticipate the implications of the following comparisons: in these small-scale face-to-face societies the learning of prosocial behaviour follows two main modes. An anxious mode in which cultural definitions of the world and of social relations as potentially dangerous are associated with the learning of shyness/fearfulness, particularly through training; and a secure mode in which supportive settings facilitate the shaping and modelling of prosocial behaviour. The secure mode societies are associated with a relatively high level of mother–child intersubjectivity, and with independent interdependence with peers which seems to encourage peer intersubjectivity.

'Peaceful societies' and 'Fierce' societies: Key settings for imputing meanings and enacting meanings

Mbuti: Growing up within the protection of the benevolent Forest.

A recurrent theme which appears in a number of accounts is the importance of the relations *between* groups for the quality of relations within the group. Turnbull's account of the Mbuti pygmies of the Congo is entitled 'The politics of non-violence'. He describes the Mbuti as a society almost without violence, yet his unusually full and sensitive study says nothing about the teaching of non-violence. How can this be? It seems that the pygmies avoid confrontation with the cultivating peoples of the forest edge. If there is a problem or dispute they simply withdraw into the depths of the forest. Among themselves they have elaborated laughter and joking as modes of dealing with conflict within the settlement. Where bad feelings remain, the dissatisfied person moves away to join a different Mbuti band. Friction of a permanent kind, as that arising from the conflicting interests of the men and the women of the band, is ritually expressed on specific occasions – i.e. allowed catharsis under strong controls. Thus the Mbuti world can be seen as consisting of a nesting set of interdependent collectivities: Negro cultivators and pygmies of the forest, the several pygmy bands, couples in relation to other couples within the band, men and women. There are several non-violent options for dealing with conflict: joking and laughter, ritualized – but stylized and controlled – expression of hostility, withdrawal. Which one is more appropriate depends partly on history and context, but also on the nature of the interdependence. Withdrawal from the Negro villagers is possible because these are marginal to pygmy life. However the men of an Mbuti band cannot leave the women without losing their basis for subsistence and domestic life; withdrawal is not an option at this level, and ritualized catharsis is used instead. How are these subtle patterns learned?

The Mbuti emphasize the provision of a benign environment for the infant, beginning even before birth, when the mother goes out alone to sing to the Forest, and to the child in her womb. This love and welcoming is extended gradually from the mother–child core to include the father's intimate nurturance within the family shelter, to settlement clearing, to the children's playspace beyond, and finally to include the interpenetration of the individual's social world with the Forest. The 'Forest' is always capitalized in Turnbull's account because the Mbuti consider it to be a strong and good Presence, which powerfully influences each life. As the child learns to crawl and walk, it comes to be able to move about within this world. Neither physical nor verbal punishment is used, and there is freedom to explore, but if the child enters the domestic space of another family group the mother will pick it up and carry it back home. The world of the children's playing ground, an area set aside on the edge of the settlement, includes all children between the ages of three to four and ten to eleven. This is the main learning environment during these years. Adults seldom go there. Within this world children do everything together in a mixed age/sex group: physical games, cooperation in role play, elaborate role playing – both instrumental adult roles and social roles – ridicule, ritual, stories. The children take responsibility for each other's emotional well-being. There are routines for managing severe teasing which restore the victim to a positive position in the group. Turnbull stresses the strong affectively positive relations between all members of this small group, of all ages and both sexes. Boys as youths sleep all tangled up, both for warmth and pleasure of physical contact. Sexuality is a growing cross-sexual expression of this closeness. Couples who marry share companionship and physical closeness as well as complementary economic and parental roles. Marriages appear to be stable.

It is as though there was an Mbuti social contract based on a premise of benevolent interdependence. This mutual responsibility extends to relations with the Forest. Hunters and pregnant mothers sing to the Forest (and themselves), telling of their enjoyment of its coolness and bounty, and praising its kind protectiveness. Within the Forest world individuals and couples move freely between bands, to see old friends, to make new ones, or when there is uneasiness in a relationship. The threatening world outside the Forest is visited temporarily, and relations there managed through joking, role play and, if necessary, disappearing back into the Forest.

Semai, Buid, Chewong: Shyness, fear and withdrawal as linked to prosocial behaviours

The Mbuti premise of benevolence contrasts with the world of the Semai for whom the forest is dangerous. The Semai are fearful, and this fear is given immediacy by the constant raiding of their Malay neighbours. Robarchek writes that for them danger is omnipresent from both the human and the natural world. They fear non-kin and Semai of other bands whose intentions cannot be trusted, and always the homicidal strangers believed to lurk along the forest paths waiting to decapitate the unwary. The Semai too respond to threat by withdrawal, but for Semai this means flight and hiding from the approach of any strangers. The natural world is also seen as full of threats from 'a vast number of malevolent situations, forces and beings nearly all of which are actively bent upon the destruction of human beings'. These dangers are met 'by a multitude of taboos that bracket even the most mundane activities in a (usually vain) attempt to forestall the dangers that lurk on all sides'. Danger is thus one central theme in the Semai world; the other central theme is dependency.

'Dependence, together with its reciprocal, nurturance, is an important structural and emotional dimension in nearly all social relationships, both between human beings and between humans and spirit helpers who come to them in dreams ... ' (p. 34). Among humans this is expressed through the ethic and practice of sharing food. The injunctions to share food and avoid violence are the central moral imperatives of the Semai, and are constantly reiterated: any public gathering, and all dispute settlements begin and end with exhortations by the elders 'stressing interdependence of the group, recalling past aid given and received by individuals, emphasizing that each is dependent on the others for survival, asserting that the band is really a group of siblings, and so on' (Robarchek in Howell & Willis, 1989, pp. 34–5).

These two themes – danger and dependency – are inextricably interwoven in individual experience and thinking. Children acquire, first, an image of themselves as helpless and dependent on a hostile and malevolent world that is largely beyond their ability to control, and secondly a set of habits that lead them to seek and expect aid and comfort from others in times of distress. The only source of security is the band of perhaps 100 related people, whose membership is permanent. In this intense social world, 'goodness' is defined as sharing and helping, and 'badness' as fighting and anger. More than anything else, the Semai say they fear getting involved in disputes; quarreling endangers the support they can claim from their band.

There are several other societies in which shyness/fearfulness is cultivated as the natural way of relating to a dangerous world. For the Chewong of Malaysia shyness and fearfulness are highly valued in individual character, although unlike the Semai, they do not fear the jungle itself. Anger, quarrelsomeness and bravery are despised and attributed to non-Chewong. Chewong children are taught to be fearful from infancy, and shyness is approved. It is largely because they are not fearful/shy that non-Chewong are seen as dangerous. Fleeing has always been the Chewong response to violence. When adults get upset with someone's behaviour they tend to withdraw into themselves, passively waiting; or they may become ill. A child's anger is ignored; adult anger is withdrawn from. When, very atypically, a jealous wife publicly broke her husband's best blowpipe several families moved away from the settlement. The Chewong turn for help especially to supernatural beings, spirit guides called 'the leaf people'. The leaf people are themselves very timid, and flee instantly from danger.

For the Buid of the Phillipines simple withdrawal is the preferred form of protection, and this is linked to fearfulness which is viewed as normal. There is also great stress on egalitarian relations among adults, so that any attempt to establish superiority is either ignored or treated with ridicule. Any assertion of dominance by force is treated as aggressive behaviour and leads to withdrawal. The importance of equality is expressed in the ethics of sharing meat. Domestic animals are killed in sacrifices to avoid serious mystical dangers. Such a sacrifice must always be shared out in exactly equal portions among all members of the community regardless of age, sex or relationship. Most meat distributions follow minor sacrifices and involve a local community of between 50 and 100; on certain occasions representatives from more distant settlements attend, many animals are killed, and the meat shared among as many as 700. Gibson writes that 'One might say that when a particular household is confronted with a mystical threat, it dissolves its individual identity into that of a larger social unit whose size is determined by the severity of the threat. It does this by sharing something closely identified with its members, the animals they have domesticated by hand, with the encompassing community in a commensal ritual' (Gibson in Howell & Willis, 1989, pp. 72–3). Animal sacrifices are necessary for protection against predatory spirits, but there is another class of spirits associated with the earth which ensures the growth and well-being of children and of crops. The earth spirits guarantee growth, partly by extending protection from malevolent forces. But this protection is withdrawn if people quarrel; hence the need to avoid conflict by respecting equality, by collective discussion, or by withdrawal.

Fore permissive nurturance and the construction of prosocial meanings: Fore prescriptive amity

The major study of the Fore by Sorenson is entitled 'The edge of the forest' because these people practise very simple agriculture on the margins of the rainforest covering the rugged highlands of New Guinea. Fore children are interdependent and cooperative but innovative. They spend nearly all their time in age/sex peer groups, especially the boys; girls more often are called to help with domestic work. Sorenson says this independence is the result of the very free exploration allowed in early childhood. He also stresses the importance of the quality of maternal care of infants and toddlers who are in constant tactile contact with the mother/another. The baby sleeps in the mother's lap while she works, and she may nurse an infant and a toddler together. Toddlers are not prevented from exploring potentially dangerous things like knives and fire, but adults are always there to help. Children use adults or older children as security figures if anxious. Toddlers are often held by older children; older children fondle, touch, and solicit attention from infants and toddlers. Three- to five-year-olds follow older children about, and are included affectionately in their play. Older children yield warmly to infants and toddlers. There is very free sharing of food among all members of the hamlet. If a younger child expresses interest in a piece of food, it is always given up by an older child or adult. There is a general cultural elaboration of sharing of meals and of feasting. There is also quick sensitivity to hints of discomfort or anger in peers – immediate response to avoid interpersonal friction of any kind. Adults and older children distract babies and toddlers from actions that might be aggressive. They systematically redefine such potentially antagonistic acts as playful, and don't let a hostile interchange develop. This is clearly documented in film sequences. Such training in the redefinition of aggression has striking similarities to the way in which joking relations are taught in other societies. For instance, in Gonja a relative of the category (grandparent/mother's brother) whose members are enjoined to joke with a young child will taunt it playfully, and then everyone present laughs and shows affection to the child. Children are at first upset and confused, but soon learn that this is 'fun' and not dangerous. A four- or five-year-old delights in joking with grandparents in this way.

Despite this premise of prescriptive amity among close associates, the Fore have a fear of strangers and people beyond the set of hamlets with which they share and cooperate. People avoid paths belonging to those individuals with whom they are not friendly to reduce the risk of arguments. They fear sorcery from 'outsiders' and believe that all deaths are caused by this sorcery. Thus the learning of benevolent responses within the known group does not ensure trust in kindness and safety from outsiders.

Sorenson has also worked with the northern Fore, where greater population density means it is no longer always possible to find new farm land by moving into an unsettled area. Here it is often necessary to defend farm boundaries, and quarrels are not infrequent. Close observation of children showed that aggression in boys' rough and tumble play may not now be redefined as benign, but instead escalates into fighting. Older boys were also tending to exclude younger ones who wanted to join their games.

The pacific Piaroa and responsible individuality

The Piaroa of Venezuela, who totally disallow physical violence, also place great value on personal autonomy and expect each person to learn to master his or her own emotions and capacities in order to be able 'to lead a tranquil and therefore moral life within a community of relationships'. The ethnographer, Joanna Overing writes that 'While there are proper ways of doing things, it is up to the individual to *choose* or *not to choose* to do them ... The Piaroa daily express to one another their right to private choice and their right to be free from domination over a wide range of matters, such as residence, work, self-development, and even marriage'. 'Coercion has no place within the domain of the social ... the notion of giving up one's rights to a "whole community" or of submitting to a decision forthcoming from the community ... would be a strange and abhorrent idea to them' (Overing in Howell & Willis, 1989, p. 89).

Nevertheless, for the Piaroa the ability to be social is the most highly valued characteristic of humans. Only humans are able to be social and live in communities; only by being social can humans acquire the capacities for creating tranquil and moral (non-competitive) relationships of cooperation which allow for the forming of a moral community. Achieving the social requires being able to prevent the establishment of (immoral) relations of dominance. This is done first by learning to control one's own emotions and powers, and then by the Piaroa political process which gives very limited authority to those who have mystical knowledge, *on condition that they refrain from domination.* Any suggestion of domination or constraint results in realignment of residential and factional groups, with the isolation of the one seen as coercive.

The skills of mystical knowledge necessary to build community are acquired gradually throughout life. Young children play together in a free-ranging small group. They never see physical violence among adults, and are never punished, so they have no model for physical confrontations. Temper tantrums are discouraged by mocking and teasing; like adults, children express strong anger by pointed silence. When they reach the age of six or seven it is time for the children to be 'tamed' through their first lessons in the knowledge of wisdom. The local 'man of knowledge' (*ruwang*) gathers together the children and teaches them about the emotions they will need to learn to master; about consciousness, will and responsibility that are necessary for managing relations with others; and about such vices as ferocity, ill-nature, cruelty, malice, arrogance, jealousy, dishonesty and vanity. However, consistent with views of personal autonomy, it is held to be the children's own decision which virtues to develop, as later it is the individual's choice what capacities and powers to learn.

'Peaceful societies' and the linking of shyness/fear, nurturance and the denial of anger

This set of societies, which emphasizes shyness and the avoidance of anger and aggression, also tends to organize social relationships so as to emphasize nurturance. The emphasis on nurturance is largely implicit, and emerges from affectively positive and physically intimate mother–child relationships; from the partial delegation of nurturance to older children, especially girls; from settings which permit close and independent inter-dependence of children's peer groups; and, in one form or another, the insistence by everyone in the child's world on 'prescriptive amity', that is the absence of adult quarreling as a model for conflict resolution, and the systematic redefinition of potentially aggressive acts as benign and playful.

Constructing and imputing meanings of interaction among 'the fierce people'

The Yanomami of Venezuela

The Indians of the Amazon and Orinoco river basins in South America are most often described not as peaceful, but as warlike raiders. Chagnon's classic account of the Yanomamo (Yanomami) is entitled 'The fierce people', and several studies of other groups show very similar patterns. People fear attack from other settlements, but since they live in very large multi-family houses near their fields they cannot usually hide or flee. The best protection is to have a reputation for fierceness so that others leave them alone. When a village is particularly afraid of another it may make a pre-emptive attack; ambush is a favoured form of warfare since it minimizes danger to the attackers. Clearly, once this pattern of relations between communities is established, the participants have little chance of surviving in the 'peaceful' mode. What are the patterns of such prosocial behaviours as nurturance and sharing in a fierce society? And how are these learned? Unfortunately there is little material on nurturance or learning for these groups. However Lizot's ethnographic picture of the Yanomami contains extremely rich descriptions and transcripts of daily events.

Sharing is formalized, and subject to strict rules. Meat must be shared to the other households in the settlement, and among the Yanomami this is enforced by the prohibition on a hunter eating meat which he himself has killed. When those from one settlement visit another they bring some food, but their hosts supply and cook most of what is shared out in the feasting. Garden produce is the product of individual labour, and used by the family which has grown it. Food other than game is not normally shared between households, nor is there a norm of older children yielding food to younger ones, even within the family:

> Humoama offers a fruit to her little boy; his older brother immediately snatches it away from him; the little one immediately breaks into tears and remains in the middle of the path, howling and rubbing his eyes, deaf to the voices telling him to move away. The older brother squats down in front of him, and mocking him, eats the banana with much noise and snickering. To calm her son, the mother detaches another banana and holds it out to him, but the child pretends not to see it. Ten times at least, she repeats:
> 'Here, take this banana, it is yours.'
> Her patience is praiseworthy. Finally, with an angry motion, the child snatches the fruit out of his mother's hand and begins to peel it without even looking at her.
>
> (Lizot, p. 158)

Yanomami children are not physically punished but they are expected to defend themselves in any confrontation, and there are strong pressures, from women and peers as often as from men, to retaliate for any injury. Lizot describes how the mother of a child hit by another gives her son a stick with which to return the blow. On another occasion a mother whose small son is bitten by a playmate places the culprit's hand in her son's mouth and tells him to bite it. In children's games play is pre-emptively defined as attack which must be reciprocated, and the pain born stoically. In one game the children of the settlement are divided into two teams, each child armed with a club of soft wood with which blows are exchanged.

> The blows are haphazard; those who are struck grit their teeth in order not to show their pain and try to trade blow for blow with their opponent. When they are finished the children plant their weapons in two parallel rows.

> (Lizot, 1985, p. 39.)

Older youths are subjected to the same pressures to retaliate. In one episode when a husband strikes his wife's would-be lover on the head with a heavy log, everyone, settlement residents and visitors, takes sides with one or the other. All the youths and men seize some kind of weapon; everyone is shouting. One of the senior men walks about telling them to calm down, but no one listens. Furious women shout at the injured man to return the blow, but he is too groggy to react. One women rushes up with a club which she thrusts into his hands; she yells into his ear:

> Avenge yourself, go on, avenge yourself! Return the blows you have received.

The two young men stand confronting each other with clubs raised. Watching youths are shouting and revelling in the confrontation. Soon the injured youth returns to his hearth, where a youth comes to attend to his scalp wound.

> Some women are still shouting, but a general fight has now been avoided.

> (pp. 155–6)

This presumption of injury which must be avenged seems to permeate all aspects of Yanomami life. All deaths are attributed to enemy sorcery, and retribution must be taken to show these evil shamans that their power is limited. The young son of the shaman, Turaewe, died a year before despite his father's attempts to cure him. The shaman is depressed and angry. He knows his son was killed by enemy shamans from Hiyomisi village to the south. They stole his soul. Finally the father announces he will avenge his son and calls fellow shamans. They inhale the hallucinogenic drugs which enable them to set out on a mystical journey of revenge. They sing that they will find a boy who is beautiful and kill him and feed his soul to the cannibal spirits. After great difficulty and suffering they bring back the soul of the enemy child and feed it to the spirits.

> May your heart be at peace, Turaewe, your child is avenged; you have finally repaid the evil and suffering that had been inflicted on you! The eaters of souls have come by; tears have erased other tears.

(p. 137)

The Yanomami provide a stark example of how institutions for protection can constrain other levels of prosocial behaviour. Social setting shapes responses as retaliation, peer groups and adults scaffold interaction so that particular routines of retaliation are learned, and mothers actually train their children to feel that they must retaliate.

Other 'fierce' societies

There are other 'fierce' societies, however, which do not shape, scaffold or teach retaliation. The Illongot headhunters of the Phillipines stress equal sharing of food within the household and equal sharing of game between households. Companionship among youths, and among young girls is highly valued; fierceness and retaliation are not emphasized. Youths want to join those who have killed and taken a head because this is evidence of their forceful maturity. The Birifor of northern Ghana were feared for their fierceness in raiding, but this is channeled through patrilineage confrontations. Daily life within the settlement and household is much more shaped by the relations between women and their daughters and daughters' children, who form the nucleus of emerging matrilineage segments.

The dynamics of prosocial behaviour in 'fierce' societies raises again the question of the nature of the link between fear, anger and nurturance. This is most explicitly treated in Jean Briggs' account of the Utkuhikhalingmiut, an Inuit group who live just north of Hudson's Bay (hereafter Utku).

Intentional training for prosocial behaviour

Most of the patterns of learning prosocial behaviour described thus far have not been the result of formal teaching. It is seldom possible to know what the adults had in mind when they shaped or scaffolded behaviour because the ethnography seldom addresses this question. However there are a few accounts which do provide this kind of information; the Utku forced their ethnographer to attend to their preoccupation with anger and with nurturance.

Utku: Training in nurturance and fear within the immediate family

Jean Briggs's account of Utku Inuit society is based on her long and intimate participation in the social life of a tundra settlement. The title of her book, 'Never in anger', reflects both Utku values and their way of managing relations with each other. Its very special contribution is to show the interplay between primary adult norms of nurturance and non-violence and the way children learn and are taught the meanings of their own and others' behaviour. Jean Briggs describes how when she first joined the tiny isolated settlement she was welcomed, and she felt that her hosts were genuinely concerned for her comfort and morale. Gradually she realized that the expression of protective concern (*naklik*) was their way of responding to a new person in their midst, and more generally to any interactions involving uncertainty and fear. The Utku defined *naklik* as 'the desire to feed someone who was hungry, warm someone who was cold, and protect someone who was in danger of physical injury' (Briggs, 1970, p. 320). People explained why they performed services for her by saying 'Because you are alone here and a woman, you are someone to be naklik'd' or 'Because you lack skill, you are someone to be naklik'd'. A mother watching her child struggle to carry a heavy kettle of water said 'She makes one feel naklik'. 'Naklik feelings are given as reasons for taking care of the ill, for adopting orphans, and for marrying widows, all categories of people who are in need of physical assistance' (Ibid., p. 321). The term was chosen by missionaries to translate the biblical concept of 'love'.

But why should the Utku react with protective concern to the strangers whom they fear? The Utku term for social fear, fear of being treated unkindly, of causing a response that is disapproving, is *ilira*. Children have *ilira* for adults, adults do not *ilira* children, but may have *ilira* towards other adults. '*Ilira* is one of the major emotional sanctions supporting [Utku] values in general and the *naklik* value in particular' (Briggs, in Montagu, 1978, p. 65). *Ilira* is uncomfortable, and also dangerous in that others are afraid of adults who express *ilira*. The Utku cure for this, says Briggs, is to reassure the *ilira* person by being protectively considerate, *naklik* of him. 'If I am kind to the person who fears me, he will stop being afraid of me – that is, will stop wanting to threaten me – and peace will be restored' (Ibid., p. 65). In short, people who are afraid, may defend themselves through attacking others. Her Utku friends told Briggs that it was dangerous to go fishing or on a trek alone with a man who feared you, because he might decide to stab you in the back as the only way of being safe from your attacking him. It seems that the experience of *ilira* from anxiety about others' unkindness or disapproval can act to make people conform with what others wish; but it is also recognized that the discomfort from *ilira* feelings may cause people to try to protect themselves by destroying the source of such feelings, the person who is unkind or disapproving. The only solution is to treat others with such obvious kindness that they do not fear you. Thus there are two ways in which '*ilira* fears motivate protective (*naklik*) behaviour: (1) *if someone is afraid of me, I'll behave protectively towards him;* and (2) *if I am afraid of another person, I'll behave protectively towards him – and of course, he will treat me protectively too, because he sees that I fear him, and because I have reassured him that he has nothing to fear.* So the circle is (ideally) complete' (Ibid., p. 66).

What is particularly striking is that Utku children could be said to undergo formal training in *naklik* and *ilira*. Infants are treated by parents, siblings and other settlement residents (all kin of some kind) with exaggerated affection and concern. (This is particularly noticeable in a society in which behaviour between adults, and towards older children is markedly reserved.) An infant or toddler is never refused food, not allowed to cry, and great efforts are made by everyone to amuse the child. Thus the infant receives *naklik* from everyone s/he meets. Further, from a very early age the infant is taught routines which are considered beguiling, and which serve to elicit affectionate *naklik* from observers. Thus the infant is trained to be an active elicitor of *naklik*. But this same infant is subject to what the ethnographer calls 'benevolent aggression' (*kiilinngu*). This occurs when the adult is in an affectionately excited state through intense hugging and kissing. It is sufficiently restrained that it does not physically injure the child, and is accompanied by affectionate smiles and laughter. Nevertheless it may be carried to the point where the child cries, withdraws, or tries to appease the aggressive adult by hugging him or offering a bit of food. Adults interpret the reaction as fear of affection and comfort the child tenderly. One woman explained her actions by saying 'A hurt child is lovable.'

The Utku hold that mature people who have understanding do not express anger or aggression against others. Their mature intelligence (*ihuma*) makes them feel loving, not hostile. For Utku parents benevolent aggressive games are expressions of love; and indeed, these routines arouse protective feelings, *naklik*, in the adult, and pacificatory (love and protection eliciting) behaviour in the frightened child. The child is taught to be afraid, and to manage this fear by eliciting *naklik*. This training is repeated on another level when the indulged infant is displaced by a new baby. Because of the extreme cold of the Arctic, Inuit infants literally live inside their mother's loose coat, snuggled against her bare back. The ethnographer considers that displacement from this physical intimacy is probably more traumatic than weaning from the breast for the Utku toddler. Not surprisingly, such toddlers tend to demand attention from their once-indulgent mothers, but they are now ignored. At the same time they are explicitly and insistently told they must be *naklik* to the new baby, and taken through routines in which they hug the infant, give it food, etc. This intruder of course is treated with the same demonstrative indulgence once directed to the toddler. And later as the baby becomes more active it is taught the beguiling routines which elicit yet more expressions of affection. If the baby wants an object being played with by an older sibling, it must be given up. The toddler may respond to these injustices with tantrums, but these are ignored. Profound withdrawal can follow, which is also ignored. Briggs remarks on the striking difference between the subdued, passive child of six and the enchanting infant of two or three years. But the child of six has learned to manage fear by giving protective concern.

This dynamic of *ilira* and *naklik*, fear and nurturance, remains the mode through which interactions within and outside the family are transacted. The Utku are dependent for survival in extreme arctic conditions on sustaining cooperative relations with other Utku. The suppression of any anger clearly is one means of avoiding the conflict which would put this cooperation at risk, and indeed an individual who expresses anger is so feared that he will be ostracized. Jean Briggs herself, following an angry outburst, was for a while excluded from social interactions, though because of her identification with government authorities she was not left to survive physically alone. But the training in expressing *naklik* means that situations which might seem likely to lead to anger are explicitly and publicly defined as benevolent.

Conclusions

'Culture' sets premises for the meaning of events – like a child's blow at another, taking food from a younger sibling, being struck by a jealous husband, death from illness. From infancy children are taught these meanings by the interventions and interpretations of the central adults in their lives. These meanings tend to form interlocking sets that reinforce each other. For the Yanomami type of 'fierce' society the set includes the need to avenge oneself for the slightest injury, the assumption that misfortunes are intentionally caused by hostile neighbours, and finally, real warfare between villages, in which death and injury must always be avenged. But other 'fierce' societies do not have such a tight set of mutually reinforcing antagonistic definitions. Both the Ilongot and the Birifor were greatly feared for their institutionalized killing. But neither society trains its children for inexorable retaliation within the community, nor sets brother against brother in competition for food. Indeed both stress the importance of sharing food within the family, and between families. And both value greatly the cooperation among youths which they see as the natural basis of adult interdependence.

The balance of cultural forms and premises in a given society is the result of many factors. Social institutions like descent groups are very important because they hold meanings about joint rights and obligations, sharing and trust. Descent groups also shape patterns of action for inheritance and living together which create settings for nurturance, sharing and protection. Both Birifor and Ilongot young women live with their mothers while they are rearing their own children. For children reared within this setting, prosocial responses are modelled by their mother and her sisters and their grandmother every day through their own warm relations. These same women shape the children's interactions with each other by their premise of trust. Where, as for the Birifor, these people are identified as sharing a special kind of close kinship, such modelling and shaping is given explicit meaning – we are 'children of one grandmother' – that further reinforces it.

In the more detailed ethnographies there is a striking linkage of prosocial behaviours, nurturance, dependence and sharing, with fear and aggression. The Utku child is taught, partly through fear, to elicit nurturance; and then taught when older to respond to fear with nurturance, and *never* with anger. The child learns that anger leads to rejection and the threat of isolation, while fear leads to nurturance, and social support. This linkage of prosocial behaviour and the denial of aggression also appears in Semai, Buid, Fore, and Mbuti. They explicitly recognize that anger will endanger support from their fellows. This pattern is in fact much more widespread. Fortes has referred to it as *the axiom of amity*, and it regularly requires sharing and prohibits hostility within the descent group in West African societies. However, the prohibition of aggression within the community is by no means always a norm. This is clear in the contrast between the Fore and Yanomami responses by older people to one child hitting another. In the Yanomami case, the mother insists that the other child retaliate; in the Fore case a child who strikes another is distracted, and the incident laughed about in a light-hearted way. These incidents show strikingly the power of mothers and older children to *define* the meaning of a young child's actions. And what is learned depends on this definition.

A stress on shyness/fearfulness as desired and normal occurs in several societies, linked with reliance on withdrawal as the main form of protection from threat. This is elaborated among the Mbuti, the Semai, the Chewong and the Utku. For the Utku, fear and nurturance continually reinforce each other. What all these societies have in common is the absence of any political or kin structures that might articulate defensive retaliation, and an ecological setting which makes separation a viable response. The Utku are perhaps the limiting case, since individuals cannot survive alone in the Arctic. Families commonly joined other settlements as a way of managing friction; however, such new families would only be tolerated if they were 'never angry'.

It seems there are at least two modes of learning prosocial behaviour in these small-scale egalitarian societies: an anxious mode (Utku, Semai) and a secure mode (Mbuti, Birifor, Fore). The anxious mode can have two components. The first is the cultural definition of human social relations and of the nature of the world as potentially dangerous. This requires the child to learn the skills of shyness and fearfulness. The other component is an insecure relation with the mother which leads to an anxious individual personality. These are *not* the same thing, though they may be related. The Semai teach shyness/fearfulness and value this behaviour in children, but they have warm close mother–child, family–child relations which do not seem to produce anxious children. The Utku dynamic is more complex; older children and adults seldom express affection towards each other, and Briggs considers some individuals to be psychologically withdrawn. However behaviour towards infants and young children is very expressive and affectionate, and within the family relations are sensitive and supportive.

Mbuti and Fore provide detailed accounts of the learning of prosocial behaviour contextually, through shaping and modelling within a strongly supportive setting. Accounts of both these societies argue for a high level of mother–child intersubjectivity. This is linked with the toddler's increasing exploration away from the security of the mother, and the child's gradual extension of reliance to older children and peers. In both these societies, as with the Birifor, children spend most of their time away from adults in interdependence with other children in mixed age peer groups. This 'independent-interdependence with peers' is a setting in which peer intersubjectivity seems to develop. Peer intersubjectivity involves sensitivity to the needs and reactions of others, and often the learning of routines for negotiating conflicts.

A critical feature of scaffolding is that the novice participates in the activity at first without the need to either understand it or be able himself to carry it out. Directly parallel with the training oriented dyadic scaffolding of the mother–child or teacher–child relationship is scaffolding through participation in activities. When the Mbuti child spends its days at the playing ground with older children whose play routines avoid hostile escalations through laughter and satire, these activities act as a framework within which the child practices increasing understanding and skill. It is here, perhaps, that societal arrangements for caretaking, socialization, and particularly peer group activities, become important. Initially intersubjectivity is a *social*-psychological process, building reciprocal attention and responsiveness on the affective bond between mother and child. But it develops very differently depending on the socio-cultural structuring of relationships. For instance, where mother–daughter–granddaughter sets are a primary context for early childhood, the warmly supporting intersubjectivity of infancy extends into childhood. This promotes reciprocal attention and responsiveness within an expanding field of social relationships, particularly among children. Conversely, where mother and child are isolated – by social structural features such as polygyny or social stratification or poverty – the intimate scaffolding of relationships is bounded. Wider relationships must be learned by applying the initial skills in a new framework which may severely limit their effectiveness, or in which they may be socially defined as inappropriate. And of course in a dangerous world they may indeed be inappropriate. The extension of the child's intersubjective relations to a peer group may, as with the Mbuti and the Fore, provide a wider sphere for exploration and support. But in highly stratified societies the peer group itself is restricted, as in modern Gonja élites and Hindu joint families.

Where the learning of prosocial behaviour is mainly through shaping, modelling and situational scaffolding, as with the Mbuti and Birifor, this seems to be largely implicit. Elsewhere, e.g. for Utku, it is elaborately explicit. Possibly this is related to utilization of the training in shyness/ fear among the Utku and Semai (anxious mode); and to a gradual extension of the sphere of secure relationships among the Mbuti, Birifor and Fore (secure mode). This raises the question of how these two modes are related to the type of learning (shaping, modelling and training), and to its explicitness. These factors in turn are probably related to the nature of bounded groups within which prosocial behaviour is practised and expected.

Finally, in most of these accounts (Mbuti, Semai, Buid, Ilongot, Yano-mami, Piaroa, Utku, Birifor) there is a clear stress on individual auton-omy. It is important to realize that this can mean socially responsible autonomy (Piaroa, Yanomami and Birifor retaliation against infringe-ment of rights). Or it can be combined with stringent training in concern for others (Uktu). The stress on autonomy seems to be related to the absence of authority roles – of any social and conceptual framework that legitimates authority of one person over another – rather than to selfish-ness. It can be linked with great sensitivity concerning the infringement of the rights of others.

In all these small-scale egalitarian societies sharing is culturally elabor-ated, and not left to individual decisions. In hierarchical societies, and those in which formal authority roles are recognized, the socio-cultural patterning of prosocial behaviour is greatly influenced by internal bound-aries defining who belongs to *our* group, and thus who deserves nurtu-rance, sharing and protection. These definitions in turn restrict individual autonomy.

It seems probable that optimal conditions for learning of prosocial behaviour through shaping, situational scaffolding and modelling occur in societies that are not threatened by competition for subsistence resources or aggressive neighbours. Societies like the southern Fore, where land is ample and neighbours not hostile, can develop a culture in which there is a premise of benign intentions. Others' acts are assumed to be friendly, and children learn that they live in a friendly world. The Mbuti provide another example, for although they manage relations with the cultivators outside the forest by withdrawal, they define their forest environment as benign, and relations among themselves also seem to follow this pattern. Within the forest there is no threat from either neighbours or subsistence crisis.

Where societies are threatened by aggressive or competitive neighbours, their protective responses may also act to model aggression as an appropriate response to conflict. This in turn may lead to pre-emptively defining interactions as hostile, and to the training of children to retaliate for even unintentional injury (Yanomami). Other threat-oriented societies have institutionalized withdrawal as the appropriate response to threat, and train their children accordingly (Semai, Buid). Where the threat is from a harsh environment rather than from neighbours, this may lead to a cultural response that stresses mutual support and the danger of anger and aggression. The Utku must survive in arctic weather with a precarious food supply. They also have elaborately explicit norms about the need to nurture others: children, strangers, widows, orphans – everyone.

How general these patterns in the learning of prosocial behaviour are must be a matter for further enquiry. Where the factors noted are similar, then one would look for similar responses, both at the level of individual behaviour and in social forms. However, such patterns are easier to discern in the small-scale societies considered here where differences in wealth and political hierarchy are minimized, and where social forms are able to adapt relatively quickly to changing external constraints. In more complex societies it might still be possible to see similar mechanisms at work, such as prosocial learning through peer intersubjectivity.

Acknowledgement

This chapter was written while the author was a Fellow at the Wissenschaftskolleg zu Berlin, and she wishes to express her thanks for the facilities and hospitality received.

References and further reading

Briggs, J. L. (1970). *Never in anger: Portrait of an Eskimo family.* Cambridge, Mass.: Harvard University Press.

Hendry, J. (1986). *Becoming Japanese: The world of the pre-school child.* Manchester: Manchester University Press.

Howell, S. and R. Willis, (eds.). (1989). *Societies at Peace: Anthropological perspectives* London: Routledge.

128 E. Goody

Chapters on: *Semai* by C. A. Robarchek
Chewong by S. Howell
Buid by T. Gibson
Piaroa by J. Overing.

LeVine, R. A. (1977). Child rearing as cultural adaptation. In P. H. Leiderman, S. R. Tulkin, and A. Rosenfeld (eds). *Culture and infancy: Variations in the human experience.* New York: Academic Press.

Lizot, J. (1985). *Tales of the Yanomami.* Cambridge: Cambridge University Press.

Lutz, C. A. (1988). *Unnatural emotions.* Chicago: University of Chicago Press.

Mead, M. (1935/1950). The mountain-dwelling Arapesh. In *Sex and temperament in three primitive societies.* New York: Mentor.

Montagu, A. (ed.), (1978). *Learning non-aggression: The experience of non-literate societies.* New York: Oxford University Press.

Chapters on: *Fore* by E. R. Sorenson
!Kung by P. Draper
Inuit/Utku by J. Briggs
Semai by R. K. Denton
Mbuti by C. M. Turnbull

Sorenson, E. R. (1976). *The edge of the forest: Land, childhood and change in a New Guinea protoagricultural society.* Washington, D.C.: Smithsonian Institution Press.

Rosaldo, M. Z. (1980). *Knowledge and passion: Ilongot notions of self and social life* Cambridge: Cambridge University Press.

Whiting, B. B. and C. P. Edwards. (1988). *Children of different worlds: The formation of social behaviour.* Cambridge, Mass.: Harvard University Press.

C.

SITUATIONAL AND PERSONALITY DETERMINANTS OF PROSOCIAL BEHAVIOUR

Editorial

Consider the following situation: a member of a youth gang in an inner-city area helps the member of a rival gang who has been hurt in a fight. His own group would regard this behaviour as violation of their gang rules and call him a traitor; but an external evaluator would possibly take him for a secret Christian or a real altruist, as there would be a low probability of any direct reward for his act. The example demonstrates four basic (interrelated) variables that play an important role in the analysis of prosocial behaviour:

> the social environment and the norms in the context of which the behaviour takes place
>
> the cost–reward relationship for the actor
>
> the labelling of the behaviour by others
>
> the (perceived) personality characteristics of the actor

In our example, the interrelation between these variables is as follows: as the social norm (gang rule) makes an immediate situational reward for the behaviour unlikely, the motive for helping is attributed either to a higher level reward system (e.g. religion) or to a specific personality trait (altruism).

Many theories of prosocial behaviour assume a reinforcement- or reward-oriented motive behind such behaviour, and regard the existence of completely unselfish altruism as unlikely. The reinforcement or reward characteristics, however, may be ordered along a scale from near-altruism to pure selfishness disguised as prosocial activity (see below). By some, behaviour is accepted as 'altruistic' only if it is completely free of *any* personal or social interest. But if one agrees that, for example, the hope for a (better) life after death, the positive emotional climate in a helping situation, or the probable survival of relatives through self-sacrifice, are in fact rewards in themselves, then it is difficult completely to exclude the reward aspect from prosocial behaviour. One way out of this dilemma might be to regard as altruistic that behaviour for which the expected costs exceed the expected rewards (see the discussion in Chapter 6) – but rewards and costs, even if they lie on the same dimension, may be too intangible to measure. Another route is to focus on the stability of prosocial behaviour across situations, and its immunity against external pressure and change, as a crucial variable. In this case, 'hedonistic' but stable helping behaviour may be more constructive than a 'morally' more appreciated unselfish but inconsistent altruistic tendency.

The application of moral categories is necessarily associated with attribution processes. One chapter in this section discusses three categories of helping, and their probable causal bases (Fultz & Cialdini, Chapter 7) and another the factors that determine whether helping behaviour elicits an attribution of altruism from an observer (Swap, Chapter 8). As these authors conclude, prosocial behaviour often arises from a mixture of different influences – experience, expectations, rewards, personal predispositions, situational factors – and can hardly ever be assessed as purely altruistic, especially as most motives are not completely open to external observation and valid evaluation. Even an actor himself/herself is usually not conscious of all factors that influence his/her behaviour: at best, he/she can offer a more or less plausible explanation/attribution for the behaviour.

Heal (Chapter 9) takes up this question of the bases of purportedly altruistic behaviour from the perspective of the philosopher. She concludes that experimental evidence for the existence of 'pure' altruism is likely to be elusive, but nevertheless the fact that we can envisage pure altruism as a possibility is in itself important: it provides a goal for the educator and the policy maker. Thus the conceptual problems do not mean that the analysis of intentions is of purely academic interest or that only outcomes count. Trying to know as much as possible about the bases of prosocial behaviour may contribute to prosocial education and increase the probability of establishing altruistic norms as societal goals. Interestingly, Heal's discussion leads to the conclusion that altruism can best be understood not as an isolated phenomenon internal to the actor but in relation to the social context and the actor's perception of that context.

The notion of altruism is related to that of justice, or what is 'fair': for example, an individual who contributes to a project more than what others consider as fair may be considered as altruistic. Justice, however, is also far from a simple concept. Clayton & Lerner (Chapter 10) show that different principles may operate at the levels of individual, relationship and society. However the fact that, despite differing visions of justice among different people, individuals and groups are concerned with unselfish goals at all is already a source for optimism.

Thus the abstract questions of whether behaviour is purely altruistic, or of whether justice has been done, yield no simple answers but lead us to try to specify the diverse factors that mitigate for and against prosocial behaviour. Briefly, what can be summarized about the interaction between personality and situation in this context? Following social exchange theory people are more likely to help if their costs are minimized and the rewards are maximized. These, costs as well as rewards, can be placed on two independent continua whose balance or net-outcome determines the probability of prosocial behaviour. Without considering additional variables, help is less on the average if the costs are very high and the rewards are evaluated as low, unless sacrifice itself is perceived as a final, perhaps religious, reward. However, costs as well as rewards may stem from very different systems. Costs can be purely economical (e.g. money), psychological (e.g. overcoming fear), social (e.g. loss of status), or physical (risk of being hurt or even killed). The relative importance of these may differ from individual to individual, making it difficult to set standards for the final costs the individual has incurred. The same is true for rewards. As we have argued earlier, it may make sense to order rewards along a scale of levels of immunity against external change or pressure: pure altruism would represent the (ideal) extreme on the scale, completely independent from any external influences. The postulation of such a possibility may help to establish an altruism norm which many would perceive as a desirable goal for their own behaviour. Pure group pressure to help without individual acceptance would represent the opposite end: as soon as the pressure faded, helping behaviour would disappear. Between these two extremes many different types of reward are possible, which alone or in combination with other influences determine prosocial motivation:

> empathy; reward: shared emotion
> caring for family, peers; reward: reciprocity, 'we-feeling'
> affection/attachment by the helped
> insight: own survival depends on others' survival
> transcendental rewards: religion, 'paradise'
> social norms: reciprocity norm, responsibility norm, ethics

modelling/diffusion of prosocial ideas, social infection
social reward: hero status; acknowledgement
symbolic rewards: medal, press coverage
fashion, habit: Zeitgeist (e.g. 'altruistic 1960s vs. materialistic 1980s')
curiosity: risk-taking and helping as experience seeking
reduction of cognitive dissonance: suffering impairs emotional climate
indolence: helping is 'cheaper' than letting someone suffer
personal guilt/debit pattern
external control/laws
fear of punishment, laws on neglect of help
payment
social pressure

Personal factors, interacting with but not necessarily depending on the different reward systems, that influence the probability of help are mood and gender and, less clearly, traits such as the 'altruistic personality', though this is often more a socialized outcome of an intersituational behavioural habit than an inherent predisposition. Happy people seem to be more willing to help than those in a neutral mood. Sad adults are unlikely to be willing, and sad children are less likely to act prosocially than their emotionally neutral peers. For gender, the surrounding situation plays an interactive role. In high risk situations, men are more likely to help, while in low risk situations women are, especially in crowd situations which intensify the pre-existing tendency to take care of others. Role socialization together with biological predispositions may be factors in this case.

Several classical studies on situational variables have demonstrated that with an increasing number of (passive) bystanders the likelihood that individuals will help decreases, whereas observation of bystanders actively helping increases the prosocial tendency (modelling).

Further reading

Batson, C. D., Bolen, M. H., Cross, J. A. & Neuringer-Benefiel, H. E. (1986). Where is the altruism in the altruistic personality? *Journal of Personality and Social Psychology*, **50**, 212–20.
Darley, J. M. & Latane, B. (1968). When will people help in a crisis? *Psychology Today*, December, 54–7, 70–1.

7

Situational and personality determinants of the quantity and quality of helping

JIM FULTZ AND ROBERT B. CIALDINI

The study of the determinants of prosocial behavior has flourished in the fields of personality and social psychology over the last two decades. Prosocial behavior researchers have greatly advanced our knowledge of the situational and personality factors leading to increased helping. Many researchers, however, have not been satisfied with simply investigating the determinants of the quantity of helping. The enquiring mind wants to know not only when more help is provided and who helps more but also why help is provided and what its purpose is, that is, the quality of helping. To structure the following review of this research, we will first outline three broad possible types or qualities of helping and then discuss the factors that influence how much of each type of helping occurs.

A conception of three qualities of helping

One way to distinguish between different types or qualities of helping is to consider that helping could result from distinct orientations toward others. We shall call these orientations: the other as object, the other as evaluator, and the other as individual.

An orientation to the other as object may strike many people as deplorable, yet it is undeniable that much human interaction is guided in this way. The helping that results is directed toward the goal of satisfying basic physiological, psychological, or safety needs for the helper. Although it may seem paradoxical that self-sacrifice by helping can lead to the self-benefit of satisfaction of basic needs, we will show that enhancing another's welfare often brings with it rewards to the helper. Helping that results from an orientation to the other as object, therefore, involves

135

the use of helping a victim as an instrumental means to attain some basic form of self-benefit.

Helping that results from an orientation to the other as evaluator also involves the use of helping as a means to attain a form of self-benefit, but the type of self-benefit is qualitatively different from that involved in the orientation to the other as object. Orientation to the other as evaluator involves at least a rudimentary 'caring' about the other, whereas orientation to the other as object does not necessarily involve 'caring.' The type of 'caring' that leads to helping guided by the orientation to the other as evaluator, however, is caring about how the other evaluates or thinks of the helper. The helping that results is directed toward the goal of satisfying higher-order psychosocial needs of belonging, acceptance, and self-esteem for the helper. Because acting prosocially is generally evaluated positively by others, helping will often bring along with it enhanced social-esteem and consequently enhanced self-esteem for the helper. Thus, helping that results from this orientation is designed to achieve the self-benefit of maintaining or enhancing social- and self-esteem.

The orientation to the other as individual stands in marked contrast to the two previous orientations. It involves recognizing the other as a person in her or his own right, a person with needs, desires, and goals of his or her own. Helping that results from this orientation would involve acting to enhance the victim's welfare as an ultimate goal, not enhancing the victim's welfare as an instrumental means to attain the ultimate goal of some form of self-benefit. The genuinely altruistic helping resulting from this orientation, however, is at present best viewed as a theoretical possibility rather than an empirically established fact. As we will show, the evidence that people help to enhance another's welfare as an ultimate goal is not, at present, conclusive.

Having delineated the three orientations to the other that result in qualitatively different types of helping, we will now examine some of the factors that lead to increased helping guided by each orientation.

Determinants of helping guided by the orientation to the other as object

That people often act socially to further their own self-interest is evident, but it is nonetheless true that no person is an island, entire of herself or himself. Another's current state can affect one's own current state. Interestingly, the capacity to be affected by another's current state appears to be inborn in humans. As any hospital nursery attendant can attest, hearing one baby cry can often lead to a symphony of crying by other babies. Research by developmental and social psychologists has shown that newborns cry as loudly or even more loudly in response to another infant's tape-recorded crying than their own tape-recorded crying (see Hoffman, 1981, for a discussion of some of this research).

That people are affected by cues indicating another's current state has profound implications for prosocial behavior. If exposure to such cues is unpleasant, ending that exposure should be rewarding. Early research by Weiss and his colleagues (1971) tested this implication by giving people the opportunity to terminate shocks (apparently) being administered to another person. Their subjects quickly learned to terminate the shocks rapidly, leading the researchers to conclude that the sight of another's pain caused distress in observers and that the opportunity to help by relieving the pain was a rewarding event. The lesson to be learned is that victims can be objects that produce distress in observers and that are manipulated, that is, helped, in order to reduce observers' distress. There is nothing selfless or altruistic about such helping: it is an instrumental means to relieve the observer's physiological or psychological distress caused by witnessing another's suffering.

The principle of distress reduction as a motivation for helping has been developed into a comprehensive theory of prosocial behavior by Piliavin and her colleagues in their book, *Emergency intervention* (1981). These researchers have amassed considerable evidence in support of their arousal/cost model of helping, which predicts that people will help faster and more vigorously the greater the arousal they experience when observing a victim in an emergency situation. There is one exception to this rule: when helping is a more costly behavior than alternatives such as putting the victim's suffering out of sight and out of mind by leaving the scene. Piliavin and her colleagues suggest that their research indicates the way to engineer a more prosocial society is to increase contact between people and reduce the costs of helping. Curiously, though, their model also would imply that people are motivated to reduce contact with others because of the arousal they would have to endure and the costs of helping they would have to incur when they are exposed to others' suffering. The prospects of building a more prosocial society based on the principle of distress reduction thus would not appear promising.

Helping that results from an orientation to the other as object, however, can be based on factors other than arousal caused by observing another's suffering. Cialdini and his colleagues (1981) have developed a negative state relief model of helping which holds that helping can be used instrumentally to relieve a different form of negative mood that springs from anther's bad fortune – sadness. The negative state relief model was originally developed in an attempt to explain the fascinating, much replicated social-psychological effect that after harming someone, harm-doers tend to become help-givers, even when the one receiving the help is not the person who has been harmed. In contrast to the more specific guilt-expiation explanation for the harm-help effect, the negative state relief explanation proposed that the mere sight of another experiencing harm might lead to a general negative affective state akin to temporary sadness or sorrow. This negative state, it was reasoned, would tend to produce helping because people learn that there is a sense of personal satisfaction that accompanies the performance of good deeds. Helping, therefore, could be used instrumentally to dispel the negative mood state caused by witnessing or causing harm to another.

Cialdini and his colleagues found support for the negative state relief analysis of the harm-helping effect by having subjects either perform or witness a transgression in which another person's neatly organized work was accidently knocked to the floor in disarray. The subjects were then given a chance to help a third party. Both types of subjects helped more than subjects who neither did nor saw any harm. Further support for the negative state relief analysis was provided by research showing that subjects who received a rewarding event designed to relieve sadness (the unexpected receipt of money or praise) between a transgression and an opportunity to help no longer exhibited the harm-helping effect. That is, it seems that if some other mood-raising event came along to lift their spirits, saddened subjects no longer needed to help. It was only those individuals who needed to help as a way of improving their moods who chose the helper's role.

It is important to note that the negative state relief model only applies to temporary sadness or sorrow that potentially could be relieved by helping. Chronically depressed individuals would not be expected to be more helpful because such individuals have lost, or never acquired, the ability either to become more saddened by exposure to another's need or to dispel their depression by helping. This expectation is supported by a wide variety of personality research showing that more helpful individuals tend to be those not suffering from chronic depression or mental illness.

A further qualification of the negative state relief model, which is also part of the Piliavin distress reduction model, is that the negative affective response to another's suffering will lead to helping only if helping is a relatively uncostly, gratifying way to dispel the negative affective state. Temporarily saddened individuals facing the prospects of large costs and small rewards for helping would not be expected to help because it would not be perceived as gratifying overall. Support for this argument comes from research assessing the effects on helping of varying the costs and benefits of acting prosocially. For example, Weyant (1978) found that negative mood subjects helped more than neutral mood subjects only when they were offered a helping opportunity involving both low costs (sitting at a donations booth) for high benefit (the American Cancer Society). Thus it appears that sadness increases helping only when it involves an overall reward value that is, on balance, sufficient to dispel the mood.

The Piliavin distress reduction and negative state relief models have generated considerable research, perhaps because both emphasize subtle, easily overlooked ways in which helping leads to some basic form of self-benefit. Other research has looked at the effects on helping of more blatant forms of self-gratification. For example, early research in the 1960s based on a behaviorist perspective found that the helping of young children could be conditioned by making helping instrumental for satisfying a 'sweet-tooth' through the receipt of a gumball.

Several common threads, however, run through all research on the effects of both subtle and blatant forms of basic self-gratification on helping. One is the assumption that personality factors play a relatively small role in affecting prosocial behavior. Because basic needs, whether to reduce aversive arousal, dispel temporary sadness, or satisfy a sweet-tooth, are seen as universal (or nearly so, except for special populations like the chronically depressed), situational factors that affect the magnitude of these needs and how they can most easily be satisfied are seen as the major determinants of helping behavior.

Another common assumption is that there is nothing inherently 'special' about helping that makes it a preferred form of behavior for satisfying basic needs. Indeed, under appropriate circumstances it is assumed that alternative forms of action, such as escaping further exposure to another's needs, dispelling sadness through less costly forms of self-gratification, or satisfying a sweet-tooth by stealing a gumball, would be preferred over helping. Thus, factors that influence the quantity of helping guided by the orientation to the other as object are seen as having effects on when and how basic physiological and psychological needs are satisfied by helping. The helper is seen as a 'comfort' seeker rather than a 'care' giver.

Determinants of helping guided by the orientation to the other as evaluator

Critics of the view that people help only in order to satisfy basic needs can find some comfort in research showing that helping can be motivated by the desire to satisfy higher-order psychological needs of belonging, acceptance, and self-esteem. We conceptualize such helping as being guided by the orientation to the other as evaluator. The research we describe in this section shows that there can be a 'kinder and gentler' view of helping because helping can be motivated by a desire to be seen by others and by oneself as a 'kind and gentle' (rather than selfish) person.

Children in many societies often learn early on that parents, teachers, and significant others want them to act prosocially. This can be communicated to children in many ways; by exhorting them to be helpful, by modelling helpfulness, and by showering them with material or social rewards such as treats or praise when they help. In the short run, some of these techniques are more successful in producing helpfulness in children than others. For example, exposing young children to models who practiced helpfulness was shown in early research by social learning theorists to be more successful than exposing children to models who preached but did not practice helpfulness. Also in the short run, the mere presence of an adult will lead children who have apparently learned the social appropriateness of orientation to be more helpful than when they are anonymous.

But in the long-run, two types of factor seem to influence whether teaching children the social value (or social norm) of helpfulness leads to their later helpfulness as adults. The first involves differences in how much people care about whether others evaluate them positively for being helpful or negatively for not. Research on the personality characteristics of criminals suggests that one indicator of antisocial behavior patterns is an indifference to what others think of one's actions. Similarly, there may be individual differences in the degree of concern about others' evaluations of oneself that are related to adult patterns of prosocial behavior. One such individual difference might be the extent to which people possess the personality trait of self-monitoring, the tendency to orient to others and attempt to make favorable self-presentations (see Snyder, 1987). Situational variables also might influence the degree of concern over other's evaluations about oneself. For example, one of the characteristics of the psychological state of deindividuation, which can be produced by feeling 'lost' in a crowd, is lessened concern about evaluations by others. At present, though, we know of no systematic research on the variables associated with lessened concern about evaluations by others. Hopefully such research will be pursued in the future.

In contrast, the second type of factor that would seem to influence whether learning norms of helpfulness establishes a tendency toward helping has received considerable research attention. This involves differences in the degree to which people adopt (or internalize) norms of helpfulness and consequent differences in the prominence of such norms in guiding their behavior. An important assumption underlying this research is that learning that helping is socially valued is not enough to produce helpful behavior; the individual must be motivated to act in accordance with learned helping norms. In the absence of external social sources of motivation to comply with social norms, the motivation must come from within. It is assumed that internalizing social norms of helpfulness provides this motivation, and that individuals differ in their degree of internalization.

Research on the relation of internalization of helping norms to helpfulness is decidedly mixed. A typical example comes from research on individual differences in the norm of social responsibility. This social norm dictates that we should help those in need, especially those who are dependent on us for help. Berkowitz & Lutterman (1968) have measured individual differences in adherence to this norm by a pencil and paper questionnaire on which respondents are asked to indicate whether they think people should act in socially responsible ways in different social dilemmas (such as when a cashier gives back too much change from a purchase). Individual differences on the personality measure are sometimes strongly related, sometimes weakly related, and sometimes unrelated to measures of helpfulness.

Critics of normative explanations for helping have pointed out several reasons why adherence to social norms is often not strongly related to helping behavior. For example, they point out that norms are often so general that they may not tell us what to do in particular situations, and that conflicting norms may seem equally applicable in certain situations (as when some situations draw attention to a norm not to meddle in other people's affairs as well as a norm to be helpful). More recently, normative explanations for helping have been given new life by Schwartz's (1977) concept of personal norms – feelings of moral obligation to act in a given way in a given situation. Schwartz's approach incorporates, among other things, several lessons learned from research on when people will act in attitudinally consistent or inconsistent ways: (1) to predict a specific behavior, the specific attitude toward that behavior must be measured, and (2) the behavioral intention to act, or feelings of moral obligation to act, in a certain way will have a more direct influence on behavior than will more general social norms. Making use of these principles, Schwartz has found that questionnaire measures of endorsements of personal norms are often useful in predicting specific prosocial actions.

In Schwartz's theorizing and research on personal norms one also can see how concern for enhancing social-esteem by being helpful can be transformed into concern for enhancing self-esteem by being helpful. Schwartz suggests that adherence to social norms of helpfulness, because they are so general, will be somewhat, but imperfectly, related to more specific personal norms. Personal norms are also influenced by other factors, such as whether previous experiences with the specific helping act have been rewarding or punishing. Personal norms, then, will be more highly related to helping than social norms, because in addition to possibly enhanced social-esteem for being helpful, the individual antici-pates enhanced self-esteem for acting in accord with personal norms and avoids guilt for failing to do so. Thus, in the effect of personal norms on helping can be seen a greater reliance on self-evaluation as a guiding principle, but social norms and concern with evaluations by others, as possible influences on personal norms, are certainly not left out of the picture.

Even if norms of helpfulness have been internalized and are relevant in a specific situation, such norms would be expected to influence helping behavior only if they are accessible and salient to the individual contem-plating help. Research by Gibbons and Wicklund (1982) and other researchers has shown that exposure to self-focusing stimuli (e.g., mirrors, one's tape-recorded voice) can lead to increased prosocial behavior. Presumably, exposure to self-focusing stimuli makes one's attitudes, including attitudes about helpfulness, more accessible, prominent, and influential in guiding one's actions. Under such conditions, people in general would be expected to be more helpful. This effect of increased self-focused attention on helping, however, is highly circumscribed. If the appropriateness of helping is not readily apparent, if self-preoccupation precludes noticing the other's needs, or if the self-focused state can be very easily terminated, then helping would not be expected to increase. So although situational factors such as self-focused attention can influence the impact of norms and attitudes on helping, the effects of such factors would not seem to be far-reaching.

To conclude this section, we reiterate that helping guided by an orientation to the other as evaluator is directed toward the goal of attaining positive evaluations from others, and by extension, from oneself. Such helping is intended to demonstrate unselfishness, but what it really demonstrates is selfishness in the unique form of concern for self-evaluation. With this conclusion we now pose the question: 'Are people ever genuinely altruistic in helping another?'

The possibility of helping guided by the orientation to the other as individual

Many of the more optimistic philosophies of human nature, and certainly many religious ideologies, propose that people can transcend selfish tendencies and truly care about others. To assess the validity of these claims, one can retire to an armchair for some serious speculation or one can attempt to observe people's behavior for any indication of genuine concern for the welfare of others. We hasten to add that there may be strengths as well as follies in the adoption of either type of inquiry, but social and personality psychologists have opted for the second approach. It may be surprising to learn that psychologists have made what we feel is definite and important progress in addressing so philo-sophical an issue as human altruism. Perhaps not so surprisingly, however, the issue is not at present resolved.

Psychologists have nominated vicarious empathic emotion (empathy) as the link to unselfish altruism. Conceptions of empathy vary greatly among psychologists, but one particular conception has emerged as most useful in the study of the possibility of genuine human altruism. Vicarious empathic emotion is conceived as an emotional response to another's suffering that is characterized by feelings of sympathy, compassion, tenderness, and the like. The suggestion that this emotional response leads to behavior directed toward the ultimate goal of benefitting another has come to be called the empathy-altruism hypothesis.

It is largely through the research efforts of Batson and his colleagues that the empathy-altruism hypothesis has advanced from a theoretical possibility to its current status as an arguably plausible explanation for some human helping behavior. (The empathy-altruism hypothesis is pro-posed as an explanation for some helping because it in no way denies that helping also can occur to satisfy basic or higher-order needs of the helper.) Batson's early efforts to test the empathy-altruism hypothesis involved testing whether empathy leads to selfishly oriented, arousal reduction motivation of the sort discussed in the Piliavin arousal/cost model of helping. These early efforts proved quite successful in demonstrating that sympathy evokes motivation to help that is qualitatively different from the motivation evoked by aversive personal distress (Batson, 1987). A typical study would involve leading subjects to experience either distress or empathy toward a victim. Subjects would subsequently be given an opportunity to help the victim under conditions in which *not* helping would allow the subjects to escape the situation (and the victim's suffering) or would not. The research consistently showed that empathy leads to high rates of helping, regardless of whether subjects could easily escape the victim's suffering without helping. In contrast, distress leads to high rates of helping only when escape is difficult. Together these findings suggest that a distressed person will help only if helping is instrumental in reducing his or her own distress, but an empathic person will help mainly for the victim's benefit.

But if empathy does not lead to selfish, arousal-reducing motivated helping, it remains possible that some other form of selfish motivation accounts for empathic helping. Fultz and his colleagues (1986) noted that helping responses in Batson's earlier research were not anonymous, so they conducted research to test whether empathy might be accompanied by increased concern about avoiding negative evaluations by others for not helping. A selfish, impression-management explanation for the empathy-helping relationship was not supported by this research; that is, empathy was found to lead to high rates of helping even under conditions in which subjects believed that they would remain fully anonymous if they declined to help.

Even more recently, Cialdini and his colleagues (1987) have proposed and tested a selfish, mood-management explanation for the empathy-helping relationship. These researchers found that empathy for a victim is normally accompanied by saddened mood; they further showed that when helping was not necessary for or instrumental to relieving this sadness, then empathy did not lead to increased helping. However, this support for a selfish, negative state relief explanation for the empathy-helping relationship has not been consistently corroborated by other researchers. Thus, although we recently seemed to be tantalizingly close to having a definitive answer about the nature of the motivation to help evoked by empathy, the empathy-altruism hypothesis apparently has been restored to its plausible status.

Is the validity of the empathy-altruism hypothesis, with its important theoretical, philosophical, and practical implications, destined to remain debatable? We hope not, but it is our feeling that further progress in testing its validity will require a new research approach. Previous research has tested the empathy-altruism hypothesis *indirectly*, that is, it has tested whether an alternative selfish form of motivation can account for the empathy-helping relationship. If the empathy-altruism hypothesis survives tests of one form of alternative selfish motivation, this merely means that still other forms of alternative selfish motivation remain viable as explanations.

What are needed, then, are *direct* tests of the empathy-altruism hypothesis, tests of predictions derived directly from the hypothesis, rather than just tests of predictions derived from alternative selfish models of helping. Even better, multiple tests of several (ideally, of course, all) alternative selfish explanations should be performed simultaneously with a direct test of the empathy-altruism hypothesis. Such research, admittedly, will be difficult to conceive and conduct, but the challenges of addressing the possibility of human altruism have not stymied researchers so far, and we doubt that they will be stymied in the future. We join with all who are interested in human nature in looking forward to the results of such future efforts.

Summary

In this chapter we have shown that advances in knowledge of the situational and personality determinants of the quantity of helping have been accompanied by advances in knowledge of the determinants of the quality of helping. There is considerable evidence that helping can be guided by the selfish orientations to the other as object and the other as evaluator. There is some inconclusive evidence that helping can be guided by the unselfish orientation to the other as individual. Additional research is needed to determine if conceptions of prosocial behavior should be broadened to include genuine human altruism.

References and further reading

Batson, C. D. (1987). Prosocial motivation: Is it ever truly altruistic? In L. Berkowitz (ed.), *Advances in experimental social psychology*, Vol. 20, pp. 65–122. Orlando: Academic Press.

Berkowitz, L. & Lutterman, K. G. (1968). The traditional socially responsible personality. *Public Opinion Quarterly*, **32**, 169–85.

Cialdini, R. B., Baumann, D. J. & Kenrick D. T. (1981). Insights from sadness: A three-step model of the development of altruism as hedonism. *Developmental Review*, 207–23.

Cialdini, R. B., Schaller, M., Houlihan, D., Arps, K., Fultz, J. & Beaman, A. L.

(1987). Empathy-based helping: Is it selflessly or selfishly motivated? *Journal of Personality and Social Psychology*, **52**, 749–58.

Fultz, J., Batson, C. D., Fortenbach, V. A., McCarthy, P. M. & Varney, L. L. (1986). Social evaluation and the empathy-altruism hypothesis. *Journal of Personality and Social Psychology*, **50**, 761–9.

Gibbons, F. X., & Wicklund, R. A. (1982). Self-focused attention and helping behavior. *Journal of Personality and Social Psychology*, **43**, 462–74.

Hoffman, M. L. (1981). Is altruism part of human nature? *Journal of Personality and Social Psychology*, **40**, 121–37.

Piliavin, J. A., Dovidio, J. F., Gaertner, S. S. & Clark, R. D. III. (1981). *Emergency intervention*. New York: Academic Press.

Schwartz, S. H. (1977). Normative influences on altruism. In L. Berkowitz (ed.), *Advances in experimental social psychology*, Vol. 10, pp. 221–79. New York: Academic Press.

Snyder, M. (1987). *Public appearances/private realities: The psychology of self-monitoring*. New York: Academic Press.

Weiss, R. F., Buchanan, W., Alstatt, L. & Lombardo, J. P. (1971). Altruism is rewarding. *Science*, **77**, 1262–3.

Weyant, J. M. (1978). Effects of mood states, costs and benefits on helping. *Journal of Personality and Social Psychology*, **36**, 1167–9.

8

Perceiving the causes of altruism

WALTER C. SWAP

The great majority of social-psychological investigations of prosocial behavior have focused on laboratory or field experiments of bystander reactions to a person in distress. These studies have generally dealt with those situational variations and cognitive processes that increase the likelihood of intervention in emergencies and other situations in which a victim requires aid from a bystander (see Fultz and Cialdini, Chapter 7). Among the factors that have consistently predicted helping or apathy are the number of other bystanders present and their reactions to the emergency, the clarity or severity of the emergency, mood states of potential helpers, and empathy toward the victim.

Very little attention has been paid, however, to how these prosocial behaviors are actually viewed by others, what interpretations or attributions observers make for helpers' actions. When are they viewed as altruistic or merely helpful? When are they praised, ignored, or even treated with contempt? Despite the absence of research attention, these judgements may assume considerable importance for a number of reasons.

First, as Krebs and Miller (1985) suggest, assessing the implicit theories of altruism held by average people is a useful approach in clarifying definitional problems. Specifically, it is important to be clear about the nature of those behaviors we are labeling 'altruistic,' and that there be some correspondence between these operationalizations and altruism's everyday connotations. Heider's (1958) call for a 'naive' psychology in which theory is mirrored in everyday perceptions is echoed here. Some of the controversy over issues such as the very possibility of truly self-less behavior may rest on differing conceptualizations of the term itself. For

147

example, must altruism require personal sacrifice? Or, if the helper gains some benefit (including psychological benefits such as tension relief), however unplanned or unanticipated, does this necessitate a descriptor other than 'altruistic'?

Second, the assignment of credit or blame for an action depends not on observation of the helper's behavior *per se*, but on the causal attributions for the behavior made by observers (Krebs, 1982). While the negativity bias in causal attributions ensures that instances of bystander apathy so frequently reported in the media will usually be condemned and attributed to essential depravity of the 'apathist,' it is unclear just when positive, prosocial behaviors will be explained by the inherent goodness of the actors. For example, even though a behavior results in considerable benefit for a victim, the helper might not be credited if the act was perceived as unintentional, fortuitous, or otherwise a product of external forces.

Third, as with other causal attributions, locating the source of the altruistic behavior either within the person or the situation helps inform us as to the actor's likely future behavior. Or, as Staub (1978) has written: 'To predict later behavior it is often necessary to understand what motivated a prosocial act; if the act was motivated by selfish intent it is less likely that the person would act prosocially under different circumstances' (p. 6). On the other hand, prosocial behavior caused by truly altruistic dispositions is likely to be viewed as predictive of similar behavior in other, perhaps quite different, contexts.

Finally, attributions made about an actor's motives for helping may influence the observer's own prosocial behavior. Hornstein's (1970) research, for example, shows that when a model's helping behavior is believed to be the result of social pressure, observers subsequently behave less prosocially themselves than when the model helped *despite* social pressures to the contrary. Such attributions provide cues as to the appropriateness of prosocial behavior in certain contexts, thereby serving as guides for one's own behavior.

If clarifying definitional issues, awarding credit, predicting future behavior, and modelling of behavior all depend not on the prosocial act itself, but on how it is perceived or attributed, then it is important to discover the implicit rules people use to interpret instances of prosocial behavior as altruistic. The remainder of this chapter is concerned with elucidating these rules. I will take as a starting point Ervin Staub's definitions of altruism, using it as a guide for discovering the rules people actually use in drawing causal inferences of altruism from prosocial behavior. The validity of these rules will be judged against empirical work, primarily my own.

For Staub (1978), '*prosocial behavior* refers to behavior benefiting others ... A prosocial act may be judged altruistic if it appears to have been intended to benefit others rather than to gain either material or social rewards' (p. 10). Prosocial behavior therefore requires no judgement of intention, no calculation of benefit or sacrifice to helper or recipient; it is sufficient that the recipient is helped. Staub's later elaborated definition of *altruism* is much more restrictive: 'behavior resulting in substantial benefit to the recipient, behavior that demands great self-sacrifice from the actor and is intended to benefit only the recipient, not the actor' (1978, p. 6). An attribution of altruism, unlike mere prosocial behavior, thus necessitates a series of judgements: Was help intended? Was it delivered? How much? Did the actor endure some costs or enjoy some benefits? Were these intended?

The requirements for earning that embodiment of altruism, the Carnegie Hero Fund Commission medal, closely reflect these judgements, inferences and attributions: The act must be voluntary, must involve an extraordinary risk, must be directed toward a non-relative, and must not be performed as part of one's occupational role (Rushton, 1982). Consider two simple examples: Mary, a professional lifeguard, rescues her son from drowning in the ocean. Good thing she was there. Doris, recovering from open-heart surgery, plunges into the surf to save the life of a man who fell from his boat. Give Doris a medal! Mary and Doris both *behaved* in an identical, prosocial manner. Many would agree that Doris also behaved altruistically; few would assign that label to Mary. (And besides, why did she let her kid get so far out in the first place?)

While psychologists (e.g. Cialdini, *et al.*, 1987) and philosophers (e.g. Rachels, 1987; Heal, Chapter 9) may debate the existence of pure altruism, it is not the purpose of this paper directly to address that issue. It is clear, however, that naive observers do use the term 'altruistic' to describe certain instances of prosocial behavior and the term 'altruist' to describe the perpetrators of those behaviors. But what are the underlying rules for such inferences? Staub (1978) discusses three reasons why people may behave prosocially: to benefit others, to benefit themselves, and to comply with personal norms. These three general reasons may serve as rough guidelines for predicting when altruistic attributions are made by observers. Many of these predictions derive from Jones and Davis's (1965) seminal analysis of *correspondent inferences*, i.e., inferring the existence of underlying traits from an observation of behavior. Most generally, 'the distinctiveness of the effects achieved [by a behavior] and the extent to which they do not represent stereotypic cultural values determine the likelihood that information about the actor will be extracted from an action' (p. 264). That is, to the extent that a behavior is extreme, surprising, out of role, or otherwise 'distinctive,' that behavior will be informative about what the person is 'really like.'

In the remainder of this chapter, I will discuss a set of four general conditions under which prosocial behavior is likely to be attributed to altruism on the part of the helper. Specifically, reflecting Staub's analysis, it should be the case that a prosocial act will be perceived as altruistically motivated when the act results in substantial benefit for the recipient, when the actor endures costs in the course of helping, and when the prosocial act occurs independently of, or contrary to, social or personal norms. Furthermore, these effects should be found only when the prosocial act is clearly intended by the helper. The discussion will be roughly guided by Jones and Davis's (1965) theory of correspondent inferences. Much of the empirical support derives from recent work by the author. A brief review of this methodology will help clarify subsequent references to the work.

Swap (unpublished manuscript) generated a set of 14 vignettes in which hypothetical people behave prosocially. Each vignette had a high- and low-altruism variation, tapping various aspects of the above four determinants of attributions of altruism. One vignette, for example, involves an elderly woman who asks someone ('Monica') to help her get home. In the high-altruism form, it takes Monica over an hour to locate the woman's apartment (high costs to helper), while in the low-altruism form, it takes but five minutes (low costs). Experimental subjects each responded to a set of high-altruism versions of seven vignettes and low-altruism versions of the remaining seven. Following each vignette was a set of rating scales measuring perceived altruism of the act and of the person, and praiseworthiness of the act. Respondents were asked to use their own definitions of altruism when making judgements, thus reflecting 'commonsense' or 'naive' (Heider, 1958) perceptions which can then be compared to more formal definitions.

Benefiting others

Of Staub's three reasons for behaving prosocially, only the first should relate positively, and perhaps monotonically, to altruism – the greater the good conferred on the recipient by the actor's behavior, the stronger the correspondent inference to an altruistic motive. As Jones and Davis (1965) argue: '[E]ffects assumed to be highly desirable are more likely to enter into attribute-effect linkages than effects assumed to be variable or neutral in desirability' (p. 227).

My research supports this notion, as people rated actors as more altruistic when their actions have clearly beneficial results. For example, when $20 is given to a panhandler, the donor is considered more altruistic when the money is used to buy clothing for a job interview than when it goes for wine; and a donor to a charity is viewed as more altruistic when the charitable organization clearly needs the money than when it has already reached its fundraising goals. Note that the prosocial *behaviors* in each case were identical. The extent to which the behavior benefited the recipient, however, strongly influenced perceptions.

Benefiting the self

Generally speaking, the greater the benefit that accrues to an actor as a result of his or her prosocial behavior, the less likely an observer will be to infer an altruistic motive. If the actor does benefit in some salient way, our naive observer would tend to 'discount' any internal altruistic motive as the cause of the behavior and would focus on the external incentives that might have been present. On the other hand, attributions of altruistic motives should be made to an actor who does not benefit from his or her action, or especially, who suffers as a result. 'Inferences concerning the intention to achieve desirable effects will increase in correspondence to the extent that costs are incurred, pain is endured, or in general, negative outcomes are involved' (Jones & Davis, 1965, p. 277). Observers would thereby be led to assume that, since the actor was willing to suffer adverse consequences, he or she must have been strongly internally motivated to aid the victim.

Batson (1987) considers the 'ultimate goal' of prosocial behavior to be crucial in differentiating egoism from altruism. If that goal is to benefit oneself, the motivation is egoistic; if it is to benefit others, the motivation is altruistic. However, it is often difficult to distinguish altruistic from egoistic motives. While Batson's elegant programmatic research argues strongly for the existence of truly altruistic, empathic motives, where the primary reason for helping is to relieve the distress of the victim, egoism can also motivate helping. If the awareness of a victim's suffering produces personal distress rather than empathy and compassion, and if the actor helps the victim in order to relieve that personal distress, then the behavior should not be labeled altruistic. Batson's 'altruistic motivation' has also been reinterpreted within an egoistic framework (e.g. Cialdini *et al.*, 1987; Dovidio, 1984). Cialdini and colleagues demonstrate in two experiments that empathy toward a victim may create sadness, and it is the egoistic desire to reduce this sadness that propels helping; ergo, it is not truly altruistic. (See chapter 7, but see Batson, 1987, for a rebuttal.)

Sociobiology would extend this analysis of self-benefit to include close relatives: a prosocial act, e.g., an organ donation to one's kin, should be viewed as less altruistic than identical behavior directed toward a stranger. Since one is, in a sense, benefiting one's 'own' genes by aiding, say, a son or daughter, such aid should be viewed as less altruistic than aid to a stranger.

In my own research, I have examined the effects of personal costs in a variety of ways. I have found, for example, that people are more likely to view an actor and his or her behavior as altruistic when the actor exerts a great deal of effort to help, has no ulterior motives for helping (e.g., the potentially grateful victim is physically unattractive as opposed to attractive), spends a great deal of time in aiding the victim (the aforementioned 'Monica' vignette), may suffer negative consequences (e.g., a poor grade in school) as a result of helping, or donates money he can ill afford.

These latter two examples also illustrate that while judgements of altruism are generally related to the praiseworthiness of the behavior, there are clearly examples where a prosocial act demands so much personal sacrifice that the altruist is judged to be foolish. For these two vignettes, there were no differences in how praiseworthy subjects judged the actors. Indeed, many volunteered in marginal comments that the actor was 'stupid' or 'crazy' to have given away lecture notes before studying them, or donated $50 to charity while on welfare himself.

Finally, the sociobiological prediction was supported by the finding that aid to one's offspring is viewed as less altruistic than identical help for a stranger.

Social and personal norms for helping

A similar explanation of this last finding can be made, of course, without recourse to sociobiology. Since parents are expected to help their children and are strongly socialized to do so, such aid should be viewed as relatively unremarkable, even expected, constrained by the parent role and therefore normative, not altruistic.

Personal freedom, or the perception of free will, has been an important concept in social psychology in such areas as cognitive dissonance theory, self-perception theory, and reactance theory. Not only is freedom of action valued in and of itself (Rokeach, 1973), but the knowledge that an individual behaved either freely or under some constraint leads to very different perceptions of that person's underlying attitudes or motives. While investigators have typically manipulated perceived freedom or constraint externally, e.g., informing an observer that an actor is either being forced (through threats or rewards) to make an unpopular speech, or is free to refuse to do so (Jones & Harris, 1967), constraints may function internally as well. The roles we choose or the values we develop constrain behavior by limiting the options with which we feel comfortable. People seem to be aware of this when forming impressions of individuals behaving under the constraints of powerful roles and social expectations. For example, Jones, Davis and Gergen (1961) demonstrated that observers are far less confident in assigning personality traits to people behaving 'in-role' than to those whose behavior violates normal role prescriptions. Their subjects were informed of the different personality characteristics deemed important for successful submariners and astronauts. They then heard job applicants reporting traits that were either congruent or inconsistent with these roles. Since reporting that one is 'sociable' is congruent with the role requirements specified for a submariner, the applicant's behavior could be due either to impression management or to actual possession of the trait. Subjects therefore expressed far greater confidence than those applicants who reported the inconsistent traits actually possessed them.

Applying this reasoning to prosocial behavior, Staub (1978) makes an important distinction between valuing others' welfare and valuing doing what is right. 'The former may lead people to focus on the welfare of others and to minimize their distress and enhance their well-being. The latter implies concern about adherence to norms and a focus on discharging one's duty' (p. 44). Presumably, prosocial behavior in service to the former would lead to an attribution of greater altruism, since one's 'duty' may be viewed both as a stronger constraint on behavior and as essentially egoistically motivated. Thus, for example, behaving prosocially in response to a specific request for aid or in reciprocation of an earlier favor should be viewed as less altruistic than similar behaviors divorced from these powerful norms.

This brief analysis suggests that an individual behaving prosocially either in a role that clearly supports prosocial behavior or in adherence to an internalized prosocial norm will be perceived as constrained by that role or norm. The observer will be more likely to explain the behavior as due to praise-seeking, outside pressure (e.g., the good bureaucrat following orders), or guilt avoidance, while true empathic concern for the victim will be 'discounted' as an explanation. The converse of this prediction is that an actor whose prosocial behavior actually defies normal role or value constraints will be perceived as strongly altruistic. Thompson, Stroebe, and Schopler (1971), in a rare use of hypothetical vignettes to study prosocial behavior, found that a student was evaluated more positively when he gave help to a janitor rather than to his professor. Help to the professor was presumably normative, hence any altruistic motive discounted, while assisting the janitor was supranormative, yielding some confidence that the student was indeed generous and considerate.

Again, my research supports this analysis. A young man clearly acting outside normal role expectations (a motorcycle gang member helping an elderly woman) is viewed as more altruistic than one whose chosen role (a boy scout) supports such behaviors. And a person who goes beyond normative prescriptions (returning tools in better condition than when they were borrowed) is viewed as more altruistic than one who merely complies with them (returning them in good condition).

Of course, the effectiveness of this attributional rule should depend on the familiarity of perceivers with the norms governing social behavior, and when those norms are being violated. For example, in the last example, female subjects were far more likely than males to view the tool borrower's behavior as altruistic when he repaired the screwdriver and sharpened the chisel, but not in the case where the tools were simply returned in good condition. Perhaps men are more familiar than women with the etiquette of borrowing tools, realizing that returning them in somewhat better condition is, like returning a borrowed car with a full tank of fuel, normative.

Intentionality

A fourth major factor influencing perceptions of altruism is intentionality. A person who intends to benefit himself or herself while helping another should be viewed as less altruistic than one who accomplishes the same result by accident or while intending some other outcome. 'It's results that count' may be true in football and the stock market, but not in person perception. Intentions to aid a victim are likely to yield greater attributions of altruism than will the same aid that occurs fortuitously. Indirect support for this hypothesis comes from research on reactions to favors, in which recipients are more likely to reciprocate when the help they received was seen as voluntary. Using a vignette approach, Tesser, Gatewood and Driver (1968) found that observers placed in the role of recipient indicated that they would feel more gratitude toward a hypothetical person who clearly intended to benefit them and who did not expect a reward.

My research clearly shows the importance of intentionality. When a young man undertakes a long bicycle trip expressly to raise money for charity, he was awarded the highest altruism ratings of all 14 protagonists in the study; if, however, he undertook the trip for purely personal reasons while, unknown to him, a group of friends arranged for people to sponsor the trip for charity, his altruism ratings were the *lowest* in the study. Somewhat more subtly, I found that a person who clearly goes out of her way to aid a victim is viewed as more altruistic than one who can escape the situation only with difficulty. In the former case, the helper shows through her actions that she intended to help; in the latter case, the situation created the opportunity. Batson has used the variable of ease of escape frequently in his studies of empathic and egoistic helping. Indeed, Batson considers help to be most altruistic when escape from a suffering victim would have been easy. Since the potential helper's own distress could have been alleviated by merely distancing himself or herself from the victim by escaping, those people who chose instead to refuse to escape and deal directly with the victim's suffering must have been behaving altruistically.

Spontaneity and 'premeditation'

One additional potential factor in judging altruism may be briefly mentioned, the extent to which the helper behaves either reflexively and spontaneously rather than premeditatedly. Some parallels with the literature on aggression might clarify this distinction. Altruism and aggression have frequently been analyzed together, perhaps because they seem to anchor a 'moving toward' – 'moving against' continuum, but also because they share many common dynamic elements. For example, paralleling the above analysis of altruism, definitions of aggression often stress intentionality and aversive consequences to the victim; an actor whose violent behavior is in response to strong provocation is often viewed as following an appropriate social norm and is not labeled as aggressive.

One factor that has been linked to perceptions of aggression is premeditation (Swap, in press), a relationship that is given considerable weight in most legal systems. The relationship between 'premeditation' or spontaneity and perceptions of altruism, however, is less clear. Severy (1974) indicates that altruistic acts must be spontaneous; otherwise one is led to believe that consideration of rewards might have guided the act. On the other hand, in his study of people who shielded or rescued European Jews during the Second World War, London (1970) seems to particularly value those rescuers who 'undertook the work deliberately and with benevolence aforethought' (p. 244). Thus, clearly planned prosocial behavior could be construed in two very different ways, as reflecting either attention to self-gain or careful consideration of how to maximize benefits for the recipient. Similarly, spontaneous help might suggest either behavior that was so automatic that no consideration of self-gain could have been intended; or behavior that was so mindless that it must have been constrained by some powerful external force. The inherently ambiguous meaning of spontaneous and premeditated prosocial behaviors therefore makes it difficult to develop general predictions about how such acts will be perceived. Indeed, in my research there were no differences in perceptions of altruism between a person immediately agreeing to donate money to charity and when she thinks it over for several days before pledging.

Conclusions

This chapter began with a discussion of four reasons why interpretations or attributions for helpers' actions may be important: Clarification of the definition of altruism, understanding the factors that result in praise or credit for prosocial acts, aiding observers in predicting a person's future behavior, and providing cues for guiding one's own future prosocial behavior. The analysis presented here is primarily focused on the first of these, resulting in a confirmation, within this person-perception context, of Staub's more formal definition of altruism. To paraphrase and augment Staub, we may consider a 'naive' perceiver's definition of altruism to be *behavior intended to, and resulting in, benefit to a needy recipient unrelated to the actor; that does not intentionally benefit the actor or, especially, that involves some sacrifice by the actor; and which occurs outside a normal helping role, or despite a help-inhibiting role.*

Armed with this conceptualization of altruism, we are in a position to interpret most instances of prosocial behavior as lying at some point along an altruistic–nonaltruistic continuum. Clearly, most such behaviors will not fall at the extremes characterized by the presence of all or none of the above attributes. The former are generally rewarded with medals, the latter with indifference. The great majority of helpful behaviors will consist of a complex mix of these factors, and our judgements of altruism will be correspondingly moderated. Further, the analysis does not indicate the differential impact of each factor: what is more important, benefit to victim or sacrifice to actor? Answers to such questions are clearly highly situation-specific.

The list of variables potentially relating to attributions of altruism is obviously not exhausted in this analysis. Additional variables that could fruitfully be examined in future work include: deservingness of the victim (is an act considered more altruistic when directed toward a victim who is more 'deserving' of aid?); and compensation (is an act viewed as less altruistic when it is an attempt to 'make up' for a previous harmful act toward the recipient?). These two examples suggest that a fifth determinant of perceived altruism, that of behavior intended to establish or restore *justice* could be well worth exploring.

Of the remaining three reasons for studying perceptions of altruism, only the first is directly addressed by the research reported here. A prosocial act is considered praiseworthy generally as a direct correlate of how altruistically it is viewed. Mother Theresa's unstinting self-denial in the service of others wins our praise and a Nobel Peace Prize. However, naive observers may draw the line somewhere prior to reaching near-saints. At some point, self-abnegation may yield judgements of foolishness or stupidity rather than praise. It may simply not seem 'normal' for a student to sacrifice a good grade to help a classmate or for a man on the dole to hand over most of his welfare check to charity. Such behaviors threaten the standards to which we hold ourselves, and may result in a defensive, even contemptuous reaction.

Finally, the clarification of perceptions of altruism helps us better to understand the social-psychological coin of the realm, social behavior. With a clear notion of how people interpret prosocial behavior as altruistic, we are in a stronger position to predict the future behavior of others, both altruists and those who observe their prosocial acts. Judging an actor's behavior as truly altruistic gives us confidence that similar behavior can be expected from that person in the future; in addition, an observation of an act we judge to be altruistic according to our definition may increase the probability of behaving altruistically ourselves.

References and further reading

Batson, C. D. (1987). Prosocial motivation: Is it ever truly altruistic? In L. Berkowitz (ed.), *Advances in experimental social psychology*, Vol. 20. New York: Academic Press.

Cialdini, R. B., Schaller, M., Houlihan, D., Arps, K., Fultz, J. & Beaman, A. L. (1987). Empathy-based helping: Is it selflessly or selfishly motivated? *Journal of Personality and Social Psychology, 52*, 749–58.

Dovidio, J. F. (1984). Helping behavior and altruism: An empirical and conceptual overview. In L. Berkowitz (ed.), *Advances in experimental social psychology*, Vol. 17. New York: Academic Press.

Heider, F. (1958). *The psychology of interpersonal relations.* New York: Wiley.

Hornstein, H. (1970). The influence of social models on helping. In J. Macaulay & L. Berkowitz, *Altruism and helping behavior.* New York: Academic Press.

Jones, E. E., Davis, K. E. & Gergen, K. J. (1961). Role playing variations and their informational value for person perception. *Journal of Abnormal and Social Psychology, 63*, 302–10.

Jones, E. E. & Davis, K. E. (1965). From acts to dispositions: The attribution process in person perception. In L. Berkowitz (ed.), *Advances in experimental social psychology*, Vol. 2. New York: Academic Press.

Jones, E. E. & Harris, V. A. (1967). The attribution of attitudes. *Journal of Experimental Social Psychology, 3*, 1–24.

Krebs, D. L. (1982). Psychological approaches to altruism: An evaluation. *Ethics, 92*, 447–58.

Krebs, D. L. & Miller, D. T. (1985). Altruism and aggression. In G. Lindzey & E. Aronson (Eds.), *Hardbook of Social Psychology, 2*. New York: Random House.

London, P. (1970). The rescuers: Motivational hypotheses about Christians who saved Jews from the Nazis. In J. Macaulay & L. Berkowitz, *Altruism and helping behavior.* New York: Academic Press.

Macaulay, J. & Berkowitz, L. (1970). *Altruism and helping behavior.* New York: Academic Press.

Rokeach, M. (1973). *The nature of human values.* New York: Free Press.

Rushton, J. P. (1982). Altruism and society: A social learning perspective. *Ethics, 92*, 425–46.

Severy, L. J. (1974). Comments and rejoinders. *Journal of Social Issues, 30*, 189–98.

Staub, E. (1978). *Positive social behavior and morality*, Vol. 1. New York: Academic Press.

Swap, W. C. (in press). Perceptions of aggression: A multivariate approach. *Journal of Social Behavior and Personality.*

Tesser, A., Gatewood, R. & Driver, M. S. (1968). Some determinants of gratitude. *Journal of Personality and Social Psychology, 38*, 291–300.

Thompson, V., Stroebe, W. & Schopler, J. (1971). Some situational determinants of the motives attributed to the person who performs a helping act. *Journal of Personality, 39*, 460–72.

9

Altruism

JANE HEAL

Introduction

Altruism is liable to seem a puzzling notion. By 'altruism' I mean disinterested concern for the welfare of others. (Here, I think, I am broadly in agreement with the accounts given elsewhere in this book.) In what way is it puzzling? On the one hand we are inclined to think of it as admirable, as something we can promote, and as something which does exist, even if to a lesser extent than we might like. On the other hand, however, sceptical thoughts keep recurring. These take different and not always compatible forms. One doubt concerns the coherence of the notion. Another suspicion is that altruism, even if logically possible, is much rarer than supposed and may not exist at all. And a third doubt focuses on whether altruism is wholly admirable; will it not, at least from the agent's point of view, be in some sense irrational? I shall try to disentangle a few of the many factors which lead to these difficulties and to put forward some of what might be said by a defender of altruism to allay the sceptical worries.

This book as a whole is about cooperation, prosocial behaviour, trust and commitment. What links these with altruism? Some kinds of cooperative and prosocial behaviour can be understood without bringing in the notion of altruism at all, as Rubin stresses in his chapter and as we also see in many of the cases of animal behaviour described by Harcourt. Very often enlightened self-interest will lead to cooperation between individuals or groups. Conflict settlement can be promoted, in self-conscious creatures like ourselves, by setting up structures (of negotiation, third party mediation etc.) to encourage the needed enlightenment in the

159

self-interested parties. Related senses of 'trust' and 'commitment' emerge in these contexts. I trust you in so far as I see that you can be relied on to cooperate and you are committed to me in that you fully intend to follow what you see is your sensible policy of cooperation. It is clear that many situations can be described in these terms.

But other kinds of prosocial behaviour do not at first sight fit this account, for example those found within families and between friends or fellow citizens. (Behaviour like this is discussed by Boon and Holmes; Lund; Fultz and Cialdini; and Swap among others in this volume.) These sorts of actions and relationships are ones that we are not, initially at least, inclined to describe as based solely on self-interest. Rather, we suppose, genuine concern for others is involved and also appreciation of these others as individuals, valuable in their own right. In these contexts, trust is seen as something evoked in me by awareness that another cares for me and commitment is thought of as something that I might undertake out of concern for another, precisely because I care for that other. But are the descriptions we are here tempted to give in any way misleading? Could it be that the 'altruism' they presuppose is really enlightened self-interest? The possibility of finding redescription in non-altruistic terms is considered in the paper by Fultz and Cialdini (Chapter 7) and is of some importance in the approach taken by Lund (Chapter 12). The idea clearly has an attractive simplicity. But does it represent the whole truth about human motivation? It is because we need an answer to this question that an understanding of altruism is central to the topics of this book.

The following discussion is aimed at clarifying the everyday notions in terms of which we describe and explain the actions of individuals. My assumption (in common with many contributors to this book) is that explanations of human behaviour in terms of conscious beliefs and desires are, very often, proper and illuminating. In particular cases we may be wrong, but I shall take it that the categories and general apparatus of commonsense thinking are in order. The issues to be discussed are ones that arise when we reflect on the everyday picture. Many of these issues have a long history and in what follows I shall be drawing substantially on points made in the philosophical tradition and by some recent writers. (A brief annotated bibliography will be found at the end of the paper.)

Further questions exist about how commonsense psychological notions are to be integrated with scientific understanding, in particular with our picture of the workings of the central nervous system and with the evolutionary story of how we came to be as we are. Both of these have seemed to threaten commonsense understanding, in whole or part, for example by leading us to reclassify conscious states as mere epiphenomena or by making it difficult to see how creatures with the thoughts and feelings we take ourselves to have could have come to exist. These issues are extremely important but are not the focus of the discussion here.

Pleasure

I shall start by considering, but not in great detail, a bold theory with a long pedigree, on the basis of which sceptical doubts about the coherence of altruism can be raised. This theory is hedonism, namely the view that we do all that we do in order to avoid pain and secure pleasure for ourselves. The theory deserves a mention because it is so venerable and attractively simple. But I shall not discuss it at length, partly because it is not a major concern of other contributors to this volume and partly because we can, to some extent at least, defuse it without entering into all the complexities of the subject.

What is pleasure? In its central use in ordinary language the word picks out that state of mind which naturally expresses itself in human beings by smiling, looking happy, saying 'I enjoyed that' and the like. By transference it is used for activities ('pleasures') which typically give rise to these expressions – eating, drinking, dancing, sports etc. Doing something for pleasure is doing it in the expectation that it will be of this kind – that one will emerge from the activity spontaneously smiling, saying 'I enjoyed that' and so on.

It is extremely implausible to suppose that we always act with a view, either short or long term, to pleasure in this sense. This sort of pleasure connects with lightheartedness and is contrasted with a variety of more weighty concerns, for example, duty, justice, compassion, pursuit of ideals and so forth. These are all thought of as things which require us to sacrifice lighthearted pleasure, both in the long and the short term, and in the name of which many people do sacrifice it. The hedonists's claim, if it is to have any credibility, should be that being just, compassionate, securing ones ideals and so forth are themselves sources of their own distinctive pleasures and that they are sought for that reason.

If this is to be an interesting and defensible idea we need to introduce some new sense of 'pleasure'. The problem, however, is to do so without rendering the hedonist's view empty. The more he insists that *every* kind of project is and *must* be undertaken with a view to pleasure, the less chance there seems to be of finding some criterion for whether pleasure is sought other than the fact that action is attempted. The danger is that the remark 'she acted to secure pleasure' may come to say no more than 'she acted'.

But even if we concede the hedonist's claim and agree to say that pleasure is the ultimate goal of every action (although the next section will suggest that we can resist the concession), the interesting problem will reappear at a different place in the system. Now the pressing questions will be 'What do people take pleasure in? Do they ever take "disinterested" pleasure, i.e. pleasure simply in the fact that another human being is flourishing? Or is it the case that they take pleasure only in states of themselves?'

So let us set the topic of pleasure aside for the moment (although it will reappear briefly) and move to consider that kind of scepticism about altruism which draws our attention to more specific self-involving goals.

Detecting disinterestedness

How would disinterested concern manifest itself in behaviour? A first, quick answer is 'in behaviour which benefits others'. But, as Fultz and Cialdini indicate, it is plausible to suppose that many cases of other-benefitting behaviour produce changes in the state of the agent, for example increased likelihood of future help for the agent, approval from observers, increased self-esteem and so forth. Hence desire for these things provides an alternative explanation for the behaviour. It is unconvincing to deny the existence of desires for such effects and their importance in our lives. There has perhaps been hypocrisy about this and some scepticism about altruism may stem from reaction against that hypocrisy. But even if we recognize the existence and importance of such motives, it does not follow that disinterested concern is entirely non-existent. The question is whether such anticipated benefits to the agent are always the sole explanation of the action.

If disinterested concern existed it might act in tandem with self-interested motives. We often do actions for which we have more than one adequate reason. But in these mixed situations concern for others would be hard to demonstrate to someone with sceptical inclinations. Altruism will show up convincingly only in the choice of actions which benefit others at some cost to the agent without providing outweighing and compensating benefits to the agent.

So is it the case that benefits to the agent (increased reputation, self-esteem etc.) always outweigh costs? Measurement is exceedingly difficult here. Yet we should avoid saying that the benefits *must* outweigh the costs because otherwise the agent would not act. This is to make non-altruism a matter of definition and so to make it empty.

One may argue that there is at least one substantial benefit to any agent which I have not mentioned, namely that he or she does not suffer the regret which would have been felt had the 'altruistic' act not been performed. Let us suppose that in some particular case of other-benefiting behaviour the agent feels no regret, would have felt regret had the alternative choice been made, and knows that this is so. Now, one may say, the agent has the satisfaction of achieving what he or she wanted and of avoiding regret. And are the satisfaction and avoidance not self-involving goals, which will be sufficient to outweigh any costs and the securing of which was the object of the action?

We can press this line of thought in any case whatsoever where an agent acts clearsightedly and does not later regret a decision. What we have come up with is a version of hedonism. The sceptical threat to altruism implicit here is that of seeing the welfare of the other as something not desired for its own sake but desired only as a means to the satisfaction of the agent. It is implausible to deny that an agent will be aware of the likelihood of achieving his or her own satisfaction by acting and it is clearly often the case that the welfare of the other is a causal means to that state of the agent. It also seems strange to deny that the state of satisfaction was an attractive prospect to an agent. How then can we attack a theory which presents it as the agent's sole goal?

Such a theory is attractively simple, yet there is something paradoxical in it: the prospect of satisfaction is being invoked in a story which downgrades the other's welfare from an end in itself to a mere means. Yet is not 'getting satisfaction in the welfare' precisely what we would expect if the welfare were wanted for its own sake? How could a person not take satisfaction in getting what he or she wants for its own sake?

There are perplexities here, having to do particularly with the notion of 'satisfaction'. But I shall not pursue them. Instead I want to look at a general methodological difficulty in defending altruism which is highlighted by the argument just sketched. I shall suggest a line of thought for countering it, which, if successful, will weaken both the claim that satisfaction is our real goal and the claim that more particular states, like good reputation, are what we seek.

The methodological difficulty arises from the fact that agents must, in general, get feedback from the world in order to be effective. We cannot operate in any extended way without being guided by awareness of the upshot of what we do as it unfolds. So, in general, any change in the world produced by me will be accompanied by some correlative change in me. So the form of a sceptical move, which names the change in me as the goal rather than the change in the world, seems always to be available.

A possible attack on the empathy-altruism hypothesis provides another example of the strategy in operation. This hypothesis attempts to defend the idea of disinterested concern. It suggests that a person may become aware of another's suffering, by empathetically sharing the other's point of view, and may be moved to act solely by desire to relieve that suffering. But, as Fultz and Cialdini point out, the hypothesis assumes that the altruist becomes sad on empathizing with the suffering of the other. So why should we not say that offering help is the supposed altruist's way of relieving his or her own sadness – i.e. benefiting him or herself? This is a move which can, it seems, undercut the claims for disinterestedness, even when the influence of concerns like monetary reward or reputation has been eliminated.

We may again feel that there is something paradoxical in the move. It would be a strange kind of caring which left the carer equally happy whether he took the other to be suffering or content. And yet this feature, which is naturally taken to be a necessary corollary of the genuineness of the disinterested concern, is being used to discredit it. The sceptic is right in saying that the helper benefits himself or herself by lessening the suffering of the other. But the initial view that the concern was disinterested makes us want to say that the help is not given for that reason or not solely for that reason. How could one defend this idea?

It is important to remember that, although feedback from the world is in general present when we act, it is not universal. We are not infallible or omniscient. What is the case and what we think to be the case can and sometimes do fall apart. Let us consider a thought experiment in which they get separated and where we know that they will do so. Imagine that you are kidnapped by some aliens. They present you with a console on which are two buttons, A and B. Your choice is to press one of the buttons. The aliens explain to you that they have your loved ones in their power. Whatever you do, you will never see them again, but their future fate rests in your hands. If you press button A they will be well looked after and live long and happy lives; you, on the other hand, will be rendered unconscious as soon as your finger touches the button and when you come round you will have had a set of delusive memories planted in your brain, according to which you betrayed them to suffering; you will be plagued with guilt and misery. If you press button B, however, your loved ones will be tortured, starved and degraded while you live happily, equipped with the delusive conviction that your nobility has preserved them. The aliens persuade you that they can do what they say. You see with your own eyes the effects of their brain manipulations on others and you become convinced of the extent of their power. The moment of choice is now; what will you do? If it were true that what we aim at is pleasurable states of the self, button B would be clearly indicated. Yet it is far from obvious that this would be the universal choice. And even when we run the thought experiment again making the potential victims into strangers rather than loved ones, intuition delivers no clear verdict of a universal choice of B.

The thought experiment as sketched bears upon the existence of an extremely strong form of altruism. But to be altruistic is not to value others' welfare more than ones own or to value it so much that one is prepared to make immense sacrifices. It is to value it disinterestedly to *some* extent, to be willing to make *some* sacrifice. It is therefore worth considering also much less drastic versions of the thought experiment. Suppose, for example, that the aliens present the following choice: if you take button A certain people will be much benefitted, but you will lose all memory of the choice and suffer bad toothache for a week; if you take button B the others will suffer, you will lose all memory of the choice and win £500 in a competition. In this case (and in many others we could devise by ringing the changes on the benefits and disbenefits) it does not seem extravagant or implausible to complete the thought experiment in one's mind with the choice of button A.

These fantastical scenarios certainly need not convince the sceptic about altruism. They are only thought experiments and our dispositions to find button A outcomes imaginable could be argued to show more about what we take ourselves to be than what we are. The sceptic may also object that the imagined conditions are just not possible and hence the whole supposed experiment is an absurdity. Even if he admits that the experiment could occur in reality and would sometimes produce the choice of button A, he is not required to admit the existence of altruism. He might speak of the agent at the moment of choosing button A anticipating feeling for an instant a pleasure so intense that it could rationally outweigh all future misery. Or, more plausibly, he could speak of the agent being unwilling or unable to face up to that shattering of his or her self-image as a caring person, which would be involved in the conscious choice of button B.

What the possibility of these latter two responses brings home is the fact that experimental evidence of behaviour in this or that situation is unlikely to provide conclusive evidence for or against altruism. Behaviour will always be open to various interpretations, which will require further experiments to discriminate them. Such experiments may become more and more difficult to devise. Scepticism about altruism needs to guard against the possibility of becoming uninteresting. After a certain point, determination to explain away proposed manifestations of altruism begins to look like *a priori* dogma rather than empirical caution.

What is the upshot, then, as far as the conceptual possibility and actual existence of altruism are concerned? The position seems to be this. We have found no reason to suppose that altruism is incoherent, but as to its actual existence we have no conclusive evidence one way or the other. The merely imagined outcomes of thought experiments (even if they were unambiguous) cannot prove anything. This is frustrating for someone moved by curiosity as to the springs of human action. But the thought experiment does indicate something interesting all the same, namely that we find altruism not merely imaginable but also intelligible and appealing. If we were told that a thought experiment situation had been made actual and that someone had chosen button A we would not say in amazement 'Whatever did she do that for?' Button A outcomes are not just scenarios we can describe without contradiction but choices we can make sense of and feel the attraction of. So we are at least capable of sympathy with and imaginative inclination to altruism, even if our possession of the thing itself remains problematic. This fact is surely of some importance to the educator and policy maker even if not to the pure scientist.

So far we have looked at worries about altruism that have to do with what goes on before and during some particular action. We have asked what situation the agent envisages as being brought about by the action and which aspects of that situation are important to the agent. I want now to step back and ask how altruistic concern, if it existed, would fit into a life as a whole. This perspective allows us to examine some further sceptical worries and to consider the interrelations of rationality, altruism and self-interest in a wider context.

Is altruism a good thing?

Is it a good thing to be altruistic? This is an unclear question. We must ask 'good from what point of view?' From the point of view of those benefitted, clearly it is. And depending on who those persons are, it may be a good thing from the point of view of having some genes replicate themselves. But is it a good thing from the point of view of the agent?

Suppose I am to some extent self-interested, both in that I have a number of particular self-concerned desires (that I should be well fed, well thought of etc.) and also in that I have a general desire to live a long, rich, fulfilled human life. Should I be glad or alarmed on my own account upon realizing that I have, or seem to have, altruistic inclinations? I reflect on the thought experiments and find that I can well imagine myself choosing button A in at least some of them. For all I can tell, I am altruistic. Should this please or worry me?

We do not have to take our tastes and concerns for granted. For example, a person with a very great liking for sweet things or a fascination with Russian roulette, might seek to eliminate or diminish it, because resisting the desire is difficult while succumbing to it may lead to loss of opportunity to satisfy other important desires. Assessment of desires can, however, be a highly controversial business. For example, someone might maintain that there were such things as objective values and that desires should be assessed by whether they were directed at the really valuable or at the merely meretricious. Let us try to keep to less contentious bases for assessment, for example, consistency with other central desires and like-lihood of contributing to a life of rich and varied satisfactions.

How does altruism fare when looked at in this way? Concentration on some of the cases discussed in the previous section may make it seem comparable to possession of a sweet tooth or fascination with Russian roulette, namely an unfortunate taste, gratification of which is likely to conflict with other important goals and may lead to disaster. This appear-ance is what comes of emphasizing the disinterestedness of altruism and at the same time taking concern for others to be just one independent desire among a bundle of desires which together help to constitute a person's character. On this view, if a person acts altruistically he or she is, to that extent, merely a temporary instrument of some impersonal benevolence. The action, as altruistic, is not seen as tying up in any intelligible way with the agent's other projects or with the possibility of him or her living a desirable life. Of course other possible elements in a situation – e.g. good reputation and self-esteem – do tie up with the agent's projects and well-being. But real altruism is supposed to operate independently of these. Hence we are inclined to think that, even if altruism is logically possible, it cannot be anything other than a quirky and risky individual taste.

Is this the right way to conceive things? Let us come at this obliquely by considering a case which does not involve altruism but which does involve, so I shall suggest, another kind of disinterestedness. In this case the relationship between the various concerns may be easier to see. Imagine a group of actors putting on a play. Acting is clearly a self-interested pursuit to a considerable extent. The actor enjoys being in the limelight, holding the stage, getting the applause etc. Yet a kind of disin-terestedness can also have a place. To have the project of acting in any full-blooded sense a person must grasp that it is the shape and balance of the play as a whole which determines what he or she is required to do as an individual. The would-be actor thus needs to grasp that other actors also have to make their contribution to the production. Someone who envisages the project of acting entirely in terms of him or herself declaim-ing centre stage is likely to be a bad actor and, more importantly, will not be capable of feeling that satisfaction in the excellence of the production as a whole which will be available to another who is capable of taking up a more detached point of view.

But this capacity for appreciating the demands and success of the pro-duction as a whole carries with it certain risks. An actor who had it might find that his part, which had at first reading looked prominent, will require to be backgrounded if the overall interpretation is to reach greatness. He may get less of that enjoyable declaiming centre stage and less of the critics' attention, than if a superficially attractive but ulti-mately unsatisfactory reading is given. Of course an actor will choose his parts hoping that this choice will not arise. But suppose it does. Would it be sensible to wish then to lose that grip on the demands of the whole which faces him with the unwelcome decision? Should he say 'Away with these artistic scruples – they are just preventing me hogging the lime-light'? This would not obviously be the sensible course. He would be wishing himself back into a state of less insight, of less complex per-ceptions, and cutting himself off from those pleasures, e.g. knowing himself to be part of a really great production, which his current insight and imagination make possible. The pleasures he could secure by losing his artistic standards are ones which, from his current perspective, are not so worth having. We are here bordering on the difficult area of 'higher' and 'lower' pleasures and the objectivity of value. But even with less ten-dentious methods of assessing tastes and desires we can make sense of his not choosing to abandon his standards. His current artistically-informed projects and satisfactions are much more various than non-artistically-informed ones. They are also less likely to lead to frustration through conflict with the projects of others. So possession of his current tastes and concerns is a benefit to the actor, not in the sense of contributing to a longer life or more progeny but in the sense of contributing to a higher quality of life, to the richness, variety, complexity and stability of his possible satisfactions.

So having artistic standards and concern for the production as a whole is something the actor is not likely to want to be without, although it carries some risks. But exactly what risks? Have we got here what I am clearly manoeuvring for, namely an analogue to the disinterested concern of the previous section? The best we can do to shed light on this is to run a relevant thought experiment. This will be one in which the choice faced by the actor is between a button A outcome in which a magnificent production occurs, although he knows nothing of it and suffers some discomfort, and a button B outcome in which he thinks, falsely, that a great production has occurred. (Again we can ring the changes on the exact nature of the outcomes as we please.) Let us not make extravagant claims here. No one thinks that an actor ought to, or is likely to be willing to, lay down his lifetime's happiness for the sake of a good production. The suggestion is the much more modest one that we can at least make sense of an actor going for the button A type of choice. To imagine someone who fully appreciates the greatness of a great production is to imagine someone for whom this choice has at least some attractiveness.

My next suggestion, following on from this, is that the pattern found in the case of the actor and the play exists in many other areas of social life. The actor stands to the play as the parent does to the flourishing of the family, the citizen to the just and prosperous state of his country and so forth. The particular projects, standards, assumptions, distribution of roles etc. will vary greatly from culture to culture. In terms of the contrast discussed in the papers by Triandis (Chapter 4) and Stevenson (Chapter 5), the point I am emphasizing can be put this way. In any recognizable human culture, even the most individualistic and 'I' centred, there will be an immense repertoire of collectivist or 'we' concepts in terms of which an individual will conceive his or her goals and structure his or her life.

The possession of these concepts and the concerns and desires that go with them is a benefit to the agent – again not in the sense of something that necessarily contributes to long life or reproductive success but in the sense of contributing to quality of life. If all goes well, in the pursuit of these cooperative and social projects, there will be a role for the individual agent which will provide a great range of rich and complex satisfactions, of a kind which are unavailable to someone whose concern is limited to his or her own states and who is interested in other things only in so far as they are means to those states. There will, again if things go well, be no serious long-term conflict between the flourishing of an agent with social concerns and the flourishing of others in the project. It is precisely a complex totality, including in it the flourishing of others, that the agent desires.

But there is no guarantee that all will go well. There is a risk of conflict between what is needed for the best progress of the whole and some or all of what the individual had hoped for his or her particular role. Altruism may now be seen as the willingness to put the requirements of the whole first (e.g. the needs of other persons whose flourishing is required for the whole), despite the possible costs for the self. The suggestion I want to make is that the intelligibility and attractiveness of altruism is an inevitable corollary of a person's desiring and appreciating such complex wholes. Thus altruism – or rather its intelligible appeal to us – is not a quirky and tacked-on extra in our make-up. It could not be removed without producing substantial alteration to how we think about our lives and the projects we pursue.

The point of trying to bring out these links between our projects and desires is to dispel some of the mystery which altruism acquires when we think of it as an isolated impulse. But it may seem that, in making altruism less mysterious and in giving it these links with social projects, we have just played back into the hands of the sceptic. Have we reduced all motivation to self-interest again? This would be so, if what I have said implied that in acting on a desire, possession of which was a benefit, a person was necessarily pursuing self-interest. But this simply does not follow. It will depend what the desire is a desire for. And if it is a disinterested desire for the welfare of others, then acting on it cannot, from the very nature of the content of the desire, be acting to secure a benefit for the self. What benefits an individual or contributes to the richness of his or her life is an ability to look at, and feel concern for, things beyond him or herself. It will not do to argue from this premise – that such outward directed attention is a good to its possessor – to the conclusion that it is not after all really outward directed but inward directed.

Another point worth bearing in mind here is this. What has been suggested does not have the implication that our outward directed concerns are acquired or maintained by us with a view to the good that having them will do for us. We do not, in general, acquire our tastes and desires for any reason; we find ourselves with them as a result of biological and social conditions. If we set out to assess them we may, in the way outlined above, find a self-interested reason not to wish our sympathy with altruism diminished. But equally we shall find altruistic reason to wish the same. If I do care for others, then the thought that *they* will be worse off if I cease to care will encourage me to retain my concern, as much as will the thought that *I* will be impoverished. The question of which, if either, of these considerations is actually having influence with me runs parallel to the question which we tried to answer earlier with the thought experiment. And we are unlikely to do any better with resolving this question than we did with that earlier one.

Certain differences between intergroup relations and interpersonal relations within groups become apparent from this perspective. When groups of strangers meet, inevitably there is not yet in place any extensive framework of assumptions, joint projects and so forth of the kind that already structures the interactions of individuals within a group. Usually it is simply not the case that the welfare of this other group is seen as having an important role in any of the projects which the members of our group already find important. The likelihood then is that intergroup dealings will develop on lines shaped by each group pursuing what it perceives as its own interests. We may add also that the perceived responsibilities of political leaders push in the same direction. I may, on meeting a total stranger from another culture, be willing personally to make myself vulnerable and to take considerable risks in an attempt to build a relationship or joint understanding. And I may well be entitled to stick my own neck out, if I so wish. But I am not entitled to stick your neck out, or put you at risk, especially if I am charged by the political and power structures of my own group, with the job of seeing to your welfare. This is not to say that there is no possibility of building structures of intergroup cooperation which go beyond the negotiations of enlightened self-interest. But (as Feger suggests) exhortations to generalized benevolence will not get us far.

In very brief summary then, I have argued that altruism – or at least its intelligibility and appeal to us – is a necessary corollary of our being able to desire and find satisfaction in certain complex social wholes. And I have also claimed that the ability to appreciate and pursue these projects is something we can see reason to retain, whether we look at things from a self-interested or from a non-self-interested point of view.

Bibliography

A classic discussion of pleasure is to be found in Aristotle. *Nicomachean Ethics*, trans. W. D. Ross. Oxford University Press, Oxford, 1925. See especially Book X.

A modern treatment, much influenced by Aristotle, is G. Ryle, 'Pleasure'. In *Dilemmas*. Cambridge University Press, Cambridge, 1954.

Discussions of the relation of altruism, desire and pleasure are given in Butler. *Fifteen Sermons*, ed. W. R. Matthews. G. Bell and Sons, London, 1969. See especially I–III, XI, and XII. Hume *An Enquiry Concerning the Principle of Morals*, ed. L. A. Selby Bigge. Clarendon Press, Oxford, 1975. See especially Appendix II.

The idea of the thought experiment in the section titled 'Detecting disinterested-ness', together with a number of other points, are drawn from B. A. O. Williams. Egoism and Altruism. In *Problems of the Self*. Cambridge University Press, Cambridge, 1973.

Two recent stimulating books on moral philosophy and moral psychology which contain much material relevant to the issues of this paper are A. MacIntyre. *After Virtue*. Duckworth, London, 1981. Martha C. Nussbaum. *The Fragility of Goodness*. Cambridge University Press, Cambridge, 1986.

10

Complications and complexity in the pursuit of justice

SUSAN D. CLAYTON AND MELVIN J. LERNER

Introduction

As this chapter was being written, the President of the United States was talking about bringing General Manuel Noriega of Panama 'to justice'. Clearly bringing someone to justice was seen as a desirable goal, sufficiently so as to excuse the U.S. invasion of a foreign country. George Bush was operating on sound psychological territory here, since the goal of justice has been shown in both experimental and observational settings to be able to override other considerations. Even that basic principle of human behavior, the maximization of one's own benefit, is sometimes put aside in favor of the maximization of a fair distribution: studies have indicated that, in addition to feeling unhappy when they get less than they think they deserve, people are actually uncomfortable when they get more than they deserve, or when other people get more or less than they deserve. Given a certain set of qualifying conditions, people will try to act in a way that seems to be fair.

This is not to say that self-interest has no role to play. In many situations, where concerns of justice are not salient, the primary and accepted rule for behavior is probably the maximization of one's own comfort and profit. More insidiously, beliefs about what is just are often influenced by what is desired: some social scientists propose that people employ rules of fairness and justice to maximize other desired outcomes, either for themselves or their group. Still, very few would question the primacy of the theme of justice in public dialogues concerning the allocation of resources. Whatever the underlying psychological motives may be, it is easy to document the singular power of the theme of justice in

173

Western societies: the appeal to justice, the most 'sacred' secular value, can legitimize, if not require, the sacrifice of all other human values and goals, including life, liberty, and the pursuit of happiness.

The goal of justice not only may supersede motivations of self-interest; it is also a difficult goal to dispute. One can disagree with the decision maker who is out to maximize his or her self-interest, and other goals, though worthy, can also be given second billing ('environmental concerns have to take a back seat to economic concerns'; 'progress may require a loss of life'). Rarely if ever, though, will someone go on the record as saying: 'We can't afford to do it the fair way'. Although the pursuit of justice may be sidetracked or put on hold in day-to-day behavior, the validity of the pursuit will almost never be denied.

The phrase 'bring to justice' makes the legitimate assumption that justice is a desirable goal. It makes a further assumption as well: that there is a consensus about what justice will involve. In this particular case, the phrase is enough of a cliché for legal procedures and punishment that the way in which 'justice' will be operationalized is fairly clear. In general, however, justice is more complex than it may first appear. To some, and at some times, justice may require giving people their 'just deserts' – a retributive focus on punishing those who have misbehaved. To others, justice may imply equity through a distribution of resources that assigns outcomes in a way that is proportionate to inputs, or it may mean mandating equality of some sort among a relevant group. Justice may prescribe the means or the end; it may focus on the individual or on the larger group. Unfortunately, it is often impossible to satisfy simultaneously all different conceptions of justice. This multiplicity of definitions for justice, combined with the sense of being in the right which is conferred by working toward any one of the visions of justice, means that the pursuit of justice, rather than uniting people in cooperation and mutual agreement, may serve instead as an additional source of conflict.

Research indicates that the desire for justice in the abstract is nearly universal, but the way in which it is brought up, if at all, in regard to a particular situation can vary widely. Some of these variations depend on the abilities and proclivities of the individual, but just as a group is more than the sum of individuals composing it, the consideration of justice at a higher level of social complexity will be influenced in other ways than just by the concerns of the people involved. At the individual level, the justice of an outcome partly predicts satisfaction, and justice allows people to feel that they have control over their fate: if this is a just world, then my outcomes will depend on how deserving I am. At the societal level, justice serves also as a criterion for how well the society is functioning; and faith in justice may serve as a mechanism for social control, whereby the disadvantaged within a society refrain from revolution. We will examine, first, at the most basic level, some factors that affect an individual as he or she constructs a personal blueprint for justice.

The individual

A sense of justice and the ability to distinguish fair from unfair situations emerge very early in the average individual's development. Even a young child will probably admit to understanding that it is not just for him or her to have a larger piece of cake than the piece given to a classmate (an equality principle) or why Jill, who worked harder, should get two cookies while Jack only gets one (an equity principle). A slightly older child may even recognize that there are alternative, equally valid, justifications for distribution of outcomes in a given situation. The consideration of justice can, however, involve relatively sophisticated cognitive processing.

Conceptualizing justice requires an ability to think hypothetically, not only to contemplate whether state X is better than state Y ('Will it be better if I give this apple to my friend or eat it myself?'; 'Is my companion's ratio of outcomes to inputs better than mine?') but also to assess the degree to which each state approaches some ideal. Thus two fairly abstract mental processes may influence a determination of justice. The first involves the ability to envision alternatives; to go beyond what is to think about what could or should be. Great revolutions in science require this ability, and so do great social revolutions: it sometimes takes a visionary to point out that the present social order is not as fair as an alternative possible society. The second process requires the ability to consider two alternatives simultaneously in order to make a comparison.

Both of these abilities are related to a general 'intelligence' factor and to the ability to engage in abstract thinking – an ability some believe not to be acquired before about age 12 by most and not at all by some. This is not to say that one must be a genius to think about justice. But people who do not have or do not use these more abstract abilities will think about justice in a way different from those who do. The former may be cognizant of alternative possibilities only when their present situation has changed from a previous one, while the latter are able to give more thought to how their present situation *could* change.

Another relevant aspect of cognitive processing, perhaps best described as part ability and part tendency, is whether an individual focuses on a single facet of the issue under consideration or integrates a number of the relevant concerns. Most issues cannot be described in black and white. What may seem clearly to be the right choice when one aspect is given priority is no longer so clear when another aspect comes to the forefront. Some individuals deal with this difficulty by concentrating on a single dimension of a multidimensional problem, perhaps because they do not possess the cognitive ability to consider the complexity of the problem and perhaps because they do not possess the desire: partisans of a particular side of the problem may defend themselves against the threatening perception that there is any injustice in their position.

Such people may show more conviction in their cause than others who are more aware of the drawbacks of any potential solution. A particular course of action is likely to seem more just to those who see the situation more simplistically. Take, for example, the issue of abortion rights in the U.S.A.: the existence of a vocal anti-choice group who insist that the only issue is human life, and a minority on the pro-choice side who refuse to acknowledge any issue but women's reproductive rights, has the political function of making the question seem more clear-cut than it actually is, and increases the acrimony of the debate.

A final influence on the thought processes of the individual who is thinking about justice will be the extent to which the analysis is dispassionate or includes some motivational or emotional aspect. Most of us do not think about issues in the same way when they are abstract and theoretical as when they are more engaging, either because we are personally involved or because the situation is vivid enough to arouse our emotions. One of us (Lerner) has argued that topics that engage our emotions are likely to elicit reasoning based on psychologically deeper rules of morality and justice than the culturally-derived standards we otherwise rely on.

The degree to which a situation arouses attention and emotion may be a function of both its significance and its relevance to the individual. The trivial may be evaluated only in terms of whether it is normal: one of us (Clayton), studying perceptions of job discrimination, found that some subjects reacted to a salary disparity of $1000/year with the judgement that the difference was too small to constitute discrimination. The same goes for an incident that is uninvolving. If one hears about a stranger or slight acquaintance who lost his or her job in the reshuffling following a corporate takeover, 'That's life,' might be the response. 'That's the way business works.' But if the same thing happened to oneself or a good friend, the response might be very different: 'That's not *fair*! I'm good at my job! What have I done to deserve this?' An emotionally arousing issue is more likely to inspire consideration of whether it is fair.

The relationship

Issues of justice may arise in situations involving only a single person: a self-employed worker deciding on a fair rate for his or her services, for example. It is more likely, however, that the situation will involve more than one person, even if one is merely the observer of the other's fate; it has even been suggested that justice concerns arise only under social conflict. There will therefore exist a relationship between the involved parties. This relationship may be based on acquaintance and past experience, or on the roles to which society and culture have assigned the individuals. Clearly, one of the most important influences on whether an issue is emotionally engaging will be the relationships between or among the relevant actors and the person who is evaluating the issue. The social relationship can affect one's sense of justice in other ways as well: by making one a partisan for a particular side, by providing rules for behavior, and by describing a context in which a particular standard of justice is implicitly given priority.

The social relationship will determine the implications that the outcomes assigned to another have for the outcome to oneself. This sort of relationship varies according to setting as well as according to person: in some situations, a parent may feel that any benefits accruing to a child are indirectly enjoyed by him- or herself; in other situations, the parent and the child may be in competition for the same resources. So the relationship must be defined in terms of what holds true for these people in this setting: what is the reaction of the observer to the outcomes received by another? If the outcome to the observer is directly tied to the outcome of the other, as in a zero-sum situation, the observer is unlikely to delight in the other's good fortune. Even this, however, may be qualified by the relationship: some people may actually be capable of satisfaction when their friend wins a prize they had both aimed for. Depending on the relationship, the assignment of resources to the other may be experienced with vicarious happiness, apathy, or resentment. The observer's satisfaction with the outcome will probably influence his or her analysis of the justice of the situation.

One way of describing the relationship, which accounts for some of the differences in reaction to others' outcomes, is to distinguish between identity, unit, or non-unit relations. If the affected party is functionally the same as the observer (i.e., all Americans if the observer is an American; the father if the outcome affects the whole family; someone with whom the observer feels an empathic bond), the observer is likely to take pleasure in the benefits to this person, even if they involve a loss of benefits to him or herself. If the other person is not the same but similar (a friend, coworker, group member), our observer may feel pleased about positive outcomes that do not diminish his or her own resources, but envious and competitive concerning those that do (as when both are competing for the same promotion). Finally, positive outcomes to someone perceived as dissimilar, or in a non-unit relation, may evoke little reaction if the observer's outcomes are unaffected and hostility if they are.

The non-unit relationship may be perceived as so extreme by one or more of the parties that justice concerns are considered irrelevant. That is, there is evidence that personal definitions of justice include some description of a community to which these standards are applied and outside which they do not apply. The most heinous examples are found in race relations, where otherwise scrupulous and fair-minded white men saw no inconsistency in owning black slaves, or citizens in Nazi Germany were heedless of the unjustified harassment, imprisonment and execution of Jews. A less obvious example is found in the feeling of some adults that children do not need to be treated fairly – with care and compassion, yes, but not with any attention to what is just. The creation of a separate juvenile justice system reflected this orientation: since the focus was on what was best for children rather than on providing them with a fair trial, an adversarial legal system was less necessary.

This distinction between identity, unit, and non-unit relationships is at the level of relationships between groups or categories of people. Finer distinctions may be made at the level of relationships between individuals: are they related, are they supposed to be looking out for each other, how long has the relationship lasted? Unwritten rules, culturally defined, exist to prescribe the normal behavior between a parent and a child, or among business associates, or between a student and a faculty advisor. Behavior that violates these norms is more likely to be considered unfair. In addition, the norms may constrain impulses toward just action: the urge to put in some hours to help a coworker who needs time off for a sick child may be suppressed in a workplace that emphasizes individual productivity. On the reverse side, the norms for a particular role may mandate a concern for justice: teachers monitoring exams, judges hearing cases, and referees at football games are all required by their jobs to evaluate the situation and do their best to ensure a fair outcome.

The social setting in which a relationship exists also defines the operative goals for the situation, and implicitly thus the type of justice that is given the highest priority. Research has indicated, for example, that equity may be more important in settings that stress economic goals, since the effect of equity is to increase individual productivity (and also because inputs and outputs are more easily quantified in economic settings). Nurturant settings, like the home or the welfare state, have been found rather to evoke considerations of need. Legalistic or competitive settings may mandate equality of opportunity and experience, although the threat of punishment in court also requires the contemplation of retribution and paying for one's sins (as does the contemplation of the afterlife in some religions, or the sight of a police car in one's rear-view mirror).

A last potential contribution of the type of social relationship is to define the timeframe within which justice may be evaluated. A colleague may be miffed if he or she is not reimbursed after paying for lunch. A friend may assume that the favor will be returned at a later point or compensated for with a different type of gesture. A parent may wait 20 years or more for a child even to show appreciation of the sacrifices made on his or her behalf. And, if two or more people will never see each other again, they may be less concerned with treating each other fairly than if they expect the relationship to continue. The amount of time we are willing to wait for justice, as well as the amount of time in which the consequences of just or unjust action can emerge, depend partly on the social relationship.

We have described the distinctions among social relationships at both the individual and the group level. Although some situations clearly call for analysis at one or the other level, others are amenable to both types of analysis; yet it is difficult to see both levels simultaneously. One of the complications here is the fact that different ideals of justice may be more salient at different levels of analysis. A microlevel analysis elicits concerns with process and with *how* an individual's outcome is determined; a macrolevel analysis focuses more on what the outcome distribution looks like. Equity is a salient principle at the individual level, while equality may seem more fair at a deindividuating group level of analysis. The size of the unit of analysis in evaluating a situation as just or unjust is also in some sense a function of the social relationships involved.

Social issues

In a world whose separate communities are becoming increasingly interdependent, some of the most significant events to evaluate with regard to justice occur not at the individual level but at the societal level. These are the events with the potential to affect the largest number of people for the longest amount of time. The evaluation of these wider social events as just or unjust, or the decision about what response to a situation will be most fair, is partly a function of the relevant social and historical context, including such factors as recent events, current social groupings and tensions, and the intentional and unintentional guidance of the media and of the people in power. Due to limits in attentional capacity, an individual or even a society cannot constantly evaluate all current situations in order to see if an injustice has occurred, or to decide if an alternative scenario might be more fair than the one which is planned. For this reason we rely on cues to tell us which events are important enough to think about, and for those that are, whether we should consider them to be fair or unfair. The evaluation of a situation is dependent on the time and place within which it occurs as well as on an objective analysis of the circumstances.

The first step of course involves whether and when people care about justice. Societies generally inform their citizens which topics the latter should be concerned about; not necessarily in a direct way, but perhaps through the amount of media coverage given. Recent events can also affect the impact of a particular situation. Something which might not be considered very important in its own right might gain new significance if it appears to be similar to another event that *was* important. This can certainly happen on the level of individual relations: one thinks of the seemingly trivial marital misunderstanding which blows up into a big fight in the light of previous incidents. It can also happen on the societal level. Here the parallels may, again, be highlighted by the media who talk about 'another Munich' or 'another Tiananmen Square'.

The similarity between a previous situation and the current one may decrease the apparent complexity of the situation as it is evaluated in terms of justice or injustice; by highlighting certain aspects of the event and minimizing others, the parallel implies that certain ways of considering the situation are more meaningful or more accurate than others. This is an example of one of the more subtle ways in which the socio-cultural context can affect the perception and operation of justice: not just by influencing the perceived importance of a topic, which affects whether or not an individual thinks about it, but by framing the issue in a particular way that affects *how* the individual thinks about it. This may be done by altering the cognitive processes that were described above. The contemplation of justice by an isolated individual may require that the person have the ability to generate alternative scenarios, and may depend on the alternative scenarios that the individual happens to generate; but at the societal level, particular alternative scenarios may be presented to the individual – again, through the mass media.

Even if the consideration of alternatives is left up to the individual, the framing of the issue may make certain alternatives more likely. Recent events and popular metaphors direct an individual's train of thought: if there is a *war* on drugs, then the use of a military approach and the suspension of certain basic civil liberties may seem fair and appropriate. Alternatively, describing drugs as a *plague* might suggest a more nurturant approach to those who are ill. Salient alternative situations, whether hypothetical outcomes of a current course of action, actual outcomes of a recent similar event, or just previous situations that have recently changed, affect the perceived justice of a current situation. We are less satisfied with an outcome if we know that a similar recent train of events (whose apparent similarity may be due to the portrayal of both events in the media) resulted in a superior outcome.

We have also discussed the way in which a person's tendency to simplify an issue or to acknowledge the complexity of it may influence that person's perspective on what is just. When an issue is described at a societal level, the complexity with which it is viewed is not entirely under the individual's control. A media portrayal which consistently focuses on only one of the myriad aspects of the problem, or alternatively one which gives space to each of the various concerns, will affect the number of separate aspects that individual members of the society are likely to consider. A simplistic analysis of a current social conflict may leave each side convinced of the justice of its position – increasing the perception of intergroup conflict by increasing perceived consensus within the group. On the other hand, people who have based their opinion on a single facet of the issue may be more vulnerable to persuasion when they hear an argument based on another factor.

The societal context surrounding an issue will also have a great effect on the degree of emotional resonance it evokes. Contextual cues may increase or decrease the apparent relevance of the issue to a person, and exaggerate or minimize the magnitude of the outcome. An emotional contagion effect may also operate, such that each person's feeling of outrage or horror may be amplified by his or her perception of the outrage or horror of others. Heightened physiological arousal, such as may result from the experience of emotion, has been shown to result in a narrowed range of attention: the individual focuses more strongly on the central concern and has less attention left over to devote to peripheral issues. The emotional response to an issue may thus feed back into the complexity with which the issue is considered, encouraging the person to take a narrow view of what could be a complex issue. Alternatively, there may be social stigma attached to taking a cautious or qualified perspective on an issue which has elicited a strong emotional response.

The social context surrounding an issue will also have a strong influence on the type of relationship seen to be involved. Social issues usually involve relatively few people directly, at least not to the extent that they are aware of their involvement. Most people will be evaluating the issue from the perspective of observers. Their feelings about and reactions to the issue will depend on the type of relation they perceive between themselves and the more directly affected parties. Will they consider the latter to be in an identity, unit, or non-unit relation? Since everyone can be described as a member of more than one category of people, the answer to this question depends partly on which social groupings are emphasized. Recent events may serve to highlight some categorizations and play down others; so too may efforts by group leaders ('It's us against them') and by the media ('This is an issue that concerns all women').

Social cues may describe the relationship in a way that suggests the appropriate terms for considering the issue: is it most appropriately considered in, for example, the economic, legal, or social domain? The answer will frame what is considered to be a just way of dealing with the problem. In a psychology and law course, one of us (Clayton) has asked students for their opinions about the law requiring car drivers to wear seat belts – a law which currently exists in some states in the U.S.A. and not others, indicating the ambivalence with which people view the issue. Students often express the opinion that the State should not infringe upon the civil liberties of its citizens by enforcing a particular behavior. When it is suggested to them that the State has an economic interest in preventing highway deaths, however, many students change their minds about the law: they see the control of behavior for economic reasons as more just than the control of behavior for reasons of personal welfare. A change in the type of relationship described between the State and its citizens, from nurturant to economic, changes the criteria used to evaluate justice.

The last of the factors described in the first two sections which is influenced by the social context is the level of analysis. Although a social issue by its very nature would seem most suited for a macrolevel analysis, the media may and often do use anecdotes about concerned individuals to bring the issue down to more of a microlevel. The intent in doing so is usually just to sell more papers or attract more viewers by making the issue more interesting and personally involving. One side effect, however, is to alter the terms of the analysis by means of which the observers make evaluations of justice or injustice. When the impact of a situation on groups is considered, an equal distribution of resources may seem like the strongest indication of a just outcome. When the focus is on the effect on a single individual or family (even, in some cases, a single community), however, there may be much more concern with whether this individual had 'a fair chance'; not whether this family got what every other family got but whether they got what they 'deserved' or needed on the basis of their merit.

Implications

Justice is invoked almost reverentially, as if reference to justice made one's position unassailable. Yet all too often the term lacks a specific definition, or at least a common one. An informal survey of the *New York Times* editorial page found 'justice' words referring to things from equity, equality, and need, to lack of bias, to accuracy in judgement, to specific contexts ('justice' often serves as a euphemism for arrest and punishment); a recent publication of the *Sierra Club* introduced the term 'environmental justice'. There is a much greater degree of consensus about the desirability of justice than about what that would involve.

Even at the individual level, perceived justice is influenced by the way in which the individual reflects on alternative scenarios, if at all, and the degree of emotional involvement or partisanship felt. As the level of social complexity at which an issue exists increases, the number of factors affecting the perceived fairness of the issue goes up exponentially. Justice theorists often describe the experience of justice as having one evaluative and one emotional component. Both of these aspects are strongly affected by the social context of the experience. Our evaluations may be influenced by the inputs and outputs that we consider, which are in turn guided by the possible scenarios that occur to us, which are in turn shaped by what we regard as the rules and norms and probabilities in a given situation. Social influences are just as responsible for our emotional reactions: just how bad (or good) is this outcome? Whose side are we on in this conflict?

Does this mean that the pursuit of justice is a red herring in the search for intergroup cooperation? We think not. Although the visions of justice held by different individuals and groups may vary, the fact that they are concerned at all with goals other than self-interest is a source for optimism. What may aid the attainment of harmony or at least a satisfactory resolution to interpersonal or intergroup conflict is a recognition that justice is not absolute; that it is not usually the case that our side wants justice and the other side does not but that two different sets of standards are operating. By articulating the reasoning behind a particular view of justice, the underlying premises may become apparent and be either explicitly recognized or discarded. In recognizing the subjectivity of justice, a more just outcome ultimately may be achieved.

Further reading

Adams, J. S. (1965). Inequity in social exchange. In L. Berkowitz (ed.), *Advances in experimental social psychology*, vol. 2, pp. 267–99. New York: Academic Press.

Brickman, P., Folger, R., Goode, E., & Schul, Y. (1981). Microjustice and macrojustice. In M. J. Lerner & S. C. Lerner (eds.), *The justice motive in social behavior*, pp. 173–202. New York: Plenum.

Cohen, R. (ed.). (1986). *Justice: Views from the social sciences*. New York: Plenum.

Crosby, F. J. (1976). A model of egoistic relative deprivation. *Psychological Review*, **83**, 85–113.

Folger, R. (1986). Rethinking equity theory: A referent cognitions model. In H. Bierhoff, R. Cohen, & J. Greenberg (eds.), *Justice in social relations*, pp. 43–63. New York: Plenum.

Gartrell, C. D. (1985). Relational and distributional models of collective justice sentiments. *Social Forces*, **64**, 64–83.

Kahneman, D., & Varey, C. (In press). Notes on the psychology of utility. In J. Roemer & J. Elster (eds.), *Interpersonal comparisons of well-being*. Cambridge: Cambridge University Press.

Lerner, M. J. (1981). The justice motive in human relations: Some thoughts on what we know and need to know about justice. In M. J. Lerner & S. C. Lerner (eds.), *The justice motive in social behavior: Adapting to times of scarcity and change*, pp. 11–35. New York: Plenum.

Markovsky, B. (In press). Prospects for a cognitive-structural justice theory. In R. Vermunt & H. Steensma (eds.), *Interpersonal justice in social relations*. New York: Plenum.

Tyler, T. (1987). Procedural justice research. *Social Justice Research*, **1**, 41–65.

Walster, E., Berscheid, E., & Walster, G. (1973). New directions in equity research. *Journal of Personality and Social Psychology*, **25**, 151–76.

D.

TRUST, COMMITMENT AND COOPERATION

Editorial

In terms of exchange theory, the partners in a long-term relationship must be willing to incur costs in the expectation of long-term gains. Each behaves prosocially to the other in the hope of reciprocation. This expectation involves trust – trust that the partner will not renege. Trust in another individual or group depends in part on the reputation of the latter: individuals may invest resources in order to augment the trust that others have in them, and dealings between groups may likewise be influenced by the perceived reputation of the other. Trust, once established in some degree, is often self-reinforcing because individuals have stronger tendencies to confirm their prior beliefs than to disprove them. Trust in another is also augmented by the perception that the other party will be punished for reneging, and this perception will be augmented by the further perception that there is an agent who can be trusted to administer the punishment: the agent may be internal (conscience) or external to the other party. In assessing the trustworthiness of another party, it is of course essential to see the situation from their point of view.

In the first chapter in this section, Boon & Holmes elaborate a model of trust which combines the developmental issues that produce a readiness to trust others with the contextual determinants of trust in a particular situation. In a dyadic relationship this will include the biases that affect each individual's perception of the other and the actual behaviour of the other over time. Boon & Holmes emphasize not only that trust implies the presence of risk of being let down, but also that it is necessary to accept that risk if trust is to develop.

Trust within a relationship is closely related to commitment: commitment involves trusting the partner or group, and accepting the risk of disappointment. Lund's (Chapter 12) contrast between 'old' and 'new' concepts of commitment in close relationships has echoes of the earlier discussion on collectivist versus individualistic orientations. The 'old' commitment depends in large measure on social pressures and accepted group norms; the 'new' commitment is more a matter of personal decision to pursue a given course of action. But she emphasizes that both forms of commitment coexist. In societies in which marriages are arranged the 'old' commitment comes first and the 'new' follows, while in societies in which partners chose each other the order is reversed.

Commitment is, of course, not limited to close personal relationships, but in such relationships it is not to be equated with love. Indeed in a study of students living together Lund assessed how much each loved the other, how committed each felt, how much each put into the relationship and how much each felt they got out of it: it was commitment and how much each felt that they had put into the relationship that best predicted whether or not the couple would stay together.

Studies like this are rare in part because 'trust' and 'commitment' are not easy to measure in the complexity of real-life relationships. Psychologists have therefore used laboratory situations to assess some of the variables that influence how people behave in relationships that involve incurring risks of costs in the hope of rewards. Boon & Holmes made use of these in their discussion of trust, and Good (Chapter 13) provides an overview of this approach. Like Boon & Holmes, Good emphasizes the complex of personality factors, aspects of the situation, and the interplay between them. One factor which he emphasizes as important for cooperation is good communication between the individuals involved: this issue was present by implication in the discussions of trust and commitment, but the experimental games approach underlines its importance. Therapists would certainly concur.

It is easy to criticize the experimental games approach, and Good acknowledges its weaknesses. It is perhaps proper that we should remind ourselves that just because human relationships are complex and the factors making for prosocial behaviour are multiple, the insight provided by constrained situations, though limited, is not to be spurned.

Rabbie (Chapter 14) also makes use of this approach in considering cooperation and competition within small groups. First, however, he considers the nature of a group, emphasizing the importance of perceived interdependence amongst the members. Using a 'Behavioural Interaction Model' he explores behaviour in two-person and multi-person dilemma tasks, concluding that a collection of people is more likely to cooperate if they see themselves as interdependent and as a distinct social unit, communicate with each other and share a common goal which contributes to their individual goals.

11

The dynamics of interpersonal trust: resolving uncertainty in the face of risk

SUSAN D. BOON AND JOHN G. HOLMES

Every day we encounter numerous situations in our lives that require us to make interpersonal decisions in the face of uncertainty. These decisions range from the mundane – whether or not to work on a project with a coworker – to those more difficult and risk-laden decisions which compel us to confront our hopes and fears about another person in an intimate relationship: if I commit myself to my partner, can I safely assume he or she will still love me as the years pass by? The uncertainty inherent in these sorts of decisions forces us to turn to our *expectations* as indicators of likely outcomes. We are forced to rely on our beliefs as a means of resolving the dilemma posed by the internal conflict between our hopes and fears: the opposing forces which simultaneously push us toward and pull us away from any particular decision to depend on another. A sense of trust – confident expectations in the benevolent intentions of another – enables the successful negotiation of such ambivalence, allowing us to quell our anxieties and take action where before we might have hesitated. Indeed, in our close personal relationships, the ability to trust that an intimate will respond to our needs over a long period of time becomes paramount. Without an established bond of mutual trust, the prospects for a relationship are at best tenuous.

The purpose of this chapter is to discuss the implications of the concept of trust as they bear upon an understanding of interpersonal relationships, particularly romantic or love relationships. The first few sections of the chapter are devoted to articulating a model of trust designed to account not only for chronic or longstanding dispositions to trust but also for more situationally-determined tendencies to take interpersonal risks. This enables discussion of trust at two levels, that of the specific relation-

190

ship, and that of the individual across time. The final sections detail the development of trust within close relationships specifically, highlighting the important dimensions underlying a sense of trust between intimates, and characterizing the experiences of individuals who feel trusting, distrusting, or uncertain about their partners' motives in the relationship.

Unfolding the nature of trust

The concepts of interdependence and risk comprise the functional core common to the majority of definitions of interpersonal trust. *Interdependence* refers to the extent to which a person's outcomes in an interaction are contingent on or determined by another's actions. In this sense, an interdependent situation is one in which the other possesses some degree of control over the outcomes that an individual may obtain. In addition, because the other may not always take the person's needs and concerns into consideration when acting, dependence upon that other for the achievement of desired outcomes also creates an element of *risk*. The extent of risk involved in a course of action is reflected in the subjective value or meaning of the outcome to the individual, and the probability or likelihood that the other will facilitate the particular outcome. The individual's estimate of these two factors is described as an expectancy.

An appreciation of interdependence is critical to an understanding of interpersonal trust because the degree of interdependence between individuals determines the relevance of trust for the interaction between them: the greater the interdependence, the more crucial is the state of trust. Deutsch (1958) stated this maxim most aptly when he suggested that there is 'no possibility for "rational" individual behavior in [interdependent situations] unless the conditions for mutual trust exist' (p. 270). Also, the compatibility of people's preferences in the interaction is important because risk is always greater to the extent that the potential outcomes of those involved result in conflict of interests (i.e., are 'non-correspondent,' Kelley, 1979). When interdependent individuals have different preferences as to what they would like to do, the likelihood that each person will act in a way that takes the needs, desires, and concerns of the other into consideration is attenuated. In these circumstances, trust in the other's good intentions is necessary in order to maximize the benefits to both parties. Trust alleviates fears of exploitation and minimizes feelings of vulnerability while those involved search for optimal solutions to the problem.

The key concepts: an example

An example may illustrate these ideas in a more vivid and compelling manner. Consider a situation in which a couple returns home from work after a long day. The wife has had a particularly tiresome and stressful day and really feels the need for a break from cooking dinner, typically her responsibility on the night in question. She has two choices. First, she may ask her husband to cook that night, effectively placing her fate in his hands. Either he will agree to cook, responding to her need to relax, or he will reject her plea and leave her need unfulfilled. Alternatively, she may choose to cook dinner anyway, a choice that does not leave her vulnerable to the whims of her partner, nor exposed to the potential rejection she might feel if he were to decide not to respond to her need.

If the wife risks asking him to cook, it is the husband's decision that will determine the value of the wife's outcome: her outcome is *dependent* upon his actions. In contrast, a decision not to ask him may be considered a safe choice in that the wife's outcome is then relatively *independent* of the behaviour of her husband (and in this case it is the outcome she expected, as it is her turn to cook). Thus, the wife may choose either a safe or risky approach to this problem.

Figure 11.1 is a symbolic representation of the choices and outcomes in this hypothetical situation. Notice first that the relationship described is characterized by a certain level of interdependence of outcomes in this interaction. On the one hand, if the wife decides to cook, her outcome does not vary much as a function of her husband's actions: whether he happens to volunteer to cook, or, as expected, allows her to take her turn, her outcome is essentially the same. However, if she chooses to ask him to cook, her outcome becomes dependent on her husband's response to the situation. Glancing at the bottom left quadrant of the matrix, it is obvious that considerable risk is implied by the choice of this alternative above the safer one. After she has expressed her need for a break, allowing herself to become *vulnerable* by placing herself in the position of being dependent on him for her outcome, she is likely to feel very dissatisfied and perhaps even rejected should her husband refuse to cook. In fact, her outcome in this cell is the poorest possible. Conversely, should her husband recognize and respond to her need and decide to cook dinner himself, making a small sacrifice in order to attend to the need she has expressed, her outcome in this event is superior to the outcomes available to her from the safe alternative. This latter point is of fundamental importance: only by *taking*

		Husband	
		wife cooks	cooks dinner
Wife	does not ask husband to cook	4 4	3 5
	asks husband to cook	4 2	3 7

Fig. 11.1. Hypothetical options and outcomes of the cooking dinner example. Numbers in the bottom left corner of each quadrant represent the outcomes for the wife, those in the top right corner represent the outcomes for the husband. Higher numbers represent more pleasant outcomes. (From J. G. Holmes, (1991). Trust and appraisal process. In W. H. Jones & D. Perlman (eds). *Advances in Personal Relationships*, vol. 2, p. 77. Greenwich, CT: JAI Press. Adapted with permission from 'A modified martyr structure'.)

the risk of placing her fate in her partner's hands – of being dependent upon him, of exposing her vulnerability – is she capable of attaining the maximum outcome. (Note that the risk involved in deciding to ask for help is indicated by the large discrepancy between the two outcomes in the lower quadrants of the matrix.) Clearly, elements of an approach–avoidance conflict are likely to exist in her decision to reveal her need.

Next consider the husband's alternatives. Notice that he has no reason to choose to cook dinner, a less favourable option, unless he has incorporated some aspect of his *wife's* needs and potential outcomes into his decision. The point is that the husband must possess some *intrinsic motivation* in order to respond to his wife's needs when such action requires sacrifice on his part; it is necessary that he consider a small loss on his part of lesser importance than the attendant gain in satisfaction for his wife. He may, in fact, experience some amount of satisfaction from having met her need, particularly if his view of the situation is relationship-oriented and incorporates *both* their needs and potential outcomes. In this case, this matrix may not present a completely accurate portrayal of his outcome in this cell because it only describes each person's *egocentric* preferences, preferences that do not take those of the other into account. The husband may not view his response to his wife's need at the level of his individual loss or sacrifice, but rather perceive it in the broader perspective of their outcomes as a couple, from the vantage point of the health of their relationship as a whole. As for the wife, she may interpret his 'sacrifice' in this context as evidence of his desire to be responsive to her needs, as confirmation of her belief that he cares for her; she may perceive his behaviour as *diagnostic* of his true feelings for her. Indeed, Kelley (1979) argues that, 'interpersonal dispositions [people's true motives] ... are both expressed and perceived in the ways in which behavior is independent of the actor's direct outcomes and geared to the total pattern of interdependence' (p. 114).

Finally, note that if the husband in this example considered cooking dinner a greater loss than presented in this particular matrix – if the partners' outcomes were more highly non-correspondent – the risk inherent in the wife's decision to ask him to trade turns with her would increase accordingly. She would be more vulnerable to the possibility that he might respond in accord with his own preferences and would require greater trust in his motives before being willing to accept the risk involved in this choice rather than opting for the safer alternative. Conversely, should the partners' outcomes be completely correspondent, no risk would be involved, and no trust required for smooth functioning. However, such a situation would provide little in the way of diagnostic information concerning the husband's motives.

A process-oriented model of interpersonal trust

The above discussion highlights certain aspects of trust that are critical to a definition of the concept. In this chapter, we choose to view trust as a state involving confident positive expectations about another's motives with respect to oneself in situations entailing risk. Central to this definition are inferences and attributions regarding the other's *motives* across situations from which *expectations* may be derived; but ultimately, the perception of *risk* in a particular situation may be the key parameter, determining whether and if the dynamics of trust come into play. Thus, this definition of a state of trust is intended to convey a concurrent summary of the various factors that might influence a person's tendency to act in a trusting way.

Chronic personality factors influencing a capacity to trust   Among the early theorists who wrote of the development of trust, both Erikson (1950) and Bowlby (1973) consider trust a chronic characteristic of personality. According to Erikson, trust is the resolution of an early inner conflict around dependency occuring during the infant's first year of life, a resolution determined in large part by the quality of the maternal relationship. In his 'architectural' model, basic trust is the first building block in a hierarchical identity structure. It is the very foundation upon which a personality is constructed, a foundation which contrains the nature of the 'person' that can be built upon it. He describes trust as the 'most fundamental prerequisite of vitality' (1968, p. 96) and as a capacity for faith. Thus, from his perspective, a capacity to trust or not to trust develops very early in life and from that point onward shapes all other aspects of personality development, laying at the very core of a person's sense of identity.

Bowlby similarly bases his discussion of trust on a developmental model emphasizing the quality of care during early childhood. According to his theory of attachment, a sense of trust is derived from feelings of security in attachment afforded an infant whose primary caregiver is consistently responsive to his or her needs and easily accessible should a need arise. Over time and through experience the child generates 'working models' (p. 203) or *representations* of self and others which incorporate this sense that others, generalized from the caregiver or 'trusted companion' (p. 201), are readily available and responsive to one's needs. These mental models are postulated to persist throughout the lifetime relatively unchanged and to influence an individual's general orientation to seek closeness with or maintain distance from others. More recent work on adult attachment (Shaver & Hazan, 1988) suggests that this claim may be especially applicable within romantic relationships.

What these perspectives have to say with respect to our definition of trust is twofold. First, patterns of caregiver responsiveness experienced during the early years of life produce more general expectations regarding the willingness and ability of others to attend to and satisfy one's needs. Second, these expectations may be anticipated not only to endure in some form throughout the lifetime, but also to play a role in determining further personality development and the way in which an individual perceives and attempts to cope with the social world. Expectations rooted in childhood, therefore, are critical bases for the development of a trusting orientation.

This emphasis on chronic, persisting expectations and a relatively enduring and stable disposition to trust differs from our conceptualization of the construct. Instead we argue that such chronic expectations may impact upon a person's *readiness* to trust, either impairing or enhancing the individual's ability to make choices that leave him or her in a vulnerable position in the face of interpersonal risk. Consequently, we see the effect of chronic expectations as laying the foundation for a 'preferred adaptation' (Holmes, 1990), a predisposition to trust given the appropriate set of circumstances. This, in conjunction with concurrent situational and relationship variables, determines whether trusting behaviour will be enacted or not. Thus, from our perspective, a *state* of trust is determined by a wide array of variables, of which a chronic disposition to trust is only one.

Situational/contextual determinants of trusting behaviour   Deutsch's (1958) pioneering exploration of trust in laboratory situations, in contrast to the theoretical work of Erikson and Bowlby, centred largely on the situational factors that relate to the operation of trust. In a series of experiments involving repeated decisions on tasks where the interests of the two participants conflicted, Deutsch equated trust with people's tendency to seek cooperative solutions to the problems they faced. In general, he concluded that context-specific variables had a profound effect on the rate of cooperative responses and that the crux of the issue for the subjects seemed to be determining the other's motives or intentions. Cooperation tended to be low when situational factors conspired to make learning about the other person difficult, for example under conditions where decisions were simultaneous or where effective communication was constrained. More specifically, cooperation was reduced most when opprtunities to appraise the other's motives were minimal, and when the decision context underlined the risk of being exploited if one were to behave in a trusting way – in other words, when the other was perceived as having incentives to succumb to temptation. Furthermore, participants' motivational set (in these studies either cooperative, competitive, or individualistic) not only influenced how they responded, but their initial perceptions of opponents exerted a reciprocal influence on their subsequent reactions. The main point to be appreciated with respect to this latter finding is that evidence relating to the particular orientation of the other in any situation involving trust will influence a person's own readiness to trust in that situation. For instance, in Deutsch's studies, when cooperative subjects encountered early signs that they faced a competitively-motivated opponent, they responded more competitively than if the opponents showed signs of also being cooperatively-oriented.

Kelley & Stahelski (1970) noted a fascinating twist on this theme. In their studies they found that competitively-oriented subjects (those selecting a competitive goal as the purpose for their decisions) responded competitively *regardless* of their opponents' orientation, whereas cooperatively-oriented participants adjusted their level of cooperation to match that of their opponents, cooperating if the opponent did, competing otherwise. (This latter effect the authors term 'behavioral assimilation', implying that the cooperators' change in tactics was a result of the competitors' dominance in the interaction). Interestingly, subjects' perceptions of their opponents' motives similarly varied as a function of their *own* orientation. Competitors frequently erred by judging cooperative partners to be competitively-motivated, whereas cooperators tended to be accurate in their judgements of partners' goals. As Kelley & Stahelski state, 'these errors on the part of the competitor suggest that he is not aware of the influence he exerts upon his partner' (p. 68). Thus, competitors appear to possess a rather homogeneous and suspicious view of the world, assuming that others are competitive like themselves, and then acting in such a manner as actually to *elicit* the competitive behaviour they expect.

This bias would seem consistent with the idea that some people hold rather rigid expectations about others' motives and that they act on their suspicions in a way that blocks any opportunity to learn that they might be wrong: as Strickland (1958) put it, ultimately, if one is to trust another, one must *first* take the risk of acting *as if* the other deserves that trust. Thus, distrusting individuals who project their own competitive motives onto others appear to have a 'theory-driven' approach to processing relevant information within social interactions. In contrast, the relatively more heterogeneous and accurate world-views of cooperators would seem more consistent with the idea that they have a 'data-driven' or evidence-based approach to social interactions, an approach more amenable to realistic evaluation of their partners' motives. They first give the partner the benefit of the doubt and take the risk of trusting another by acting cooperatively. The behavioural response of trusting individuals is thus a contingent one, adjusted in an adaptive manner to suit both the features of the situation and the characteristics of the other. In this manner, the role of chronic expectations and stable dispositions to trust (or, in this case, not to trust) would likely be greater for competitive persons than for cooperative persons, for whom the expectations primed by the context of an encounter with either a competitively- or cooperatively-motivated other would probably carry greatest weight. Thus the commonly held assumption that trusting individuals are 'naively optimistic' is quite unwarranted.

In terms of our conceptualization of trust, the research of Deutsch and Kelley & Stahelski points to the relevance of considering more acute situational determinants of the expectations people may hold regarding others' motives. The key dimension is the amount of risk involved in choosing to trust as a consequence of the specific contextual features inherent in the situation. As Deutsch argues, 'when the fulfillment of trust is not certain, the individual will be exposed to conflicting tendencies to engage in and to avoid engaging in trusting behaviour' (p. 208). In other words, when features of the situation and the interaction evoke feelings of vulnerability – when the risk of potential exploitation is perceived to be high – the dilemma surrounding the decision to trust or not is likely to be exacerbated.

Relationship history   The particular history of a relationship may be considered a contextual variable of fundamental importance as it imparts a refined and perhaps unique quality to the expectations those involved possess about each other. In addition, the extent of interdependence between the partners in specific domains of their lives, and consequently, the amount of risk involved and relevance of trust in particular inter-actions, are also heavily determined by the various patterns of experience which combine to form the relationship's history.

Within close relationships, trust provides a secure platform from which important relationship issues may be approached (Holmes, 1990). The *prospective* nature of trust implicates the importance of subjective fore-casts of what the future holds. Patterns of responsiveness and validation that have characterized the relationship in the past lay the foundation for these forecasts. When a partner has proven over the course of time to be consistently responsive to one's needs and accepting of one's self-definitions, a sense of security or confidence in the partner is likely to develop. Even then, however, pockets of vulnerability may sometimes remain, tagged to situations that have raised doubts in the past.

Thus, the particular history of the relationship – the times when needs were met and when needs were left unsatisfied – may profoundly influence what types of expectations will be primed in a given situation and therefore a person's readiness to engage in trusting behaviours in the face of risk. From this perspective, a concern with the dynamics specific to the given relationship is an integral component worthy of consideration in an analysis of trust.

In established relationships, then, the particular expectations that are foremost in people's minds will depend on the conjoint influence of their chronic dispositions, their relationship history, and the features of the particular situation they face. In our earlier example, the conflict of interest in the *situation* made salient the issue of personal risk for the wife if she were to ask that her needs in the domain of household duties be met. In addition, the risk that she perceives, and the attendant anxiety she might feel, would also depend on the *couple's history*: if the husband had responded unselfishly in the past to such requests for help, then her decision may involve little or no sense of trepidation. On the other hand, household duties might be a hot issue in an otherwise trusting relationship, resulting in a reluctance on her part to expose herself to what she sees as his fossilized attitude toward the male role. Furthermore, if her *chronic disposition* was one of suspicion and distrust about the willingness of men (or perhaps people in general) to respond to her needs, the *conjunction* of risks she experiences would probably result in her taking a self-sufficient stance in which she avoids depending on her husband. Alternatively, the selective but temporary focus on this negative aspect of the husband might be realigned or balanced by a chronic readiness to trust combined with her general sense of faith in his caring for her, reducing the impact of any negative feelings she might have.

On seeing and believing: The effects of perceptual distortion on trust   An element that further amplifies the conjunction of risks in situations involving trust is the tendency for people's expectations to distort their assessments of others' behaviour in particular situations. A large and well-documented literature (see Darley & Fazio, 1980, and Miller & Turnbull, 1986, for reviews) supports the notion that social perception is a constructive process heavily based on our initial expectations. We are not the veridical scientists, objectively and accurately processing information, that we might like to think we are. Instead, a variety of biases and perceptual distortions often provide us with a view of the world that looks the way we want or expect it to, rather than the way it really is.

One of these biases is the *self-fulfilling prophecy* in which an expectation leads people to act in ways that elicit expectation-confirming behaviour on the part of another. For instance, consider what might happen if the wife in our example were to assume, in advance of any decision to ask her husband for help, that he would refuse to meet her need. If she were to enter the interaction with this negative expectation in mind, it is quite possible that her behaviour toward her husband would be tempered by this expectation; she may be especially quiet and reserved, appear upset, or even act in a cold and distant way toward him. His response to behaviour of this sort would probably also be negative; at the very least, such behaviour on the wife's part is unlikely to *increase* the likelihood that the husband will respond to any need she might express. Thus, the husband's response will most probably reinforce her initial expectation that he cannot, in fact, be relied upon to meet her needs. It is also unfortunate that people in interdependent relationships generally tend to underestimate the extent to which their *own* behaviour shapes that of others. As a consequence, it is highly unlikely that the wife would be aware of the causal influence her behaviour exerted on his response. This aspect of the self-fulfilling prophecy implies an especially pernicious bias: the target may in essence be 'blamed' for behaviour which is largely or in part the 'fault' of the perceiver.

What perceivers 'see' may also be less contingent on objective evidence than on original expectancies. There are several ways in which people may distort evidence that is inconsistent with their assumptions or 'theories' about another. First, they may ignore such behaviour and selectively attend only to behaviours that are expectancy-confirming. Or they may simply interpret ambiguous behaviour as consistent with their beliefs. Finally, they may attribute expectation-inconsistent behaviour to temporary or situational causes, explaining it away and discounting its value, rather than accepting it as evidence of what the other is really like. Again, consider the wife in our earlier example. Even if her husband somehow recognizes that she would like to rest instead of cooking dinner and takes on this responsibility himself, negative expectations might colour her perception of his actions such that she fails to give him credit for meeting her needs. She might explain away his response by attributing it to some ulterior motive she believes he might have (e.g. 'He's just doing this so that I won't be angry that he forgot to clean up after breakfast'), or to some rather fleeting whim of his to treat her nicely because of his good mood that day. Essentially, people are inclined to project their own expectations and feelings onto others and to resist modifying these projections on the basis of relevant social evidence.

Given the focal role of expectations in our perspective on trust, then, the potential for perceptual distortions and self-fulfilling prophecies has major implications for our model. Bear in mind the dual reality that not only do expectations derive from experience, but also experience derives from expectations. Thus, trust may be based on confident positive expectations, and such expectations may also breed trust.

A portrait of trust in close relationships

The weaving together of two lives is a complex process, fraught with the potential for conflict of interest and bounded on all sides by people's vulnerabilities and fears. The capacity to trust is an essential ingredient in fulfilling the promise of such relationships.

The birth and maturation of a capacity to trust

Trust within a close relationship derives much of its flavour from the major issues in focus at any given period of relationship development. Current concerns such as the rewards to be gained and costs to be paid, compatibility of needs, a partner's core motives, and so on, set the stage on which a bond of mutual trust is established. The form this sense of trust takes is also in part a reflection of the particular type of '*appraisal process*' – the search for and evaluation of evidence concerning a partner's motives – activated by the current concerns of those involved. As the relationship grows and matures, these two determinants often change in correspondence with the changing dynamics of the intimate relations between partners.

The romantic love stage   The early periods of a successfully progressing romantic relationship are typically characterized by a profusion of positive feelings and idealization of the partner. Consistent with the surplus of positive emotion at this stage, the focus is largely on the rewards that the relationship provides.

During this time of intense but rather superficial emotional involvement, a person's expectations about a partner are both fuelled and protected by a process of projection. By projecting his or her own strong feelings onto the partner, a person develops and defends 'theories' speculating that the partner cares just as he or she does, theories lent apparent support by the partner's displays of affection. Although more objective evidence is sought to corroborate these budding positive expectations – the partner's behaviour is likely to be monitored for indications that he or she truly cares – unless there is a considerable imbalance of involvement in the relationship, optimistic assumptions about the partner are not likely to be questioned. As a consequence, trust and love tend to be strongly related at this early stage, even though a fragile expression of hope may be the only basis for this sense of trust. Any doubts or fears regarding trust tend simply to be denied importance, drowned amidst a sea of positive feelings.

The evaluative stage   As the depth of interaction and interdependence increases, the partner's 'imperfections' are gradually revealed and the person becomes motivated to reduce a newly emerging sense of uncertainty regarding the prospects for the relationship. Establishing the worth of the partner in broader terms that do not deny the implications of more negative aspects of the relationship becomes the dominant goal at this point. However, as the constraints of reality begin to intrude on perceptions of the partner, projection is no longer an adequate means of assessing the partner's feelings and motives, and a more data-based approach to this task must be implemented. Along with this more active evaluative stance toward the partner, a sense of risk originates with the realization that the increasing depth and intimacy afforded by the discovery of the partner's real nature amplifies the potential losses accruing from dissolution of the relationship. Furthermore, in some cases a more realistic approach to determining the merits of the relationship may force the partners to acknowledge the unpleasant possibility that the relationship may not work.

Thus, the somewhat 'blissful' perceptions of the very positive romantic love stage eventually give way to a greater concern with the underlying realities, motives, intentions, and dispositions of the partner. The earlier focus on rewards recedes in importance, replaced by a concern with the overall pattern of a partner's behaviour and what it symbolizes. Viewed as a coherent whole, does the behaviour of the partner support the assumption that he or she truly cares? Does the evidence indicate that he or she is dependable, responsive to needs, kind, considerate, and so on? The search is for stable, unified expectations regarding the partner that enable more accurate forecasts about the course of the relationship and the benefits to be derived from it. The emergence of a sense of risk during this period in fact enhances the opportunity to make such dispositional inferences. As we suggested earlier, only when the possibility of exploitation by the partner exists can inferences regarding his or her intrinsic motivation and trustworthiness be at all conclusive.

Entry into this stage provides the first opportunity for a real sense of trust to take root. The growth process is a conjunctive responsibility, determined by the actions of both partners. When a rough equivalence or balance in perceived affection exists, it serves as a safe base from which partners may take the risk of expressing their feelings and making moves designed to increase the level of intimacy in the relationship. Ideally this process of self-disclosure is reciprocal: each partner both initiates actions that reveal his or her feelings and motives and responds in kind to the revealing actions of the other. The growth of trust is thus a cyclical process, escalating and expanding at each step.

Because trust becomes a real issue for couples at this point, they become highly motivated to determine the nature and bases of each other's responsiveness. The key question to be answered is whether such behaviour is intrinsically or selfishly motivated. To the extent that responsiveness involves costs or self-sacrifices on the part of a partner, it is particularly likely to be perceived as diagnostic of caring. However, these are also precisely the conditions under which the perceived risk of exploitation is greatest: noncorrespondence of interests and considerable interdependence. In any event, if consistent evidence of unselfishly motivated caring is obtained, the resultant gain in mutual attachment allows fears and concerns about security in the relationship to diminish in importance, enhancing the partners' ability to respond directly to each other's needs. Potentially, as a healthy capacity to trust continues to grow, people may shift from an emphasis on assuring that their own needs are appropriately met to a concern with maximizing the communal welfare of the couple as a whole.

The accommodation stage The evaluative stage paves the way for a period of 'accommodation' involving the negotiation of conflicting needs and preferences. Dealing with the latent incompatibilities exposed during the preceding evaluative phase provides an especially valuable arena for partners to solidify their state of trust.

At this point in the development of their relationship, partners may become almost painfully aware of the need to 'fit' together more closely on a variety of dimensions and styles of relating. Ambivalence and an increase in the incidence of conflict often result from this enhanced awareness of the need for further compromise and adjustment. In contrast to what couples may fear, however, these signs of negativity do not typically auger poorly for the relationship or suggest that love is waning. Rather, a partner's responses during episodes of conflict may provide important diagnostic evidence concerning the extent to which he or she values the person's needs and preferences and is prepared to work to make the relationship successful. Of course, some couples may discover that their differences are irreconcilable or that they are too uncomfortable dealing with conflicts directly. If issues are avoided or either partner withdraws from problem-solving interactions, the opportunity to use episodes of conflict to evaluate further a partner's caring and responsiveness is lost. Furthermore, over time partners are likely to become entrenched in these dysfunctional patterns of conflict resolution.

Achieving an understanding of the partner's inner self, his or her motives and dispositions, remains the primary goal as the accommodation process begins. As in the evaluative stage, behaviours are monitored only in so far as they speak to their underlying causes. However, the special focus unique to this later period of relationship development concerns the achievement of a sense of psychological closure or *confidence* regarding the partner's ultimate designs and the viability of the relationship. This can be a difficult objective to accomplish because partners inevitably realize that the evidence upon which attainment of psychological certainty rests remains rather inconclusive. It is necessary for them to come to terms with the fact they will never amass sufficient information to remove all elements of risk in making a judgement. Acknowledgement of this state of affairs necessitates a 'leap of faith', an emotionally charged conclusion about the partner that goes beyond what the evidence permits. Therefore, trust becomes a necessary *construction* permitting an illusion of control, a resolution of the uncertainty continually promoting a sense of vulnerability.

For some, this leap of faith is more successful than for others. As a consequence of the consolidation process enabled by faith in the partner, trusting individuals possess well-integrated, confident attitudes concerning their partners. Rather than existing as isolated and compartmentalized entities, the positive and negative elements of these attitude structures have been reconciled with each other, in line with the positive conclusions about the partner, and organized into a consistent, coherent whole. Consequently, faith in the partner's benevolence enables trusting persons either to refute the implications of discrepant information regarding their partners or to assimilate it with their attitudes. For instance, trusting persons may discredit or downplay the significance of a partner's expectation-inconsistent acts, relegating such behaviour to more peripheral areas of concern. Alternatively, the wealth of positive attitudinal elements stored alongside negative elements may serve to *buffer* the impact of any negative information obtained. The positive elements may simply drown out any activation of negative elements.

In contrast, an individual whose faith lacks the strength to eradicate all his or her fears and doubts, whose trust therefore remains somewhat tentative or uncertain, is likely to possess a less integrated attitude structure. This leaves the person vulnerable to the possibility that a partner's negative behaviour may activate broader concerns without *also* activating the positive elements that could moderate the effects of the sources of worry. In this case, the individual's ability to refute the implications of negative acts is substantially diminished.

A trusting orientation: Bases of confidence

A decision to trust an intimate partner requires obtaining acceptable answers to several questions regarding the deservingness of the partner and the prospects for the relationship. The research of Holmes (1990) and others points toward four main issues that must be addressed if a capacity to trust is to develop to its fullest extent.

Dependability   At the most basic level an individual is concerned with establishing the dependability of the partner, the extent to which he or she is the type of person who can be relied on to act in a benevolent, cooperative, and honest manner over time and in different situations. Trust may be invested in a dependable partner on the basis of a sense of certainty that the partner will behave in a reliable, consistent manner. He or she is the sort of individual who can safely be counted on.

Responsiveness   The partner needs to be more than just dependable, however. Feelings of security in the relationship are not based solely on the other's general character, but also on an assessment of the feelings and attitudes that the partner has toward the person *in particular*. He or she must demonstrate a true concern and desire for meeting the person's needs, and for considering the person's preferences and wishes in addition to, and sometimes rather than, his or her own. In essence, this sort of behaviour provides evidence that the partner truly accepts the person as he or she is, *validating* his or her needs, aspirations, and fears. Such responsiveness lies at the core of the process that promotes a sense of mutual attachment between partners, the process which forms, maintains, and intensifies the affectional bond between them. As Reis & Shaver (1988) suggest, 'appropriate responses enhance feelings of connectedness' (p. 379), a crucial building block in the foundation of a willingness to trust.

Confronting conflict   Partners must also be confident that they can successfully engage issues of disagreement within their relationship and develop solutions that consider each of their concerns. A feeling of efficacy in terms of the ability to resolve conflict is integral to the evolution of a trusting orientation. If conflict is not dealt with in an open and constructive manner, it tends to fester and resurface in other areas. Furthermore, failed attempts at problem-solving may breed a 'terminal hypothesis', a belief that the couple is in fact incapable of negotiating effective solutions to their problems and hence the relationship is on a downward course that they cannot control (Holmes & Boon, 1990). Because engaging issues not only involves placing oneself in a position of vulnerability by opening the problem to discussion – thereby facing down the spectre of rejection or disappointment – but also usually requires some degree of self-sacrifice on the part of one or both parties for resolution to occur, trust may be greatly intensified through successful negotiation of such risk-laden terrain.

Faith   Ultimately, trust cannot flourish unless a person takes the step of actively placing faith in the partner. At some level trust is a positive illusion developed to curtail feelings of uncertainty once commitments have been made. It is a construction, a working model of the partner and the relationship that functions to allow one to act *as if* the partner is trustworthy and to feel secure in a sometimes capricious world. The fact that such a conclusion cannot be fully warranted by the evidence must become secondary to the *assumption* that the partner is worthy of trust. Among those who are able more or less to consolidate their positive and negative beliefs about their partners, this transformation of hypothesis into 'fact' enables them to relax their efforts at appraising the partner and relinquish their fears and insecurities regarding trust. Thus, faith is the capstone of an orientation to trust.

Calm seas or crises of confidence: The dynamics of trust in established relationships

The above depiction of the factors that contribute to a state of trust in a relationship makes it clear that a sense of security can be undermined on a variety of fronts. Most relationships start on a relatively trusting note, but few survive without at least some challenges or crises of confidence. In this section, we will first describe the fabric of people's experiences in a trusting relationship and then portray the sort of dilemmas that people face if the issue of trust has again become an open question in their relationship. What then will be the nature of their current concerns?

Trusting relationships   Trusting individuals feel no need to question their partner's motives and commitment. Nor do the exigencies of avoiding disappointment and hurt cause them to feel pressured to engage in active monitoring or appraisal of their partner's behaviour. Rather, stable and confident expectations that their partners will consistently act in positive and responsive ways enable them to conquer any fears or vulnerabilities about trust they may yet harbour. Theirs is a state of intimacy free from the threats that might upset their sense of assurance in the partner.

As a consequence of this well-established faith in the partner's trustworthiness, trusting persons exhibit a *charitable* orientation towards their partners. Considerable latitude is granted in what is interpreted as acceptable behaviour, and the partner is often given the benefit of the doubt even when his or her behaviour is discrepant with trusting expectations. For example, a lack of thoughtfulness may be readily attributed to the partner's being preoccupied, stressed by his or her job, or being in a bad mood. These types of explanations serve to *preserve* feelings of trust because they do not involve negative inferences about the partner that would have broader implications for the relationship. In fact, negative acts by the partner have often been discounted *in advance*.

When major disagreements or especially discrepant behaviours do activate an analysis of the situation, however, trusting individuals tend to frame the issue in an abstract manner which enables them to interpret the evidence with much poetic licence. They may 'see' what they want to see, regardless of the facts. Furthermore, there is some support (Holmes, 1990) for the notion that confronting such people with evidence of clearly discrepant behaviours in fact serves to amplify the positivity of their feelings toward their partner and their views of the partner's motives. Thus, they are so liberal in their interpretations of even negative events that they appear able to find the silver lining behind almost any cloud. Apparently, the act of considering the implications of negative behaviour also primes their positive attitudes and thus may cause them to draw even more positive conclusions about the partner than would be true had the partner acted in a positive, expectation-consistent manner. Not surprisingly, the end result of this very benevolent, lenient way of 'asking the questions' regarding trust is a positive confirmatory bias affecting how the evidence is perceived and translated into meanings for the relationship.

Overall, individuals in trusting relationships tend to process their partner's behaviour in a rather automatic and positive way. A negative consequence of this manner of processing is that sometimes particular acts of caring may be taken for granted, paling in significance against the backdrop of other 'positive' events. This also occurs because trusting persons tend to assess the meaning and impact of events and partner actions within a relatively long-term perspective. This approach to relationship accounting has a stabilizing or levelling effect as the impact of any particular event or behaviour is moderated by aggregating it with others in the broader scheme of things (part of the explanation behind taking positive acts for granted). Thus, the effects of infrequent negative events would tend to be cancelled out by the effects of more positive ones. All things considered, then, these various processes clearly serve to maintain and stabilize the attitudes of trusting individuals; it probably requires serious challenges to the integrity of their beliefs before they are called into question.

States of uncertainty   Unfortunately, for some people the history of their relationship has failed to instil or sustain a sense of confidence in the partner as an object of trust. They sense continual challenges to whatever level of confidence they have mustered. The insecurity that results from this state of uncertainty leads them to continue actively to test the hypothesis that the partner can be trusted. Unfortunately, there is a tendency among uncertain persons to frame this test in a negative way, typically resulting in confirmation of the more pessimistic side of their expectations. They tend to be vigilant, constantly monitoring the partner's behaviour for signs diagnostic of his or her motives and intentions. Furthermore, the cumulative hurts from their present or past relationships accentuate their feelings of vulnerability and constrain their hopes that the partner is worthy of trust. As a result, they tend to be extremely conservative in the inferences they draw about the partner's behaviour – very ready to assign blame for negative acts, but rather reluctant to grant credit for positive acts.

In essence, uncertain individuals are '*risk averse*' (Holmes, 1990): unwilling to accept the risk of being hurt from drawing an unwarranted positive conclusion about the partner's trustworthiness, they instead adopt a sceptical stance, delaying any final conclusions in favour of collecting yet more evidence. Research indicates that, in fact, uncertain spouses are particularly suspicious of behaviours that appear too positive. This suggests that as they begin to approach the point of drawing a positive attribution about the partner, uncertain individuals' fears and doubts activate strong desires to protect the self from the potential of being hurt. Sadly, they are caught in a tug-of-war between their desire to believe the partner is motivated to be responsive, and their fears that they could be wrong. The safe position is to wait until the evidence is sufficiently compelling that it 'speaks for itself' – unhappily, it seldom does.

Because uncertain individuals actually test the hypothesis that their partner cares, they are drawn into a limited, short-term accounting process that may further amplify their feelings of vulnerability. Their restricted perspective prevents them from being able to contain the damage caused by localized negative events, in contrast to evaluating them within a broader context that would include a more representative sample of positive events. Furthermore, because they persist in tracking relationship events and adding symbolic meaning to them, they are more emotionally reactive or volatile, unable to downplay the implications of a negative act. They focus more exclusively on the moment-by-moment exchanges and transactions within their relationships, responding to any one event in relative isolation from any other. They are tossed around in a sea of often changing and contradictory impressions of the partner.

Clearly, then, the dilemma for uncertain individuals is dealing with the risk inherent in trusting on the basis of imperfect and perhaps partly equivocal evidence. These individuals, in contrast to those low in trust, still harbour hopes that the partner really cares and that the relationship will succeed. But their fears of exploitation and rejection are hard to dispel, constraining them from integrating the negative aspects of the relationship into a coherent, balanced account which might allow them a sense of faith in the partner. Instead, their worries are readily activated by occurrences that recapitulate their larger concerns and they remain trapped between their hopes and their fears.

A state of distrust An accumulation of violated expectations regarding a partner's responsiveness may at some point exceed a person's threshold of tolerance. The negative evidence may reach some criterion level after which a person concludes with reasonable certainty that the partner cannot be trusted and ought to be viewed with considerable suspicion.

In direct contrast to a trusting person, a distrusting person is likely to distort *negatively* a partner's behaviour, perceiving it to be consistent with his or her negative expectations. A type of defensive pessimism is invoked as a mechanism for coping with this crisis of confidence. Such individuals typically adopt a perspective from which they assume the worst in any instance because of serious concerns about their own psychological safety if they were again to venture the risk of depending on the partner. This self-protective tendency results in a vigilant orientation which is likely to enhance the implications of negative events, and diminish the significance of positive ones. Thus, a vicious circle is established in which a lack of trust promotes a pessimistic and defensive appraisal of the partner's behaviours, which in turn leads to corroboration of negative expectations and further erosion of any state of trust in the relationship.

Conclusion

It seems rather paradoxical that in order to establish a state of trust it is first necessary to take the risk of trusting. In cases where we are dependent on another, when the other's interests are not always our own, the only protection from our fears is the hope that we will not be exploited in our decision to trust. When there is evidence suggesting that such a seed of hope is valid, this dilemma may not seem insurmoutable. When such evidence is not forthcoming, the dilemma may well become paralyzing.

The dynamics of this synopsis are nicely illustrated by the challenge that trust poses for people in relationships where issues regarding trust remain unresolved. Uncertain individuals cannot relax their guard – cannot overcome their vulnerability – for fear that the partner might not truly be worthy of trust. They flounder amidst doubts and suspicions, unable to reach the safety and security of a positive conclusion about their partner's motives.

Despite the hurdles, some people do achieve this safety. Some do confront the risk of trusting and successfully emerge on the other side with a confidence born in defiance of their apprehensions. Their success is a reminder that it is possible to trust and that our fears need not restrict our opportunities to learn to depend on those who merit our confidence.

Acknowledgements

Thanks to Lisa Barham and Jacquie Vorauer for their helpful comments on an earlier version of this manuscript.

Preparation of this manuscript was facilitated by a Social Science and Humanities Research Council (SSHRC) Doctoral Fellowship to the first author and an SSHRC Research Grant to the second author.

References and further reading

Bowlby, J. (1973). *Attachment and loss, Vol. 2. Separation: Anxiety and anger.* London: Hogarth Press and Institute for Psychoanalysis.

Darley, J. M. & Fazio, R. H. (1980). Expectancy confirmation processes arising in the social interaction sequence. *American Psychologist*, **35**, 867–81.

Deutsch, M. (1958). Trust and suspicion. *Journal of Conflict Resolution* **2**, 265–79.

Erikson, E. H. (1950). *Childhood and society.* New York: Norton.

Erikson, E. H. (1968). *Identity, youth, and crisis.* New York: Norton.

Holmes, J. G. (1991). Trust and the appraisal process. In W. H. Jones & D. Perlman (eds.), *Advances in personal relationships, Vol. 2.* pp. 57–104 Greenwich, CT: JAI Press.

Holmes, J. G. & Boon, S. D. (1990). Developments in the field of close relationships: Creating foundations for intervention strategies. *Personality and Social Psychology Bulletin*, **16**, 23–41.

Kelley, H. H. & Stahelski, A. J. (1970). Social interaction basis of cooperators' and competitors' beliefs about others. *Journal of Personality and Social Psychology*, **16**, 66–91.

Miller, D. T. & Turnbull, W. (1986). Expectancies and interpersonal processes. *Annual Review of Psychology*, **37**, 233–56.

Reis, H. T. & Shaver, P. R. (1988). Intimacy as an interpersonal process. In S. W. Duck (ed.), *Handbook of personal relationships*, pp. 367–89. New York: Wiley.

Shaver, P. R. & Hazan, C. (1988). A biased overview of the study of love. *Journal of Social and Personal Relationships*, **5**, 473–501.

Strickland, L. H. (1958). Surveillance and trust. *Journal of Personality*, **26**, 200–15.

Webb, W. M. & Worchel, P. (1986). Trust and distrust. In S. Worchel & W. G. Austin (eds.), *The psychology of intergroup relations*, pp. 213–28. Chicago: Nelson-Hall.

12

Commitment old and new: social pressure and individual choice in making relationships last

MARY LUND

What is commitment to an interpersonal relationship? Is it an obligation, the effort to be a good spouse or friend, or is it just another word for the indistinguishable good feelings that keep relationships going? Questions about commitment assume greater importance as relationships become more fragile due to the stresses of a mobile society and the emphasis on individual growth and rewards. Whereas a person in a small town a century ago could count on most relationships continuing, today most people have a choice about whether to maintain relationships and how much energy is put into maintaining them. The executive in a corporation is expected to put commitment to the organization above commitment to family and friends. The saying at IBM is that the corporation's initials stand for 'I've been moved.' How does this hypothetical executive develop and maintain committed relationships with spouse, family, and friends when pressures from the organization compete for his or her time and energy?

The most dramatic evidence of the fragility of relationships in the twentieth century has been the rising divorce rate. Since any serious discussion of commitment should examine what factors were associated with this change, this chapter will focus primarily on commitment in love relationships. While divorce was possible in the early part of the century, the emotional and legal barriers to getting one made it relatively rare. Going into the last decade of this century, roughly one third of marriages in industrialized countries will end in divorce. Although the simple explanation for the increase in divorce may be the liberalization of divorce laws, the more complicated explanation includes changes in how people view commitment in love relationships and marriage.

212

The discussion of commitment in this chapter is based on the notion that we are living in a period of transition from one type of commitment in interpersonal relationships to another. The 'old commitment' in more stable, traditional cultures depended on maintaining a role in relationships, putting the good of the group above the good of the self, and avoiding punishment from the group for deviating from social expectations. The 'new commitment' depends more on the individual's decision-making about a given relationship in which rewards, investments, and alternatives to the relationship are weighed. The 'new commitment' is experienced more by the individual as coming from within and not from societal pressure. Although this discussion will contrast the 'new' and 'old' commitment, the forces from each coexist in modern relationships. This blend of commitment results in an individual's motivation to maintain relationships not just for the social good but also for the personal integrity that comes from following through on building a relationship.

Commitment defined

Commitment is used popularly to mean sticking with an activity or staying in a relationship. There is general agreement among authors on commitment in relationships that the focus is on the duration of relationships. Committed relationships are expected to endure and withstand periods of conflict or low rewards to the individuals in them. Beyond agreeing that commitment has something to do with the duration and stability of relationships, authors vary widely in their ideas of how people become committed or how commitment works to keep people in relationships.

Variations in definitions

Commitment has been a topic of theory and research in psychology, sociology and anthropology. The following definitions of commitment from a sample of prominent authors on the subject show the differences in how commitment has been conceptualized:

The sociologist Georg Simmel wrote about the related concept of faithfulness, 'a specific psychic and sociological state, which insures the continuance of a relationship beyond the forces which first brought it about; which survives these forces with the same synthesizing effect they themselves originally had.' In his definition of faithfulness, Simmel distinguishes between what draws people to a relationship and what keeps them in it over time.

Relationship theorist Robert Hinde states, 'the term commitment is used here in a general sense to refer to situations in which one or both parties either accept their relationship as continuing indefinitely or direct their behavior towards ensuring its continuance or optimizing its properties.' In this definition, Hinde refers to the private pledge a couple make to each other as opposed to the public pledge which makes commitment enforced by social punishment for leaving the relationship.

Zick Rubin, who has done extensive research on how love develops in relationships, defined commitment as, 'the pledging of oneself to a line of action, whether it be the fight for a political ideology or the struggle for intimacy with another person.' Rubin's definition is very close to the one used in the psychology of commitment to attitudes in which behavior consistent with an attitude (signing a petition against nuclear arms) strengthens the cognition (belief that nuclear arms should be eliminated) and leads to further behavior consistent with the cognition (marching in an anti-nuclear arms demonstration.).

In her work on commitment to religious communes, Rosabeth Kanter defined commitment as, 'the willingness of people to do what will help maintain the group because it provides what they need ... commitment links self-interest to social requirements ... it forms the connection between self-interest and group interest.' This definition of commitment holds the process of commitment responsible for a shift in orientation from individual rewards to rewards that come from group connection.

These authors have gone beyond describing the phenomenon of commitment, that is the observation that relationships do endure despite immediate problems, and begun to address the process of commitment. The definitions of commitment are the starting points for the theory presented here of the 'new' commitment. A process of commitment, separate from initial attraction, begins when a person accepts that the relationship will continue and acts accordingly. The acts themselves bind the person to the relationship and create a strengthened attitude about the importance of the relationship continuing. The motivation to continue the relationship comes both from the tendency to act consistently with the belief that the relationship will last and from the shift in the perception of rewards from self-interest to what is in the interest of the relationship.

Commitment vs. love

Part of defining commitment is to distinguish it from other aspects of interpersonal relationships. Most attention has been paid to commitment in a love relationship. Love in a close relationship has been defined by Rubin as encompassing feelings of needing another person (attachment), wanting to take care of another person (caring), and having revealed one's self to another person (intimacy). Commitment differs from love because it is possible to love someone without being committed, as in a brief extramarital affair. It is also possible to be committed without being in love, as in a marriage of convenience in which there is little attachment, caring or intimacy. The ideal in our society is that commitment and love increase along parallel lines in a relationship, but the exceptions point out how there may be something unique about commitment and its development.

Types of relationships

A person may be committed to many types of relationships that involve many different feelings. A man may be committed to a long-term partnership with a business colleague whom he likes but with whom he shares little social connection. A soldier may be committed to friends in his platoon and close to them during war but not intend to continue a relationship during peacetime. Parents expect a lifelong commitment to children, but the relationship will vary in contact and intimacy as children grow older.

Commitment in some relationships implies exclusivity. Generally, a love relationship moves toward a monogamous commitment. Commitment to religious groups and some work groups may also demand foresaking outside involvements. However, commitment to friendships, family relationships, and many other types of groups allow for other commitments. Hence, commitment alone does not convey the depth of a person's feelings in a relationship, the degree of interdependence, nor the exclusivity of the relationship. Commitment in these different types of relationships means acting in the interest of continuing the relationship as the relationship is defined.

The 'old commitment'

Since this chapter focuses on the dramatic changes in commitment to love relationships in the twentieth century, it is important to discuss the sources of commitment in traditional western societies before the mid-twentieth century. Instead of being a process, commitment to marriage was an event, the public pledge. There may have been some individual choice in picking the partner, but once married, society regulated the relationship. Whatever the feelings about the partner, the relationship was guaranteed to continue because of *external* pressure. Avoiding social censure provided a motivation to stay married. (It is interesting that one leading demographer has noted that there was as much marital disruption in the 19th century as in the 20th because of the higher mortality rate and that increase in divorce may be due to unhappy spouses not being able to contemplate the end of a marriage through the partner's death!)

Until recently, courts maintained control over the decision about ending marriages and what would happen to the family's financial assets and children. Lenore Weitzman traces how in the 1970s, many of the United States changed divorce laws to allow for more individual decision about continuing a marriage in the form of no-fault divorce. Before that time, the courts had more power to grant a divorce based on the conduct of the participants. The guilty party, the one whose conduct caused the end of the marriage, was usually punished by the loss of money or contact with children.

The premise underlying the laws was that marriage was an economic union with prescribed roles and obligations for husband and wife. Men were obligated to provide financially and women to care for children, and those roles continued in some fashion after divorce. The financial dependence of the wife was unquestioned, although she could be punished for moral misconduct by the court's withholding spousal support.

With the advent of no-fault divorce came the legal concepts of 'private ordering', 'joint custody,' and 'rehabilitative spousal support' for women. These concepts grew from a change in how society viewed marriage to one of an equal partnership between a man and woman, both of whom could be financially independent and both of whom were important in parenting their children. A marriage could now end when one party desired it without the other's consent. Weitzman questions whether the change in divorce laws truly reflects the current situation since women are not financially equal with men and usually suffer more economically in divorce. Still, the majority of divorce petitions are filed by women in the U.S.A. and the United Kingdom.

If the changes in divorce laws are accurate evidence of the changes in the process of commitment, the 'old commitment' consists mainly of adhering to a role in society. Society has a stake in people performing the roles of husband and wife to create stable marriages and families if that is the way wealth is distributed and children are reared. The 'old commitment' works as a force society has over the individual. Certainly there are positive reasons for the individual to adhere to a social role such as pride in being a good husband or wife. But, as the increase in divorce suggests, the 'old commitment' may no longer be operating in an era when individual happiness is given as high a value as performing a social role.

The 'new commitment'

To discuss commitment in modern relationships requires starting with the promise that it is up to the individuals in them to remain involved. 'Till death do us part' may still be part of the majority of marriage ceremonies, but all present have some awareness that the public pledge is subservient to whatever ongoing private pledge of commitment that the husband and wife will make in the relationship. Does the 'new commitment' mean staying in the relationship as long as there is love, as long as it is rewarding to the individual, or as long as no better potential partner becomes available?

Evaluating rewards in relationships

Some psychological theories of relationships would suggest that there is a simple, quasi-economic basis for relationships continuing. People bring rewarding properties to relationships and do rewarding things to which psychologists can assign some value and use to predict attraction and satisfaction in relationships. Physical attractiveness, social status, earning potential, and some personality characteristics are some of these rewards. The value of these rewards is slightly different for men and women. George Bernard Shaw highlighted these differences when he defined marital incompatibility as, 'When he has no income and she is not pattable.'

The simple economic theories of relationships, that people stay in relationships on the basis of rewards, cannot account for the observation that people can and do stay in relationships when they are very unrewarding, at least for some period of time. All relationships go through periods of conflict and most people encounter other attractive potential partners. Despite the high divorce rate, the evidence is that the majority of people actually stay married. Something besides the motivation for individual rewards must account for relationships continuing in the era of no-fault divorce.

A more complicated economic theory for explaining why relationships continue has been researched, primarily by Caryl Rusbult. In a series of studies, she has examined how people evaluate relationships by rewards and costs to them (which combined form the level of satisfaction in the relationship), attraction to alternative relationships (including being alone), and investment in the relationships. For example, a male university student may find that his girlfriend is rewarding due to her good looks and her tendency to flatter him and has the minor cost of living a 30-minute drive from where he lives. He is satisfied overall with the relationship. He may also find himself attracted to a girl in one of his classes. However, he does not act on this attraction partly due to his satisfaction with his girlfriend and partly due to his investment of his time and energy for the last six months in the relationship.

Investments

It is the concept of investments in a relationship that takes the theory of the 'new commitment' from a quasi-economic evaluation of relationships to a new level. People do not stay in relationships just because they are rewarding in the moment. The level of satisfaction can dip so low or the alternative can be so attractive that they may act to leave. But the weight of the cumulative investment in a relationship and the effect of that investment on beliefs about the importance of the relationship usually tip the scales in favor of continuing to invest in that relationship. At some point the investment is so large that to leave the relationship can only be done at a large personal cost.

An unlikely source for understanding the theoretical basis of the impact of investments on commitment comes from Rosabeth Kanter's study of 19th century religious communes in the United States. In some sense, membership in these communes does parallel the involvement in modern love relationships. These were volunteer societies in which there were no prescribed social roles as in traditional cultures. Members were free to decide to leave, and there was no pressure from outside the commune to remain. Kanter's research showed that the communes that survived longest made most of what she termed 'commitment-building mechanisms,' which increased the link between self-interest and group interest. Classes of these commitment building mechanisms were sacrifice, investment, renunciation, communion, mortification, and transcendence. For example, communes which required total investment of material possessions were more long-lasting than those which did not.

Investments in a love relationship can be small and hardly noticed or large and consciously contemplated. One of the most important is the amount of time spent together that could be spent on other pursuits. In the beginning of relationships, giving gifts is a measure of financial investment. The traditional engagement ring could be thought of as a large, irretrievable, financial investment in a relationship. Later, merging finances or becoming economically interdependent increases the investment. Limiting outside friendships, doing favors, trying to change some behavior on request are forms of investment. Making plans for the future together on a short or long-term basis, acting as a couple socially, and publicly announcing the relationship in some fashion also constitute investments in a relationship.

If the development of commitment is viewed partly as a series of increasing investments in a relationship, then there is no one event or one discrete phase during which the individuals become permanently committed. The marriage ceremony itself may be a large investment because of the time, effort, and money involved and because of the public announcement. Once a wedding has been planned, the couple will feel pressure to follow through with it. However, according to this model of commitment, a married couple who keep separate bank accounts and spend little time together may not be as committed as an unmarried couple who have completely merged their finances and spend most of their time together. If the two relationships were equally stressed, the unmarried couple may feel they have more invested in their relationship that they do not want to lose than the married couple.

Although there is no way to determine the exact point, there does come a time in many relationships in which the cumulative investment is so great that a person feels no choice but to continue a relationship. There is a joke that goes, 'What is the difference in what a chicken and a pig contribute to your breakfast?' The answer, 'The chicken is only invested. The pig is committed.' The threat of loss of an irretrievable investment may be every bit as devastating as any outside social punishment and provides part of the motivation to stick with a relationship.

Decisions and cognitions of commitment

The model of the 'new commitment' assumes that there is an ongoing decision, the private pledge, that people make about continuing their relationships, although there may not be much conscious thought given to that decision at all times. Decision theorists who write about commitment to relationships call this style of decision-making 'muddling through.' Each small change is selected as 'good enough' because it is better than leaving the situation unchanged. It is easier to become slightly more involved in the relationship than to make a decision to leave it altogether.

The person who keeps pleading or deciding to invest in a relationship – to put into it time, effort, and money and to act as a couple in social situations – builds up a belief or cognition about the importance and the future of the relationship. In psychology, cognitions are said to be strengthened through consistent behavior. This strengthened belief about the importance and future of the relationship acts as an *internal barrier* to ending the relationship. According to the model of 'new commitment,' the person who has invested in a relationship stays in it during periods of stress because of the self-perception of treating the relationship as important, having put effort into it, and having investments that would be irretrievably lost if the relationship ended. This self-perception influences the person to put effort into staying in the relationship until the stress subsides or to 'make the relationship work,' for example by trying to make changes so that satisfaction is increased.

The pacing of investments in a relationship is important to the buildup of the beliefs or cognitions that form the internal barrier to ending a relationship. The strongest self-perception of commitment comes from having consciously decided to invest gradually in a relationship. Hence, couples who marry after knowing each other only a short period of time may have a weaker self-perception of commitment than those who have a longer courtship. Levinger and Rubin discuss the importance of pacing intimacy and involvement in relationships. People feel threatened with too intimate a self-disclosure from an acquaintance and prefer to match level of self-disclosure. When the level of intimacy increases more rapidly than the level of commitment (a man on a first date reveals that he has never felt attractive to women), it may threaten the relationship. Likewise, investment and commitment in relationships have a rhythm that must feel comfortable.

The theory of the 'new commitment' presented here makes what a person *thinks* about a relationship as important in determining whether the relationship will last as what a person *feels*. Cognitive therapist Aaron Beck stresses the importance of a person's thoughts or beliefs about a relationship. Writing generally about relationship stability, he highlights the undermining influence of 'irrational beliefs' on a relationship such as, 'If I do not feel constant happiness in a relationship, it is not right for me,' or 'If my partner really loved me, he would know how to make me happy without my having to tell him.' Writing more specifically about commitment, he writes that helping couples to review evidence of a relationship's durability can strengthen their belief that the relationship will last and motivate them to risk changes and find new ways to resolve conflict.

A change to couple rewards

Along with the belief that the relationship is going to last comes a change in what the individual perceives to be rewarding in the relationship. As with the members of religious communes that Kanter writes about, the person in a committed love relationship begins to link self-interest with what is in the interest of the couple. In the beginning of a relationship, the person focuses on how rewarding the partner is, how attractive, rich, humorous, etc. As commitment increases, the focus turns to other kinds of rewards such as the reward of feeling trust and security. Making the partner happy may be experienced as rewarding since it is tied with the rewards of security in the relationship. To some extent, identities merge and there is less a focus on the self and more on the unit of the couple or family. The continuity of the relationship becomes rewarding in and of itself over time.

A blend of 'old' and 'new' commitment

The marriages of today endure both because of the external pressures from the 'old commitment,' avoiding the social cost of ending a marriage, and because of the internal pressures of the 'new commitment,' following through on decisions made to build the relationship. Which commitment has more force probably reflects an individual's values. A person who highly values a traditional role and social approval may stay married because of the 'old commitment.' A person who values individual choice may be more motivated by the 'new commitment.'

In its most positive form, the 'old commitment' provides social support for the couple who weathers bad times. Friends, relatives, and professional helpers such as marriage counsellors or clergy can share the investment made in a relationship and encourage the partners to put effort into making the relationship work. Research on homosexual relationships shows that the lack of social support for their commitment undermines the stability of these love relationships. Society can encourage relationships continuing, not just punish people for not fulfilling their social obligations in a marriage.

The 'new commitment' puts pressure on the individual to choose voluntarily to maintain a relationship. To be committed is to stay in a relationship from the perspective of one's own decision and not from a position of dependence or obligation. Individual happiness and growth are valued. However, part of what makes the person happy can be the involvement in a secure and trusting relationship. Doing what is good for the partner and the relationship, balanced with what is good for the autonomous self, can be rewarding in and of itself. This kind of commitment still depends on the avoidance of pain but not on social punishment. The pain of ending a relationship comes from the loss of the personal investment and the disappointment in the effort not being effective at making the relationship more personally rewarding.

The 'old' and 'new' commitments also blend in other types of relationships that have been stressed by social changes. The parent–child bond can be a casualty of divorce. A father in a traditional marriage could secure the connection to his child by being the bread winner while the mother did the direct parenting. In a divorce, the father must make a direct investment in the child or the relationship is likely to dwindle. While the 'old commitment' works to keep mothers involved with their children after divorce because of social disapproval for a mother leaving her children, the same is not necessarily true for fathers. Court orders have been found largely ineffective for getting fathers to pay child support or visit their children if they have not been doing so. What has been found to be effective incorporates elements of the 'new commitment.' Research shows that when disputes over child support or visitation have been mediated, so that fathers voluntarily agree on issues concerning children, they are much more likely to follow through. Helping fathers to learn to invest in their children after divorce can increase commitment in those relationships.

Much attention has been given to commitment in the workplace. Direct personal investment in a job setting increases commitment. Commitment to getting a task done versus commitment to social needs are often put in conflict in a work setting. Yet, the most effective work groups have been shown to be ones in which the time is taken to get input and voluntary agreement on the task. When workers' input on the quality and the efficiency of the task is solicited, productivity increases and mistakes decrease. Commitment to a work organization, as measured by turnover and absenteeism, increases with worker's perception of good communication and value for their individual initiative on the job.

In many types of relationships – marriage, friendship, work – we are moving from an era when people felt they *had* to stay connected to an era when they *want* to stay connected. Highlighting individual choice in commitment today does not mean that the focus is always on the self. People will work to get involved in relationships and it is that very work that transforms the 'I' into the 'we.'

Commitment may not be the most glamorous part of a relationship. Endurance and stability are not words found in many popular songs. But most people dream of growing old with someone they love. We want relationships to be rewarding but we also want them to last. Commitment provides the frame in which other parts of a relationship can grow. Part of the frame comes from living in the social order, which pressures us not to stray outside the structure of the relationship. Part comes from our decisions and effort in building the frame in the first place.

Further reading

Beck, A. T. (1988). *Love is never enough*. New York: Harper & Row.

Hendrick, C. & Henrick, S. (1983). *Liking, loving and relating*. Monterey, California: Brooks/Cole.

Hinde, R. A. (1979). *Towards understanding relationships*. New York: Academic Press.

Kanter, R. M. (1972). *Commitment and community*. Cambridge, Massachusetts: Harvard University Press.

Kelley, H. H. (1979). *Personal relationships: Their structure and process*. Hillsdale, New Jersey: Erlbaum.

Levinger, G. & Rausch, H. L. (1977). *Close relationships: Perspectives on the meaning of intimacy*. Amherst, Massachusetts: University of Massachusetts Press.

Rubin, Z. (1973). *Liking and loving*. New York: Holt, Rinehart & Winston.

Simmel, G. (1950). *The sociology of Georg Simmel*, Kurt Woolff (ed.). New York: The Free Press.

Weitzman, L. J. (1985). *The divorce revolution: The unexpected social and economic consequences for women and children in America*. New York: The Free Press.

13

Cooperation in a microcosm: lessons from laboratory games

D. A. GOOD

Introduction

Imagine that you and your best woman friend have been taken into custody by the police so that they can interrogate you about a robbery which you did in fact commit together. In the police station, you have been separated from one another, and are being interviewed about the crime individually. Subsequent to the crime, it was clear to both of you that the police would be unable to prosecute successfully unless one of you confessed. You both heard that the insurance company has offered a £20 000 reward for a successful conviction. Also, your legal advisor has just reminded you that helping the police would ensure that the court would sentence you to no more than a suspended prison sentence, and thus you would remain free. Of course, you are certain that your colleague has received the same advice.

In the interrogation room, therefore, you are faced with a dilemma. If both you and your friend do not confess, then both of you will go free. However, if you can be sure that she will not confess, then you could exploit the situation by confessing. The insurance company would give you the reward for indicting your partner who would get at least a five year sentence, and you would still go free. Of course, you know that the same thought about exploiting the situation will be going through her mind. If you think that she might succumb to the temptation and confess in the hope of claiming the reward, then you would be well advised to confess anyway. The court will recognise your confession as an act of contrition, and even if it is of no value for prosecuting your partner, you will be rewarded by being given a prison sentence of no more than three years. So, should you remain silent and adopt what is often known as a *cooperative strategy*, or confess in the hope that she will not and adopt the *competitive strategy*?

This dilemma, known as the *Prisoner's Dilemma* (henceforth PDG), is one of a type which is extremely popular as a parlour game with many people. It is obviously restricted to societies with the appropriate judicial experience, but games where different players cooperate and compete to achieve different outcomes seem to be present in most, if not all, human societies. There are undoubtedly many reasons why they are attractive to so many, but one must surely be the fact that they seemingly provide a microcosm of everyday life.

As a microcosm, the PDG and its ilk have also provided the kind of idealized situation which social and economic philosophers like to ponder in their analysis of real-life human conduct. It has a structure which pervades many of the choices which confront us in the real world in that it captures the conflict between self and collective interest, where achieving the former through a competitive strategy and achieving the latter through a cooperative strategy cannot be pursued simultaneously.

Given this background, it is no surprise to discover that psychologists interested in experimental investigations of human cooperation and competition have chosen to study the well-defined and well-articulated environment provided by these games. To experiment, it is necessary to generate a sample of the behaviour required in an environment where all extraneous factors which might provide alternative explanations of what is observed have been controlled. Ordinarily, our social relations and actions depend upon a multitude of complex, hard to chart, and at times ephemeral factors. This makes the experimental study of social life extremely difficult because it is hard to know what should be controlled. Games by their very nature make this control much easier to achieve by both restricting and specifying what is relevant. Consequently, games like the PDG have been subject to extensive experimental investigation.

As a laboratory exercise or parlour game, the prison sentences and financial rewards will obviously be replaced by minor forfeits or prizes, but the structure of the game is preserved. The *payoffs*, as they are often known, can be varied so as to emphasize the different consequences of the two strategies, or so that the strategies merge. The games may also be extended to more than two players thereby producing a more complicated set of possible outcomes. These games may also be played repeatedly within a short space of time. With these variations possible, it might be thought that such experiments would be extremely informative about competitive and cooperative behaviour in the real world. Unfortunately, this has not proved to be straightforwardly so.

Experimental social psychology has been riven in the past two and a half decades by various disputes about the viability and value of experimentation. Much of the debate has focused on the relationship between what happens in the laboratory and the mundane social world. The more extreme critics have argued that there is precious little relationship, and that the laboratory is a unique environment. Experimental studies of strategic interaction in games have not been immune from these charges. Some authors have argued that they can only provide us with information about how people behave when playing these games in the laboratory and we should expect no more. Not everyone agrees with this extreme and pessimistic conclusion, but even the optimists agree that to understand the relevance of the findings from this literature to the study of real-life cooperative behaviour, it is necessary to contextualize them by reintroducing the various factors which the laboratory eliminates. In this paper, I will do this by offering a brief review of the major experimental findings concerning cooperativity, and a consideration of some of the more important factors which are hard to capture fully in the laboratory and which are relevant to an understanding of this literature.

Personal characteristics

When first considering the topic of cooperativity, it is tempting to think that there are some people who are naturally cooperative and some who are not. Thus, cooperativity could be seen as a personality trait or as related to some other personal characteristic such as gender, class or race, for example. This idea would appear to receive some support from the finding that subjects will differ radically in their responses to an unremittingly cooperative partner in the PDG. Such partners can be continually exploited for maximum personal advantage. While some subjects do this, some respond by being totally cooperative, and others only exploit the situation to a limited extent. Such unremitting cooperation is rare though in everyday life, and other attempts to support this idea have not been overwhelmingly successful.

The idea, which is central to many all-encompassing theories of personality, that individuals are consistent in their behaviour across a wide variety of situations has been strongly challenged since the late 1960s. The most extreme criticism was offered by Mischel who argued that all our social behaviour is dictated by the situations in which we find ourselves, and that there are no interesting contributions to it from what had been termed our personality. He subsequently retreated from this position, but his criticism left its mark. It is now fashionable to study person by situation interactions instead, and the explanatory ambitions of most personality theories have been greatly reduced as a result. One personality theory has, in part, explored cooperativity as a trait, and the experimental studies of its claims have strongly supported the importance of the person by situation interaction. This is the theory of Machiavellianism offered by Christie & Geis (1970).

Christie & Geis took the central propositions of Machiavelli's *The Prince*, and constructed a questionnaire where the respondents were asked if they agreed or disagreed with these propositions. Machiavelli's precise words were not used, but propositions such as 'Anyone who completely trusts anyone else', or 'Most people who get ahead in the world lead clean and moral lives' were offered instead. The level of agreement or disagreement with these propositions permitted the respondents to be classified as 'High' or 'Low Machs'. By definition the Low Machs would be unlikely to exploit a situation for personal advantage and would be more likely to act cooperatively.

They attempted to validate this questionnaire using a variety of laboratory games one of which is of particular interest here. This is known as *The Con Game*. This is a three-person board game where each player moves forward along a path according to the value of a dice throw multiplied by the value of one of the cards from the hand which he or she was dealt at the start. Players may form coalitions which by using the highest value cards from the combined hands of the coalition enable a pair to move faster than an individual. Coalitions may be made or broken prior to any move.

In their experiments using this game, Christie & Geis varied the players according to their Mach scores, and the knowledge of each player about the others' hands. Generally, those who scored highly on the Machiavellianism index were most able to manipulate the situation to their own advantage when each player's hand was hidden from view. When the hands were open, this advantage disappeared, and coalitions were more stable. Thus, the cooperative stance adopted by the low Machiavellian players in forming coalitions was best supported by a high degree of shared knowledge.

In a different vein, there are compelling arguments that a substantial component of our individual identity is derived from our group membership. This component of our individual identity is derived from our group membership. This component is established through our under-standing of both intra- and inter-group comparisons. In terms of the latter, an important factor for the individual is group distinctiveness, and this has clear consequences for his or her willingness to cooperate. Generally, we are far more willing to cooperate with in-group members than out-group members, and non-cooperation can increase inter-group hostility thereby making cooperation even less likely.

Furthermore, if competition maximizes the relative difference between the groups this will then be desired for more than just the intrinsic value of the payoff. Indeed, some studies have even shown that subjects will seek to maximize the difference between their own group and another even if this results in members of their own group receiving smaller rewards.

This inter-group effect depends upon how we perceive the group membership of the other players in a game. Similarly, it has been found with the PDG game that a subject's perceptions of the intentions of the other player as basically competitive or cooperative depends upon their own outlook. In repeated plays of the game where one subject plays according to a pre-programmed sequence of competitive and cooperative choices, the other subject saw this behaviour as reflecting an overall attitude towards the game which was similar to their own cooperative or competitive disposition. More recent work has confirmed this, and has also suggested that those with a cooperative or competitive outlook tend to see the opposite in others as being due to weakness and idiocy.

This position is also supported by Rotter's related work on trustworthiness. He and his coworkers examined the characteristics of individuals who are willing to trust others on a wide range of issues, and in a number of different spheres. They found that those who are more willing to trust other people are likely to be trustworthy themselves in that they are less likely to cheat, steal or lie. It is also worth noting that they are usually happier individuals, and are typically more liked by their friends and colleagues than non-trusting persons.

The only other personal characteristic to be examined using these games and for which there is any consistency in the results has been gender. The results of these experiments are surprising to most people in that they contradict the common sex-role stereotypes, and the findings from other studies, of men as competitive and women as cooperative. Typically, in the PDG and similar games, men choose cooperative strate-gies more often than women do.

This finding could, of course, lead us to conclude that the stereotype is wrong, but, as Colman (1982) has argued, it might be short-sighted to see the confession strategy in this game as always reflecting competitiveness in the pursuit of advantage. He argued that women might choose the confession strategy as a way of avoiding comparative disadvantage. The only certain thing in the PDG is that if a player always chooses the competitive strategy, he or she will never do worse than their opponent. This may well lead to that player receiving a smaller net benefit than he or she might otherwise have done, but the comparison with a hypothetical outcome for self may matter less than an interpersonal comparison with the actual outcome for the other player.

Finally, it should be recognized that since most, if not all, players will assume that most, if not all, other players will not adopt a self-destructive strategy, then choosing the competitive response could be construed, somewhat paradoxically, as affiliative. In this case, this choice would guarantee that the other player is left in exactly the same position as self.

Payoffs

Whether Colman is right or not in his suggestions as to why women choose the competitive strategy more often in the PDG, his observations do point to the importance of understanding the full value and meaning of the payoffs to those subjects taking part in experimental games. If the payoffs are construed merely in terms of the abstracted financial value, then factors such as a concern with inter-group distinctiveness or the minimization of individual disadvantage will not be recognized. The financial rewards which can be offered in the laboratory are rarely of significant substance, and so the likelihood of these other, possibly idiosyncratic, meanings of a payoff affecting decisions is high. Consequently, it is not surprising that experimental manipulations of the size of the rewards and penalties have failed, for the most part, to demonstrate any consistent effect. A number of studies have found that both a high and a low absolute value of the reward for cooperation will make it more likely. There is, however, one exception to this rather dismal picture.

Deutsch & Krauss (1960) developed what is known as *The Trucking Game* in the late 1950s to examine cooperation. Structurally, it has much in common with the PDG, but the form of it allows some interesting variations in the nature of the play, some of which I will return to below. Initially, this was a two-person game played on the map given in Figure 13.1. Both Acme and Bolt are manufacturers who must deliver their goods to opposite ends of the board. Each player has a private, but

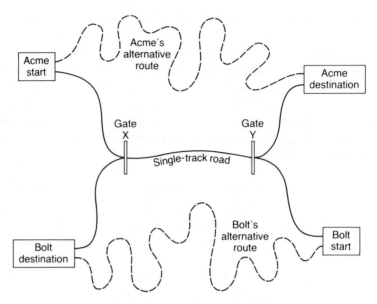

Fig. 13.1. The Trucking Game. (Adapted from Deutsch & Kraus, 1960)

lengthy and tortuous route by which their goods can be delivered. There is also a short route which they share, but which is only one truck wide. If they both try to use it at the same time, a stalemate results. Since time is money in this game, both players benefit from a coordinated use of the short route. The payoffs in this game can be made greater or smaller, but the interesting finding is that the players are more likely to cooperate in the production of high personal and aggregate wealth when the rewards for cooperation are initially small, and then gradually increased.

By itself, this finding might be attacked in the way that other experimental findings on the significance of rewards have been, but its significance is supported by two other observations. First, is the finding that the salesman's 'foot-in-the-door' technique can be demonstrated in a controlled setting. Freedman & Fraser had an executive from a road safety campaign ask householders if they would mind having a large and unsightly billboard promoting safe driving erected in their front garden. Only one sixth of this group said that they would. However, a second group to whom this request was made had been asked several days before if they would display a similar small sign in their window. Most agreed to display the smaller sign, and, when asked to accept the billboard, three quarters said they would.

Second, is a particular criminal activity which exploits the essential proposition that gradualism is the best way to build cooperation and trust. This fraud, reputedly used by the Kray brothers in London in the 1960s and widely used elsewhere, involves the establishment of a fake sales operation which begins with small orders to suppliers who are promptly paid. The size of these orders is gradually increased so that the effective credit being drawn from the supplier gradually mounts. In this situation, suppliers are happy to do this to support an increasingly important customer. The sting comes with the last order which is inevitably the largest of all. The goods supplied are immediately sold on at cut-price to an unsuspecting third party, and when the supplier sends the final demand for payment it is discovered that the sales operation has folded, and no trace can be found of its employees or directors.

The temporal factor which is central to this fraud and the related experimental finding is reflected in other studies on temporal variation which may be found in this literature.

Time

The capacity to explore in the laboratory the impact of the potential length of a relationship on the occurrence of cooperativity is severely limited. Most experiments in the laboratory last one or two hours at most, and even Zimbardo's noted attempt to simulate a prison environment lasted no more than one week. This time-scale is far shorter than is usual in most of our relationships with other persons or institutions, and is importantly different in that it is predictably shorter. While we may have passing short-term relations of many sorts, there is always the possibility that these could be extended. Therefore, our actions in being more or less cooperative are constrained by an uncertain temporal horizon. In the laboratory, a subject knows that the other subjects may be never seen again, and that the behaviour there is somehow circumscribed.

This point has been at the heart of some of the major claims that the laboratory provides a totally untypical and thus uninformative environment, and while the general force of it cannot be ignored, it does not mean that all studies of temporal variation are thereby dismissed. Indeed, one interesting finding is based on this very criticism. Pruitt & Kimmel (1976) note that subjects who believe that they will interact with one another after the experiment is over, and may thus come to know one another in a more open-ended way are more likely to be cooperative in the experiment itself.

A second important finding in this sphere is that the probability of cooperative behaviour being present or absent fluctuates across time if a game is repeated a large number of times. The general picture with the PDG is that players have a slight preference for the cooperative strategy on the first play. This preference rapidly disappears to be followed by both players choosing the competitive option repeatedly for a number of trials. Then, cooperative choices tend to reappear so that by the end of several hundred runs of the game, the cooperative choices predominate once more (Rapoport & Chammah, 1965).

There are several ways of construing this fluctuation, but it is very tempting to see it as akin to the responses of subjects to a pre-programmed, tit-for-tat strategy. In this, a subject, who believes that the other player is a real subject, is confronted in each play of the game by a choice by the other player which is the same as the subject's own choice on the previous play of the game. So, choosing the cooperative strategy on one play will lead the dummy opponent to choose that strategy in the next play. The result of this programmed strategy is a comparatively rapid move on the part of the real subject to a high and stable level of cooperative play.

The player using the tit-for-tat strategy would seem to be using the sequence of plays to offer the opponent a response which is contingent upon that opponent's own actions. It is conceivable that this is what is happening also, be it consciously or not, in the long repeated plays of the PDG. If we accept this, then there are two components to the bargaining which effectively takes place.[1] Necessarily, there is a communicative part, in that the sequence of plays contains the message, 'If you act cooperatively, I will too', and also, it would seem, a punitive part. I will turn to these next.

Communication

The importance of communication to the establishment of cooperation seems intuitively correct, but as was noted above when the issue of Machiavellianism was discussed, a channel of communication can also provide a medium of exploitation for those who are so inclined when

---

[1] This position assumes that the true player still believes that he or she is not operating against a program. If the truth is discovered, then the program's responses will be seen as mechanically contingent upon the subject's actions, and thus under his or her control. If this were so, then the same behaviour may result, but the game becomes psychologically uninteresting.

the basis of the negotiation is ambiguous. Of course, for many games, the situation is not ambiguous, and, in these cases, a large number of studies have reported that increasing the amount of communication does increase the level of cooperation. For example, Wichman (see Wrightsman, O'Connor & Baker (1972)) reported least cooperation between female subjects when they only had access to the other's choices in the game, and most when they were in close proximity, and could both see and hear one another.

The notion 'amount of communication' is far from simple, as is the means by which it has an impact on the level of cooperation. It has often been observed that when individuals have little information about or access to one another, their actions become less sociable and caring, and they often behave in a more instrumental and aggressive fashion towards one another. This was seen in a most striking fashion in Stanley Milgram's famous study of obedience. In this, subjects believed they were giving ever greater electric shocks to another person as part of a training exercise. When the trainee was in another room where the subject could not see the effect of the shocks he was administering, and could barely hear the trainee's protestations, the average level of shock which was administered was far higher than when the trainee was sitting next to the subject. Indeed, nearly two-thirds of the subjects continued to give shocks past the point at which the trainee might be thought to be seriously harmed or even dead. This suggests that the range of communicative channels, both vocal and non-vocal is an important matter, so that the quality and texture of the communication becomes important too.

The importance of the range of cues is supported by Derek Rutter's work on 'cuelessness'. In a series of experiments, he has shown that subjects whose access to one another is limited by, for example, having to interact through a screen or over the telephone deal with one another in a more instrumental and impersonal fashion. Seeing the other person's facial expressions will also increase the degree of affective communication, as will the greater sense of empathy made possible by the clearer demonstration of convergence in speech style and body movement.

It should be also be borne in mind that for successful communication, a high level of mutual knowledge is required. Anything which contributes to this will enhance our belief that we are clear about what the other intends. As confidence tricksters know only too well, many superficial aspects of our personal presentation can lead us to feel that we know one another on the basis of very little substantive information. We draw all sorts of inferences about the other's beliefs and sentiments on the basis of their age, gender, ethnicity, social class, etc., even though it is hard to support the conclusions in any objective way. As Baruch Fischoff and his colleagues have shown, increasing our access to apparently relevant information will increase our confidence as to our judgements in any sphere, even if we know that this is irrational. The net result of all this is that we may feel more sure in interpreting a cooperative overture when there is a greater amount of communication and, when face-to-face, such overtures may be more frequent.

Having greater access to the other player will also make more evident to each individual a possible level of social embarrassment and loss of face which would arise if competitive strategies were adopted. Of course, some games are played so that there can only be one winner, and only competition is possible, but in those like the PDG, where there is a cooperative option, to compete is positively to reject sociable cooperation, an implication which does not follow in the former case. In part, this suggests that in these cases, more information also increases the non-specific rewards and penalties associated with each choice.

Coercion

The success of the tit-for-tat strategy seems to suggest that effectively punishing one's opponent for non-cooperative behaviour by responding to like with like (the 'an eye for an eye, a tooth for a tooth' policy), is a sensible move. However, it would be wrong to overestimate the influence of having a retributive element on the basis of the PDG. Each player can recognize the defensive element of choosing the competitive strategy when the opponent is doing likewise, and the rewards in the game are derived through the same actions as the punishment is administered. This is not so in a version of the trucking game mentioned on p. 230.

In one variant of the game, two gates were introduced at either end of the road at positions X and Y in Figure 13.1. There were three conditions studied. In one, a mutual threat condition, each player had control of one gate and so could block the road if the other player was perceived to be acting unfairly. In a second, a unilateral threat condition, only one player could block the road in this fashion. In a third, neither could do so. At the end of a large number of trials, it was discovered that it was in the no-threat condition that there was most cooperation, and indeed it was only in this condition that either player made a 'profit' on their operations at all. Least cooperation was found in the 'mutual threat' condition.

Cognitive constraints

So far, various findings from laboratory games have been considered, as have been some of the criticisms which have arisen because of their laboratory setting. One of the features of these games which made them so attractive was the clarity with which they depicted choices which might face a social actor, but this clarity is also their major weakness. The real-life choices which individuals must make are rarely clear, nor are the settings in which they are located. As I have argued elsewhere, (see Gambetta, 1988) this lack of clarity can mean that conditions of cooperation and trust, or non-cooperation and distrust can be far more resistant to change than might be expected on the basis of these experimental findings. There are many reasons why this might be so, but there are three which I would like briefly to note here.

First, given that the full range of rewards and benefits facing someone is ill-specified, which ones they perceive will be a function of which ones are thought to be possibly present. Our perception in all domains, be they sensory or social, can be knowledge-driven to a large extent, and there is every reason to believe that this is true in this one too. Also, given that the perception of the costs and benefits of an action occur both before and after that action, and that there is a strong tendency to engage in post-hoc rationalizations which justify decisions which have been made, this tendency to perceive what we wish to perceive could be exaggerated. I have already noted that cooperative individuals are disposed to seeing others as cooperative, and that is part of this phenomenon.

Second, there is the common observation that all humans are poor at seeking disconfirmation of their ideas, and this provides the converse of the first point. Essentially, we are as poor at looking for reasons to undermine our views as we are good at looking for reasons to support them. Thus, if I believe someone is generally a cooperative person, I will ignore cases where they act uncooperatively, or explain them away as instances of that person acting out of character.

Third, when we have established a way of behaving with respect to some other person or group of persons we are reluctant to change it. This fixedness of mind has been well documented in many areas of our intellectual activity, and again there is no reason to believe it is not operative with respect to our decisions about being cooperative too. Making any decision, or deriving a solution to any problem requires a measure of cognitive work, and it is no surprise to find that individuals are unwilling to do this if they believe that nothing of great significance is at stake.

Together, these particular cognitive foibles, and others, will make cooperation, once established, more durable. Unfortunately, the same will be true for non-cooperation too.

Conclusion

In this brief essay, I have offered certain of the findings from the experimental gaming literature. To summarize, these suggest that greater levels of cooperation can be established when there is an absence of threat, and a greater facility for communication. Those who are cooperating also need to believe that the social, strategic or economic interactions in which they are taking part are potentially open-ended. Those who have a cooperative outlook will also be more likely to construe others in the same way, and thereby establish cooperative relations more quickly.

I hope I have also shown that in some respects these studies can be informative and in others provocative. While they must be taken with caution, I believe it would be wrong totally to reject them. They may not often serve the hypothesis-testing function for which experimentation is designed, but they can guide our thoughts in this area in a useful way. At the very least, they serve to demonstrate that actual performance in these ideal situations can be very variable and unpredictable, and that believing that we can investigate human behaviour by simply reflecting on them is a dangerous move. Experimentalists have been criticized for the distance between the laboratory and the real world. It is tempting to suggest that the distance between that world and the theorists' *Gedankenexperiment* is even greater.

Further reading

Christie, R. & Geis, F. (1970). *Studies in Machiavellianism*. New York: Academic Press.

Colman, A. M. (1982). *Game Theory and Experimental Games*. Oxford: Pergamon.

Colman, A. M. (ed). (1982). *Cooperation and Competition in Humans and Animals*. Wokingham: Van Nostrand.

Deutsch, M. & Krauss, R. M. (1980). The effects of size of conflict and sex of experimenter on interpersonal bargaining. *Journal of Experimental Social Psychology*, **61**, 181–9.

Eiser, J. R. (1986). *Social Psychology*. Cambridge: Cambridge University Press.

Gambetta, D. (ed). (1988). *Trust*. Oxford: Blackwell.

Pruitt, D. G. & Kimmel, M. J. (1976). Twenty years of experimental gaming. *Annual Review of Psychology*, **28**, 363–92.

Rapoport, A. & Chammah, A. M. (1965). *Prisoner's Dilemma: A Study in Conflict and Cooperation*. Ann Arbor: University of Michigan Press.

Wrightsman, L. S., O'Connor, J. & Baker, N. J. (eds). (1972). *Cooperation and Competition*. Belmont: Brooks Cole.

14

Determinants of instrumental intra-group cooperation

JACOB M. RABBIE

Introduction

In this chapter factors will be identified which facilitate the development of cooperation rather than competition among members in social groups. In cooperative relationships people work with one another toward a common goal in an effort to resolve a conflict of interest among themselves. In competitive relationships people work against each other in an attempt to gain more than the other individual or party. Our main thesis is that perceived membership in a social group facilitates intra-group cooperation.

Social group and social categories

There is no agreement in the literature how a social group should be defined. According to our Behavioral Interaction Model (BIM) (Rabbie, 1987; Rabbie, Schot and Visser, 1989), groups are viewed as social entities of various sizes and clarity of boundaries, ranging from couples, family units, teams and political parties to national movements. Consistent with the interdependence perspective of Lewin (1948), a group is conceptualized as a social system or 'dynamic whole,' ranging from a 'compact' social unit to a 'loose mass' whose members are defined, not by their similarity to each other but by their perceived positive goal inter-dependence with each other and with the group as a whole (Horwitz and Rabbie, 1989). A group becomes a compact 'we-group' or 'social group' to the extent that individuals are subjected to the experience of a common fate, perceive themselves to be interdependent with respect to their common goals and means to attain those goals, view themselves (and are also considered by others) as a distinctive social unit, can directly communicate with one another and engage in cooperative face-to-face interactions in an effort to achieve a group product or a common outcome which contributes in some way to the desired outcomes of each of the individual members and of the group as a whole. If individuals see better ways of achieving their desired outcomes like safety, prestige, status, interpersonal attraction, material rewards and task accomplishments, they will try to leave their own group for another group, provided they are not prevented from doing do by internal restraints (e.g., out of a sense of duty to stay in the group or nation); by external restraints (e.g., when one is forbidden by the authorities to travel abroad or when the desired group erects impassible barriers to entry, such as refusing illegal immigrants). This also happens when members of underprivileged groups are not allowed to 'pass' the intergroup boundaries between themselves and the more powerful majority groups in our society (Lewin, 1948).

In trying to achieve their common group goal, members are usually willing to sacrifice some of their individual interests in the service of attaining the collective interests of the group. This may reflect the operation of normative orientations induced by ingroup norms (e.g., norms that one ought to give more weight to the interests and desires of ingroup members than of outgroup members). The restraints on selfish individualistic behavior will be stronger the more the contributions of members can be personally identified. In this way they can be held accountable (and eventually rewarded or punished) for their actions. In the course of direct social interaction, members of a we-group create a common past and future together which form the basis for the development of a group identity: this way members view their group as a distinctive and bounded social unit which differs in some unique way from other 'they' or outgroups. Sharing a group identity leads to a social identification with the ingroup which is defined here as a form of interdependence whereby a group member is satisfied with the positive outcomes of other members and with the ingroup as a whole.

In contrast with the Social Identity Theory of Tajfel and Turner (1986) and its recent extension in the Self-Categorization Theory (SCT) of Turner and his associates (Turner, Hogg, Oakes, Reicher and Wetherell, 1987), we have made a conceptual distinction between a *social category* and a *social group*. As we have argued, a social group is defined as a 'dynamic whole' or social system, characterized by the perceived interdependence among its members, whereas a social category is viewed as a collection of individuals who share at least one attribute in common: for instance, they may be classified at random as members of Blue or Green groups, or as categories of people who prefer paintings of Klee or Kandinski (Tajfel, Billig, Bundy and Flament, 1971). Existing social categories or 'sociological categories' based on similarities in attributes like skin colour, gender, age, sex and occupation are likely to experience 'an interdependence of fate' (Lewin, 1948) – and become a social group in the process – since individuals in these sociological categories are often treated differentially, by outsiders and themselves, solely as a function of their membership in these types of social categories. In our view, the experience of a common fate (or 'interdependence of fate'; Lewin, 1948) is a crucial precondition for the formation of a social group.

Advocates of the Social Identity approach e.g., Tajfel and Turner, 1986, and Turner et al., 1987) reject perceived interdependence or a 'common fate' as an important motivational factor in group formation and view a group mainly as a cognitive 'perceptual category' from which other group characteristics may follow. Our central hypothesis is that the more a collection of people begins to share the features of a compact social group, as we have defined it, the more mutual intra-group cooperation will occur. Evidence for this hypothesis will be discussed in the context of social dilemma and other laboratory tasks which will be described in some detail.

Social dilemma tasks

Many human decisions involve a conflict between individual and collective interests. For example, it is to each individual's immediate benefit to take a free ride in a train, to drive a car (and pollute one's environment), to dodge taxes, to make illegal copies of computer software, to refuse to conserve water or to save energy, but the long-term collective effects of these selfish decisions are considerably worse for everyone than if each individual had exercised some personal restraint. These decision situations are referred to as social dilemmas.

Social dilemmas are characterized by two essential properties: '(1) each person has an individually rational choice that, when made by all members of the group, (2) provides a poorer outcome than that which the members would have received if no members made the rational choice' (Messick and Brewer, 1983, p. 15). Thus social dilemmas are conceptualized as a conflict of interest in which each person has a choice either to cooperate or to defect, to save scarce resources (and possibly to contribute to the common good) or to take a free ride i.e., to enjoy a public good without paying the costs for it.

In mixed motive situations the interests or preferences of the actors or players are neither diametrically opposed to each other (as in strictly competitive zero-sum games in which one party can only win at the cost of the other) nor are their preferences completely identical with each other (as in pure coordination games). Individuals in these 'mixed-motive games' are motivated partly to cooperate and partly to compete with each other and have therefore to contend with *intra*-personal conflicts arising from this clash of cooperative and competitive motives in addition to the *inter*personal conflict that is built into the game (Colman, 1982).

Social dilemmas have been studied in two types of mixed motive games: (1) the two person, two choice Prisoner's Dilemma Game (PDG) as described by Good in Chapter 13 and (2) the multi or N-person Prinsoner's Dilemma or NPD in which many persons are involved. The players or parties in these two-dilemma tasks are interdependent for achieving their outcomes. No one can achieve outcomes by his or her own unilateral actions but each needs the cooperation of the other(s) to obtain desirable outcomes or to avoid undesirable ones. These games have been characterized by McGrath (1984) as 'dilemma tasks' in which two or more parties have to make cooperative 'C' choices or non-cooperative (defecting) 'D' choices – usually simultaneously or at least independently from one another – which jointly determine the outcomes of each of them so that each one's payoffs depends on his/her own *and* the other's choices (i.e., the parties are outcome interdependent). The interdependence structure in these dilemma tasks is such that each party is motivated not to cooperate and yet, paradoxically, both prefer mutual cooperation to mutual non-cooperation or competition. In this way individual rationality leads to collective irrationality in the sense that the collective interest will suffer if each individual pursues only his/her selfish interests (Pruitt and Kimmel, 1977). In these dilemma tasks people are usually socially isolated from each other. They cannot directly interact or communicate with each other, verbally or non-verbally. Their interaction is restricted to making strategic cooperative C or non-cooperative (defecting) D choices in an effort to increase their own outcomes through others. The absence of direct communication and interaction between the players and the highly restricted choices they can make prevent the players in these laboratory tasks from becoming a social group as we have defined it. This is the reason why two or more individuals participating in these dilemma tasks have been characterized as 'quasi-groups' (McGrath, 1984). By

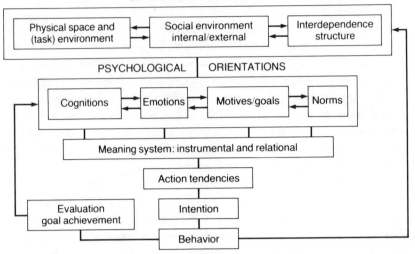

Fig. 14.1. The Behavioral Interaction Model.

studying the minimal conditions which seem to facilitate interpersonal cooperation in these 'quasi-groups' we may get some idea how fully-fledged social groups are formed.

The interpersonal cooperation in social dilemma tasks will be discussed from the perspective of a Behavioral Interaction Model (BIM) in which an attempt is made to integrate a variety of social psychological theories in the area of intra- and inter-group cooperation and competition (Rabbie, 1987; Rabbie *et al.*, 1989).

In the first section we will give a brief sketch of the model, in the second and third sections research will be reviewed with the aim of identifying the variables that facilitate or hamper cooperation in two-person (PDG) and multi-person (NPD) dilemma tasks. In the fourth section research on intra-group cooperation is discussed in minimal group situations which were designed to study the conditions in which ingroup cooperation or outgroup competition are likely to occur.

A Behavioral Interaction Model

In our Behavioral Interaction Model, depicted in Fig. 14.1, it is assumed that behavior, including the cooperative or defecting behavior of subjects in dilemma tasks, is a function of the external environment and the cognitive, emotional, motivational and normative orientations which are in part elicited by the external task environment and in part acquired by individuals in the course of their development (see Figure 14.1). The main function of these psychologicial orientations is to reduce the uncertainty in the external environment to such a level that it enables individuals, groups or organizations to cope effectively with the environment in an effort to achieve desirable and to avoid undesirable outcomes. The external environment consists of three components: (1) a physical (task) environment; (2) an internal and an external social environment: (i.e., the behavior of other people within and external to the group or social system); and (3) a positive or a negative interdependence structure between the parties which may help or hinder the parties in reaching their goals. The goal interdependence between the parties may be loosely or tightly coupled and may be symmetrical or asymmetrical with regard to the power relations between the parties. (In asymmetric power relationships party A depends more on party B for the satisfaction of his or her outcomes than vice versa.) These different psychological orientations produce a meaning system about the situation which in turn generates various action tendencies in the actor or party (an individual or a group).

Although many types of meaning systems may exist, we have focused our attention on instrumental and relational orientations which combine with the different cognitive, emotional, motivational and normative orientations which have been distinguished in the literature (Deutsch, 1982). For example, two kinds of cooperation and competition can be distinguished. In instrumental cooperation (or competition) the aim is to cooperate (or to compete) with each other to attain economic or other tangible outcomes. In social or relational cooperation the aim is to achieve a mutually satisfying relationship with the other as an end in and of itself rather than as an instrument to reach an external goal. The goal is to attain a relational understanding with the other in an attempt to explore whether the relation may grow in depth or will stay at a more superficial level (Rabbie *et al.*, 1989). The goal of social or relational competition is aimed at differentiating oneself, one's own group, organization or nation from similar others in an effort to achieve prestige, status, recognition or a 'positive social identity', i.e., that aspect of one's self-concept which is derived from one's membership in a group or social category (Tajfel and Turner, 1986).

It is assumed that among competing action tendencies and available strategies, those actions and alternatives will be chosen that promise, with a high probability of success, to attain the most valued goals or profitable outcomes, whereby the gains seem to exceed the costs of achieving them.

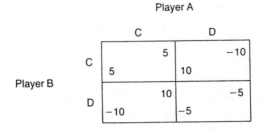

Fig. 14.2. Pay-off matrix in the Prisoner's Dilemma Game.

Thus, it can be expected that people will choose to cooperate rather than compete with each other when mutual (instrumental) cooperation seems to be more profitable to them than mutual (instrumental) competition in an effort to achieve their individual and collective outcomes.

In the next section, research on cooperative and competitive behaviour in the strategic environment of a two-person dilemma game which provides some evidence for this hypothesis will be reviewed.

Cooperation in the Prisoner's Dilemma Game

One of the most interesting mixed-motive games, often used in experimental social psychology, is the two-person, two-choice Prisoner's Dilemma Game (PDG). A typical pay-off matrix of the PDG is presented in Fig. 14.2.

Each player, A or B, in Figure 14.2 is faced with the dilemma of making a cooperative C choice or a non-cooperative (defecting) D choice. From an individual point of view, each player is tempted to make a D choice since in that case one can gain the best outcome for oneself (+ 10), at the costs of the other player (− 10). However, if the other player also makes a D choice, neither wins and both players end up in the DD cell in which both parties suffer a loss (− 5, − 5). From a collective point of view, both players will be better off if they cooperate with each other by making a CC choice so that both can win (+ 5, + 5) instead of making a mutual DD choice which leads to a mutual loss.

The PDG has been used to model the arms race between the super-powers as a series of decisions by two parties to increase or decrease the production of conventional or nuclear arms (Colman, 1982). The payoff matrix in Figure 14.2 indicates that each power is tempted to make a D choice in an effort to gain a military advantage over the other. However, if both powers increase their production simultaneously, they are worse off by making a mutual DD choice, which may lead to a dangerous escalation of the conflict, than if they had decided to make a mutual cooperative CC choice by limiting their arms production. By allowing mutual inspection of the reduction in weapons (or even their destruction), as was agreed in the INF accord between President Reagan and Secretary Gorbachev in the Washington talks, for example, the likelihood of mutual trust and commitment of both parties to a peaceful cooperative settlement between the superpowers is enhanced.

In terms of our model, the task-environment of the PDG is a very impersonal and strategic one which is likely to elicit mainly instrumental and utilitarian orientations and behaviour aimed at maximizing one's own economic and other tangible outcomes. There is little or no opportunity or incentive to engage in relational or social cooperation since the internal and external social environment of the PDG is rather anonymous and depersonalized. Most of the time subjects cannot see or communicate with the other player apart from the choices they exchange in a multi-trial game by means of written messages or mechanical devices.

The external social environment of player A (which we call the actor) consists of the variations in the (cooperative) C or (defecting) D choices of the other player (B) over the different trials of the game. The choices of B may represent the decisions of a 'real' player or are pre-programmed by the experimenter. In either case the strategy of B may have a contingent, responsive character in which the other appears to take the choice behavior of the actor into account, or follows a non-contingent, responsive strategy, independent of what the actor does. Generally a contingent responsive strategy in B elicits more mutual cooperation in a PDG situation than a non-contingent strategy. A contingent tit-for-tat (TFT) strategy involves matching the recent C or D behavior of the subject: making a C choice when he does and D choice when he fails to cooperate (see also Good, Chapter 13).

The external physical and social environment of the PDG elicits particular cognitive, emotional, motivational and normative orientations in the subjects who participate in the PD situation. *Cognitive orientations* and related concepts such as scripts, schemas, hypotheses and attributions refer to the structure of expectations and organized beliefs people have about themselves and others as they interact, for example, with the other player in the PDG (Deutsch, 1982). In a cooperative relationship a cognitive schema will emerge in which the players realize that the fates of the parties are linked together so that both players gain (positive interdependence) or lose together (negative interdependence). In this situation players intend to cooperate with the other party and expect the other to cooperate as well (Deutsch, 1982). *Emotional orientations* will be aroused when one expects the other to cooperate but in fact the other party makes a defecting or non-cooperative D choice. This kind of exploitative behavior will evoke feelings of indignation and anger especially in conditions in which the other has promised to make a cooperative choice. This norm violation of the other may evoke angry aggressive reactions when subjects have the opportunity to express them. Positive feelings are aroused when the other acts in accordance with one's cooperative expectations. In a cooperative relationship each tends to elicit a positive *motivational orientation* to the other, to trust the other, to be favorably disposed to him or her, to feel responsible for the other's welfare and to expect the other to be similar in these respects. In competitive relationships the opposite motivational orientations will be activated (Deutsch, 1982). It should be noted that these kinds of positive motivational orientations to each other will be much stronger in relational than in instrumental cooperation where the other is seen mainly as an instrument to maximize one's own economic and other material outcomes. *Normative orientations* linked with cooperation tend to foster egalitarian relationships, while competitive norms sanction inequality and superiority in outcomes of one party over the other (Deutsch, 1982).

People may also have various psychological orientations about the interdependence structure of the PDG (Pruitt and Kimmel, 1977). Cognitively, the nature of the goal interdependence is rather difficult to understand for most naive subjects. It takes some time and experience with the game and with the actions of the other player to gain some insight into the interdependence structure of the game. People with an instrumental, utilitarian predisposition are likely to get more emotionally involved in the strategic task-environment of the PDG since it will be more challenging to them in trying to maximize their economic gains than to subjects who are more relationally oriented and who see no opportunity in the PDG situation to satisfy their relational or communal needs (Lodewijkx, 1989). Motivationally, the PDG choices are rather ambiguous: a C choice may reflect cooperation (maximizing joint gain) or a desire to maximize individual outcomes (max. own) without taking the interest of the other player into account. A D choice is even more ambiguous; it may reflect competition (maximize relative gain), defensiveness (minimizing a maximum loss) or individualism (max. own). In view of these ambiguities, a motive questionnaire in our PDG research is administered in which people are asked about the reasons for their choices. The responses to the motive questionnaire give a more reliable and sensitive measure of their motivational orientations than can be inferred from their behavior alone (Rabbie, 1987; Lodewijkx, 1989). Finally, the symmetric interdependence structure of the PDG seems to induce a norm of reciprocity which may elicit mutual cooperation or mutual competition. In most 'normal' PDG situations parties gravitate toward mutual competition (McGrath, 1984; Colman, 1982).

What kinds of conditions facilitate mutual cooperation rather then mutual competition in a PDG? In their Goal Expectation Theory, Pruitt and Kimmel (1977) had proposed that cooperative behavior in an iterated (repeated) PDG (which lasts for several trials), usually result from the long-range goal of establishing (or maintaining) continued mutual cooperation. Three perceptions contribute to the development of this goal (1) perceived positive dependence on the other, i.e., a recognition of the importance of the other's cooperation; (2) pessimism about the likelihood that the other can be exploited (achieving DC), i.e., that he or she will cooperate unilaterally for any period of time; and (3) cognitive insight into the necessity of cooperating with the other in order to achieve mutual cooperation (CC). The latter two perceptions amount to a recognition that the dyad in the two-person PDG must choose between mutual cooperation (CC) and mutual non-cooperation (DD) and that the former is preferable to the latter.

When the two players in a PDG engage in mutual cooperation rather than in mutual competition we may observe the beginnings of a social group. As will be remembered our central hypothesis is that the more a collection of people have the features of a compact social group, the more mutual cooperation is likely to occur. There is some evidence for this hypothesis.

In Table 14.1 the conditions are presented, summarized by Pruitt and Kimmel (1977), which favor the development of the long-range goal of mutual cooperation (1) and the development of the expectation that the other can be trusted to cooperate in the future (2). As can be seen in Table 14.1, even the past history and the expectation of future interaction in the short time-span of the PDG in a laboratory setting, seem important factors in inducing the goal of mutual cooperation. In real life situations this effect would be even stronger. The initial experience that mutual competition does not work (a) and the anticipation for continued inter-action (k), particularly when A knows that B has cooperated in the past (1), encourages intra-group cooperation. Many conditions (a,b,c) make parties aware of the strategic nature of the interdependence structure of

Table 14.1. *Factors in PDG that favor the development of the goal of mutual cooperation and expectations that other will cooperate*[a]

1: Goal mutual cooperation	i. When A and B can communicate with each other
a. Initial experience of mutual non-cooperation in PDG	j. A sees oneself as weaker or more dependent on B
b. Time to think about the pay-offs in the matrix, e.g., by filling out questionnaires about motives; by having intra-group discussions; by making provisional decisions before having to make binding decisions	k. A and B anticipate continued interaction with each other
	l. When high aspirations of A seem to require CC choices.
c. When PDG is shown in a 'decomposed' form enhancing perceptions of mutual inter-dependence between A and B	*2: Expectation of cooperation*
	1. When A knows that B has recently or consistently cooperated in the past
d. When mutual cooperation (CC) yields high outcomes	2. A has received a message from B requesting or assuring cooperation
e. When mutual competition (DD) yields low outcomes	3. When A knows that B's incentives (pay-offs) or instructions favor C choices
f. When CC choices yield equal or equitable outcomes	4. When A sees B as weaker or dependent on oneself
g. When decisions can be reversed if A or B are dissatisfied	5. When B uses a tit-for-tat or another contingent strategy
h. When B uses a tit-for-tat or other responsive strategy	6. When A sees B as similar to oneself or as a friend

[a] This summary is based on Pruitt and Kimmel (1977).

the parties in the PDG. This leads to the cognitive insight that mutual cooperation would be preferable over mutual competition if the other party can be expected or trusted to cooperate as well. When an asymmetric interdependence structure is perceived, in which an actor views her- or himself to be more dependent (or less powerful) than the other, the weaker party is expected to cooperate (j,4). Other conditions pertain to elements of the task environment of the game, particularly the kind of incentives or payoffs at stake which favor mutual cooperation rather than mutual competition (d,e,f,l,3). The most important factors in encouraging mutual cooperation are changes in the social environment over time: the behavior of the partner over repeated trials or the (cooperative) instructions of the experimenter (h,i,k,l, 2,3,5, 6). While unconditional cooperation might invite exploitation by the other party, a contingent TFT strategy is the most effective way of inducing mutual cooperation. Experiments have shown that direct task-related communication between the players, which is usually not permitted in the PDG, leads to an increase in the proportion of mutual cooperation, particularly when the announced intentions to cooperate are backed up by cooperative actions. In this case two (cooperative) deeds speak louder than words!

The proportion of cooperation is higher when subjects are paired with someone they like, than with someone they dislike. Their reactions to a stranger are in between. When the liked partner responds cooperatively to a C choice of the actor the subjects are likely to reciprocate in kind; when the liked partner makes a competitive response however, subjects become much more competitive. That does not occur when a stranger or a disliked partner makes a competitive response. Apparently, one expects a friend to be similar to oneself, while a stranger and especially a disliked opponent is perceived to be different from oneself. We expect more cooperation from someone who is similar than from someone who is dissimilar. A competitive response from a dissimilar other is to be expected and does not arouse strong reactions, especially if an actor has behaved competitively himself. A competitive strategy of a friend is seen as an exploitative response which violates the norms of reciprocity, trust and equality among friends and requires a sharp reaction. It is likely that in our relationships with friends social or relational cooperation (and competition) becomes much more important than instrumental cooperation (and competition) in the usual PDG situation. In other research we have found that the norm violation of a similar partner in a PDG evokes a stronger aggressive reaction than a breach of promise by someone who is viewed as dissimilar from oneself (Rabbie and Lodewijkx, 1987). Similarity is also an important factor in man–machine situations. When people are told that they have to play a PDG with a computer rather than with another (responsive) person, like themselves, they will tend to make more non-cooperative than cooperative choices. It is likely that all these factors which seem to induce mutual cooperation in the minimal 'quasi-group' conditions of the PDG, will enhance the probability of intra-group cooperation in 'real-life' groups as well.

Individuals differ in their propensity to make cooperative or competitive choices in the PDG. Kelley and Stahelski (1970) have proposed, in their triangle hypothesis, that individuals who describe their intentions as competitive expect the other to be competitive as well, while individuals who describe their intentions as cooperative expect others to be either competitive or cooperative. Cooperators appeared to be more flexible and responsive to the choices of the other than competitors.

In our research we have found that men are likely to follow a more contingent cooperative TFT strategy than women. They act more like cooperators, making cooperative choices when the other can be trusted to cooperate, but more non-cooperative choices when the other seems to follow a competitive or non-contingent strategy. Females, on the other hand, tend to make more competitive choices and seem less responsive to the cooperative or non-cooperative behaviour of the other player. In contrast to men, their prior behavior appears to be a better predictor for their future choices than the actions of the other players. One possible explanation for these sex differences might be that there is a greater congruence between the strategic environment of the PDG and the more instrumental orientation of men. Men will maximize their economic outcomes either by instrumental cooperation or by instrumental competition dependent on the strategy of the other player. The more relational women are less interested in maximizing their economic gains and appear to be more motivated by social competition than instrumental cooperation in making their choices in a PDG (Lodewijkx, 1989). (For these intriguing sex differences see Good, Chapter 13).

In our research we have compared the game behavior of individuals and groups with one another (Rabbie, Visser and van Oostrum, 1982). Groups who have to make collective decisions in a PDG are likely to follow a more cooperative TFT strategy than individuals, especially when they consist of males rather than of females. This finding contrasts sharply with the contention of Tajfel and Turner (1986) that groups have 'essentially' a competitive relation to each other. Consistent with the goal-expectation theory of Pruitt and Kimmel (1977), intra-group discussions about what choices to make in a two-party PDG seem to enhance cognitive insight into the nature of the interdependence structure of the game and the necessity for mutual cooperation when the other dyad can be trusted to cooperate. When individuals, dyads and triads are paired with 'groups' of the same size it is found that dyads make more defecting D choices in a PDG than individuals or triads, especially when the other party appears to exploit them. The dyad is the smallest possible group. If the members of a dyad cannot come to an agreement about the desired course of action in a PDG, the dyad cannot function any more as an effective decision-making group. The pressures to conform and to avoid conflicts among themselves are therefore stronger in dyads than in triads, in which decisions can be made by a majority vote. An internal conflict among members in the dyad about what kind of choice to make in a PDG is more easily resolved by making a D than by making a C choice. If the other dyad chooses for cooperation one's own party wins at the cost of the other. If the other makes a D choice, a defective D choice of the own party is the best way to minimize a maximum possible loss. This finding illustrates, consistent with our model, that the nature of the internal and the external social environment has to be taken into account to predict whether groups will compete or cooperate with each other.

McGrath (1984) has pointed out that situational factors, for example, payoff structures, experimental instructions, the strategy or behavior of the other party or the opportunity to communicate or make visual contact with the other party have much more powerful effects than do personal factors (e.g., gender, status, personality characteristics, or even pre-game preferences for competition or cooperation). Moreover, there is not a high correspondence between how cooperative a person is in the PDG and in other mixed-motive games. A recent meta-analysis of PDG research provides some evidence for this conclusion (Lodewijkx, 1990). In this analysis it was found that most of the variance in the proportion of cooperative choices in the PDG was accounted for by the perception that the other party was a machine or a computer rather than another person (a person elicited more cooperation than a machine). Cooperation increased when the other followed a responsive TFT strategy rather than a non-contingent strategy. Cooperative behavior is enhanced as a function of the incentive structure of the PDG: (1) the higher the payoffs for cooperation, (2) the lower the payoffs for defection and (3) the greater the punishment for mutual defection, the more cooperative choices will be made. The opportunity to communicate with the other and the nature of the incentives – when the game is played for money rather than for points – encouraged more cooperative than non-cooperative behavior although the effects of these variables were rather small. Although the sex variable was included in this meta-analysis, it does not seem to make much difference to the proportion of cooperative choices in the PDG.

There is some (conflicting) evidence that intra-group cooperation tends to decline with increasing group size. Different explanations have been suggested. Large groups foster anonymity of the group members which makes it more difficult to apply positive or negative sanctions to them. Moreover, the weaker the sense of community and group solidarity the lower the impact of group norms favoring intra-group cooperation. Since individuals in larger groups feel more anonymous and less identifiable, the social responsibility for attaining the group goal is diffused and consequently individuals are less motivated to cooperate with one another for the group's benefit. Members of larger groups may also feel that their personal contributions are less effective or efficacious in affecting the fate of the group than is experienced by members of smaller groups.

Obviously, issues of anonymity, identifiability, impact of group norms and personal efficacy can better be studied in the context of a multi- or N-person Prisoner's dilemmas (NPD) than in a two-person PDG.

Cooperation in multi-person dilemma tasks

Social dilemmas such as taking a free ride on a train or a failure to save energy, which were mentioned earlier, have often been studied by means of NPD. As noted by Dawes (1980) the two-person PDG differs from a NPD in three ways. First, the harm of defection in a two-party PDG is visited completely on the other player. In the NPD the gain for the defecting choice accrues directly to the individual actor while the costs or harm are spread out over several others. Second, when the actor makes a defective D choice in PDG, the other knows what the actor has done. In the larger group of the NDP the defecting player(s) remain anonymous and cannot be held accountable and sanctioned for their defecting behavior. Finally, each person in a two-person PDG has total reinforcement control over the behavior of the other. An actor can 'punish' the other for his prior defecting behavior by making a D choice and can 'reward' the other's previous cooperative (C) behavior by making a C choice too.

In their thoughtful review, Messick and Brewer (1983) have made a distinction between solutions to social dilemmas that derive from independent changes in individual behavior and the solutions that come about through coordinated group action. They illustrate the distinction between individual and structural solutions by referring to the responses which were made in California to the droughts of 1976 and 1977. Individual solutions were stimulated by urging citizens to use less water in washing their cars, flushing toilets and watering vegetation, etc. The appeals for individual conservation were quite successful. In some areas the water consumption was reduced to about 30% per capita. Structural solutions were instituted by placing a moratorium on new water connections which reduced the total number of users of water in some areas.

Obviously, that is a measure which can only be taken by an existing organization such as a regulatory board, but not by individual consumers.

Since we are interested in the question of what kind of conditions facilitate cooperation and ingroup formation among interdependent (anonymous) individuals, we restrict our discussion to the factors Messick and Brewer (1983) have mentioned which promote individual solutions to social dilemmas. A brief summary of their review on this issue is presented in Table 14.2

All factors listed in Table 14.2 provide strong evidence for our hypothesis that the more a collection of people are characterized by the features of a compact social group, as we have defined it, the more intra-group cooperation will occur. The opportunity to communicate, the information that others have cooperated in the past, the trust in other group members, shared social values and responsibility for the group's welfare and obviously the awareness of a common group identity encourage the arousal of cooperative orientations and consequent behavior.

As Messick and Brewer (1983) have pointed out, information about the cooperative or defective behavior of the others in the group may have conflicting effects. Learning that others have cooperated introduces normative pressures to behave cooperatively, but on the other hand, it relieves pressure to cooperate since the others have done it already. Similarly, information that others have made selfish or defecting choices leads to conformity in favour of self-interests but also activates the need for individual restraint in the service of the collective interest. It will depend on additional factors, for instance, the opportunity to communicate or perceptions that individuals can make an effective contribution toward the group goal, or whether information about the decisions of others will move them more to cooperative (C) rather than to defecting (D) choices.

In listing the effects of ingroup identity in Table 14.2, Messick and Brewer (1983) have noted that 'effects of category membership are enhanced when they are subjected to a common fate'. This statement seems to imply the traditional view that membership in a perceptual social category is a precursor rather than a consequence of experiencing a common fate. In the next section we will present some evidence which suggests that the perception of a common fate, rather than membership in a 'perceptual' social category motivates people to cooperate with each other.

Table 14.2. *Factors that facilitate individual cooperation or restraints on selfish behavior in social dilemma tasks (PDN)*

1. Communication
Prior discussion of a social dilemma increases the probability for cooperation for the following reasons: (a) it may provide information on the willingness of others to cooperate; (b) introduces beliefs that others are committed to cooperate which enhances mutual trust and reduces the perceived risk to cooperate; (c) provides opportunity for moral persuasion and communication of values that support collective goals; (d) may create a sense of group identity and cohesion

2. Information about other's choices
Feedback that others have cooperated tends to increase cooperation or restraints on selfish behavior, while information about defecting others has the opposite effect

3. Trust in other group members
The probability of cooperative behavior is increased by the expectation or trust that others will cooperate as well. Trust in others will be enhanced by: (a) their past (cooperative) behavior; (b) by 'depersonalized trust' based on the presumption that others in the group share common values, attitudes and goals: (c) individual differences in prior beliefs about other's trustworthiness in dilemma situations; (d) by the development of beliefs and group norms that defective or selfish behavior will be punished; (e) the degree to which individual decisions are personally identifiable

4. Social values and responsibility
Individuals may choose to respond cooperatively because of the value they place on behavior that serves the collective welfare above self-interest. Cooperative choices are more likely to the extent that these values are made salient by: (a) making individuals aware of their social responsibility for the long-term effects of their actions; (b) when social dilemmas are presented in terms of moral issues involving ethics, group benefit and exploitation; (c) receiving feedback that their individual contributions are effective in achieving collective goals; (d) the realization that in small groups, as compared with larger groups, each individual's failure to contribute may spoil the outcomes of the group as a whole; (e) the perceived seriousness of the collective problem which makes the exercise of personal restraints more urgent and compelling

5. Ingroup identity
A heightened sense of membership in a common group or social category probably enhances all of the factors discussed above that influence individuals' willingness to exercise personal restraints and the tendency to make cooperative rather than defective choices in social dilemmas. This heightened ingroup identification may lead to: (a) stronger pressures in cohesive groups to conform to cooperative models in the group; (b) strengthening the tendency to see members in the ingroup as more trustworthy, honest and cooperative then members from outgroups; (c) enhancement of the effectiveness of normative pressures that non-cooperators will suffer negative sanctions; (d) diminution of the sharp distinctions between individual and group welfare; (e) increase of perceived effectiveness of individual contributions. Effects of category membership are enhanced when members perceive that they are subjected to a common fate

Cooperation in the minimal group paradigm

In our definition of a social group we have stressed the importance of perceived positive outcome interdependence among members of a social unit, rather than similarity, as a crucial precondition for the development of intra-group cooperation and group formation. Turner *et al.* (1987), in their Self-Categorization Theory (SCT) have a different view. They reject 'interpersonal interdependence for need satisfaction as the basis for group formation' and stress the importance of the 'perceptual identity of people in the sense of their forming a cognitive unit or perceptual category' (p. 52). The main evidence for their position is based on the allocation behavior in the Minimal Group Paradigm (MGP) developed by Tajfel *et al.* (1971). Turner (1982) describes the experimental procedure of these researchers as follows: 'These investigators randomly classified their subjects into two distinct categories (e.g., Group X or Y) in isolation of all the other variables normally associated with group membership. Group membership was anonymous and *there was no goal interdependence, social interdependence, social interaction or other basis for cohesive relations.* Nevertheless, the subjects discriminated against anonymous outgroup and in favour of anonymous ingroup members in the distribution of monetary rewards *under conditions where they could not benefit from this strategy*' (p. 22; emphasis added). In standard MGP experiments two main allocation strategies have been found: (1) 'ingroup favoritism', the strategy to allocate more money to members of one's own social category than to members of the other social category, and (2) the strategy of 'fairness', to give members in the other social category about as much as to members in their own social category, with a slight but significant preference for members in the own category. Turner (1982) seems to argue that since subjects cannot directly allocate money to themselves, the favoritism to members in their own social category in the standard MGP experiments should be viewed as an instance of social competition by which the subjects strive to increase or maintain their self-esteem as category members or to achieve 'a positive social identity' (Tajfel and Turner, 1986).

We take a different position. In the usual instructions of the MGP, the subjects in the Tajfel *et al.* (1971) experiment were told that 'They would always allot money to others [i.e. always members of the ingroup or outgroup] . . . At the end of the task each would receive the amount of money that the others had rewarded him' (p. 156). These instructions imply that the individual outcomes of subjects in the Tajfel *et al.* (1971)

experiment are dependent upon the allocation decisions of members of their own and other social category, but not on their own decisions. Thus, their 'outcome dependence' is two-sided: on members in their own category and the *other* category. Although subjects in the usual MGP cannot directly allocate money to themselves, they can do it indirectly on the reasonable assumption that the members in their own social category will do the same to them. By giving more to members in their own category than to the members in the other category – in the expectation that the members of their own category will reciprocate this cooperative action – they will increase the chances of maximizing their own outcomes. Thus, although subjects in the MGP may be seen as acting independently from each other, they *perceive* themselves to be interdependent with respect to reaching their goal to maximize their (joint) outcomes. When the subjects act on these perceptions, a tacit coordination is achieved, which appears to help them to gain as much as they can in the MGP. It should be noted that subjects in the MGP do not receive any feedback about the effects of their actions from the recipients of their allocation decisions. The 'tacit coordination or cooperation' is not affected by feedback about the immediate consequences of the actual actions of the other subjects in the MGP. Thus the subjects in the standard MGP react to a constructed rather than to an 'objective' social reality. In our view, contrary to the opinion of Turner (1982), there is a perceived 'goal interdependence' in the standard MGP and the subjects may use strategies which they view as instrumental in providing them with valuable economic 'benefits'. The so called 'ingroup favoritism' shown to members in the own category, which is often found in these MGP experiments (Tajfel and Turner, 1986) should, in our opinion, be reinterpreted as a form of (indirect) instrumental cooperation with anonymous members of one's own social category and with members of the other social category intended to maximize one's own individualistic outcomes. The potential competition with members of the other category will be tempered by the perceived interdependence of members in the other category as well, leading to equal or 'fair' allocations to both categories, as is often found in the standard MGP and which seems instrumental in maximizing one's own outcomes. Since it is difficult to accommodate the strategy of fairness to social identity theory, our interdependence hypothesis gives a more parsimonious explanation of *all* the findings obtained in the standard MGP than that proposed by social identity theory. This reinterpretation of the allocation behavior in the standard MGP of Tajfel *et al.* (1971) challenges the basic assumption of Tafjel and Turner (1986) that 'the mere perception of belonging to two distinct groups – that is, social categorization *per se* is sufficient to provoke intergroup competitive or discriminatory responses on the part of the in-group' (p. 13).

In our view, social categorization in the MGP does not so much stimulate social competition and discriminatory *intergroup* behavior aimed at achieving a 'positive social identity', as Tajfel and Turner (1986) have asserted, but can be better understood as primarily instrumental and *interpersonal* behavior aimed at cooperating with those individuals in social categories which one perceives oneself to be the most dependent upon in trying to maximize one's own economic self-interests, at least when money has to be allocated (Rabbie and Schot, 1989a). This view implies, to put it very bluntly, that there is no 'ingroup bias' or 'ingroup favoritism' in the standard MGP, but there are only self-interested individuals in social categories, who use the category labels mainly as instruments to maximize their individual outcomes.

This interdependence hypothesis was tested in a recent experiment (Rabbie *et al.*, 1989). In this study subjects were classified into two social categories, allegedly on the basis of their preference for two different series of paintings. The perceived interdependence structure of the MGP was varied in the following ways: in an own social category condition the subjects, males and females, were made to believe that they were only dependent on the allocation behaviour of the members in their own social category (D.own) In a two-sided dependence or control condition the subjects received the same instructions as were used in the original Tajfel *et al.* (1971) experiment, which implied that they perceived themselves to be dependent on individuals in *both* social categories for maximizing their own outcomes (D.both). In a third dependence condition, the subjects were made to believe that they were only dependent upon the allocation decisions of individuals in the other social category (D.other), who had expressed a different preference for the paintings from themselves. To induce symmetric perceptions and reciprocal expectations, the subjects in each of the three dependence conditions were told that the other members in the own and other social categories were in the same position as themselves.

Contrary to the assertions of Tajfel and Turner (1986), there seems to be a rational link between economic self-interests and the strategy of 'ingroup favoritism' i.e., the tendency to give more to those individuals one perceives oneself to be dependent upon in the MGP. The more people perceived an interdependence of fate with individuals in their own social category (D.own), in the other social category (D.other) or in both social categories (D.both), the more money they allocated to those individuals, in the expectation that the allocations would be reciprocated, leading to a maximization of their economic self interests. Thus, the greater the perceived dependence on individuals in their own social category (D.own) the greater the 'ingroup favoritism' (or intra-category cooperation), while the greater the perceived interdependence on the individuals in the other social category (D.other), the greater the 'outgroup favoritism' (or cooperation with individuals in the other social category). In the standard, two-sided dependence condition, the expected intermediate level of 'ingroup favoritism' (or intra-category cooperation) was obtained since people were motivated to maximize their outcomes by allocating money to individuals in their own as well as in the other social category. Moreover, consistent with our interdependence hypothesis, the (instrumental) fairness, the tendency to give individuals in both social categories about as much money in the two-sided standard (MGP (D.both), was greater than in the two one-sided interdependence conditions. The 'outgroup favoritism', the strategy to cooperate more with individuals in the other social category than with individuals in their own social category, in the expectation that this cooperative behavior would be reciprocated in the same way, provides strong evidence for our hypothesis that perceived positive interdependence is a more important condition for inducing mutual cooperation than perceiving oneself as being part of the same 'perceptual category' as Turner et al., (1987) and others have claimed.

However, the outgroup favoritism or cooperation with members in the other social category (D.other), may be affected by the kind of incentives which are involved. It was hypothesized that the allocation of symbolic points in the MGP would arouse more social competition than instrumental cooperation with those groups one perceives oneself to be dependent upon. Consistent with this hypothesis it has been found that if economic incentives (monetary points) are involved, subjects show the expected 'outgroup favoritism' in the D.other condition; i.e., instrumental cooperation with individuals in the other social category. When symbolic points have to be allocated, the usual ingroup favoritism or cooperation with individuals in their own social category is obtained, even in the D.other condition, although this trend is much weaker in the D.other than in the D.both and in the D.own conditions (Rabbie and Schot, 1989a). Apparently, the norm to give more weight to the desires of ingroup members than to outgroup members is considerably weakened when economic rather than when symbolic incentives are at stake in the MGP. In these experimental games people can afford to show more ingroup favoritism if it does not cost them that much. Thus there are circumstances in the MGP when social competition, as proposed by Tajfel and Turner (1986), may be a more important strategy in the standard MGP than instrumental cooperation.

We have argued that a social group should be considered as a 'dynamic whole' or social system, whose members are not so much defined by similarity as by their perceived positive goal interdependence with each other and with the group as a whole. A social category was defined as a collection of people who have at least one attribute in common. In our view, members in a social category remain self-interested individuals unless they experience a 'common fate' (Cartwright and Zander, 1986, pp. 56–7) or a 'common predicament' which makes them aware that membership of a social category has positive or negative consequences for them, in comparison with individuals in other social categories. Only when they positively identify with the interests of the other members in their own social category do they become a social group as we have defined it. As we have indicated, positive identification is viewed as a form of perceived interdependence where one is satisfied with the good outcomes of other ingroup members and the group as a whole and dissatisfied by their bad outcomes.

In view of these considerations, our general hypothesis is that mutual cooperation is more likely to occur in social groups than in social categories. To obtain evidence for this hypothesis, subjects in a standard MGP were categorized at random in the usual way on the basis of their preferences for two series of paintings. The individuals in these two social categories were instructed in a Group condition to maximize their group interests as well as they could. In an Individual condition they were urged to maximize only their individualistic self-interests while in a Control condition only the usual standard instructions of Tajfel *et al.* (1971) were given. Since we assume that the allocations in the standard MGP are mainly motivated by attempts to maximize individual outcomes rather than group outcomes, as Tajfel and Turner (1986) have claimed, no differences in allocations between the Individual and Control conditions were expected, but in both conditions allocations would be significantly different from the ones in the Group condition (Rabbie and Schot, 1989b). Consistent with our general hypothesis, more tacit intra-group cooperation (and more intergroup competition) did occur in the Group than in the Individual and Control conditions (which did not differ in allocation behavior among themselves). Moreover, the greater willingness to cooperate with each other in the Group condition occurred even though it implied some individual loss. Again it seems that shared perceived interdependence rather than membership in a perceptual social category is the main impetus for encouraging instrumental intra-group cooperation.

Conclusion

The evidence reported in this paper provides some support for the hypothesis that the more a collection of people resembles a compact social group the more intra-group cooperation will occur. Moreover, a perceived positive goal interdependence among members of a social group seems to facilitate more instrumental intra-group cooperation than among individuals who perceive themselves as members of a cognitive or 'perceptual' social category (cf. Turner *et al.* 1987).

Of course, this conclusion should not be over generalized. Most of the evidence is based on the game behavior of socially isolated individuals in situations in which there are severe restrictions in opportunities to communicate with each other, either verbally or non-verbally, where only a few choice strategies are available, and where the resources to be exchanged are limited to symbolic or monetary points. There is little temporal continuity in these *ad hoc* laboratory groups and less need for patterned interpersonal activities (McGrath, 1984). Consequently, in these strategic games there is much less room for the development of social or relational cooperation aimed at achieving a mutually satisfying relationship with others as happens among friends, couples, families and in small working groups in organizations. Although strategic experimental games may provide some insight into why individuals are motivated to cooperate with each other, primarily in an instrumental, utilitarian fashion, the study of the development of relational intra-group cooperation would require the use of different research strategies, which have not been discussed here.

Acknowledgements

The author is indebted to Murray Horwitz, Hein Lodewijkx, Lieuwe Visser and Jan Schot for their contributions to this chapter. The research reported in this paper was supported by the grants 57–7; 57–97 and 560–270–012 from the Netherlands Organization for the Advancement of Research (NWO).

References

Cartwright, D. & Zander, A. (1986). The nature of group cohesiveness. In: D. Cartwright and A. Zander (eds). *Group Dynamics*, 3rd edn. London: Tavistock.

Colman, A. (1982). *Game Theory and Experimental Games. The Study of Strategic Interaction.* Oxford: Pergamon.

Dawes, R. M. (1980). Social dilemmas. *Annual Review of Psychology*, **31**, 169–93.

Deutsch, M. (1982). Interdependence and psychological orientation. In: V. J. Derlega & J. Grzelak (eds). *Cooperation and Helping Behavior.* New York: Academic Press.

Horwitz, H. & Rabbie, J. M. (1989). Stereotypes of groups, group members and individuals in categories: A differential analysis of different phenomena. In: D. Bar-Tal, C. F. Graumann, A. W. Kruglanski, W. Stroebe (eds). *Stereotyping and Prejudice: Changing Perceptions.* Berlin: Springer-Verlag.

Kelley, H. H. & Stahelski, A. (1970) The social interaction basis of cooperators' and competitors' beliefs about others. *Journal of Personality and Social Psychology*, **16**, 66–91.

Lewin, K. (1948). *Resolving Social Conflicts.* New York: Harper and Row.

Lodewijkx, H. F. M. (1989) Aggression between individuals and groups. Dissertation, University of Utrecht (In Dutch).

Lodewijkx, H. F. M. (1990) Factors including cooperation in the Prisoner's Dilemma Game: a meta-analysis (in Dutch) I.S.P., Utrecht.

McGrath, J. E. (1984). *Groups: Interaction and Performance.* Englewood Cliffs, N. J.: Prentice-Hall.

Messick, D. M. & Brewer, M. B. (1983). Solving social dilemmas: A review. In L. Wheeler and P. Shaver (eds). *Review of Personality and Social Psychology*, vol. 4. Beverley Hills, CA. Sage, pp. 11–44.

Pruitt, D. G. & Kimmel, M. J. (1977). Twenty years of experimental gaming: critique, synthesis, and suggestions for the future. *Annual Review of Psychology*, **28**, 363–92.

Rabbie, J. M. (1987) Armed conflicts: Toward a behavioral interaction model. In J. von Wright, K. Helkama, A. M. Pirtilla-Backman, (eds). *European Psychologists for Peace. Proceedings of the Congress in Helsinki, 1986.*

Rabbie, J. M. & Lodewijkx, H. (1987). Individual and group aggression. *Current Research on Peace and Violence*, **2–3**, 91–101.

Rabbie, J. M. & Schot, J. C. (1989). Instrumental and Relational Behavior in the Minimal Group Paradigm. Paper presented to the 1st Congress of Psychology, Amsterdam, July, 2–7, 1989.

Rabbie, J. M. & Schot, J. C. (1990). Group Allocations in the Minimal Group Paradigm: Fact or Fiction? In P. J. D. Drenth, J. A. Sergeant and R. J. Takeys (eds.), *European Perspectives in Psychology*, **3**, pp. 251–63. Chichester: Wiley.

Rabbie, J. M., Schot, J. C., Visser, L. (1989). Social identity theory: a conceptual and empirical critique from the perspective of a Behavioural Interaction Model. *European Journal of Social Psychology*, **19**, 171–202.

Tajfel, H., Billig, M. G., Bundy, R. P. Flament, C. I. (1971) Social categorization and intergroup behaviour. *European Journal of Social Psychology*, **1**, 149–78.

Tajfel, H. & Turner, J. C. (1986). The social identity theory of intergroup behavior. In S. Worchel and W. G. Austin (eds). *Psychology of intergroup relations*. Chicago: Nelson-Hall, pp. 7–24.

Turner, J. C. (1982). Towards a cognitive redefinition of the social group. In: H. Tajfel (ed.). *Social identity and intergroup relations*. Cambridge: Cambridge University Press.

Turner, J. C., Hogg, M. A., Oakes, P. J., Reicher, S. D., Wetherell, M. S. (1987). *Rediscovering the Social Group: Self-Categorization Theory*. Oxford: Basil Blackwell.

E.

COOPERATION BETWEEN GROUPS

Editorial

The development of cooperation between individuals is, as we have seen, a complex enough process. When we consider cooperation between groups, societies or nations, the complexity is even greater. The propensity of a group to cooperate with another group, or to show prosocial behaviour, is not a simple consequence of the prosocial tendencies of its members; additional variables enter in. For example, the group context can affect the way in which individuals evaluate a situation, and thus their behaviour.

Within most groups, cooperation with other group members is vital. Professional groups, political parties, pressure groups cooperate to achieve their several goals – manufacturing a product, providing a service or achieving influence. Individual members may differ in their willingness to cooperate and the extent of their contributions, but group norms and often group pressure create a within-group cooperative tendency, and often it is difficult to determine whether such a cooperative tendency, e.g. on a religious background, can be traced back to a real individual belief, to the expected, e.g. transcendal, reward or to strict group pressure. In fact, many of the prosocial normative systems, like religion, humanistic and/or enlightment ideologies and socio-political structures may be a mixture of such different influences to guarantee a long-term balanced prosocial belief system/norm.

If the goal of the group is to help outsiders or other groups, as with charity organizations, the leadership role will usually demand a personality with strong prosocial propensities. The leader has to facilitate and exemplify prosocial behavior. However, while the leader of such a group usually has these qualities, they are not essential. An authoritarian leader could 'order' cooperation. In any case, as with cooperation between individuals, individual and situational factors enter in: group A may cooperate with B over task X, but not with C or over task Y. Only those groups which have goals of social responsibility towards others are likely to show prosocial behaviour consistently.

In any case most groups find that, from time to time, their goals clash with those of other groups. In such cases the forces that operate to promote in-group cooperation, discussed in the preceding chapters, may also lead to denigration of the outgroup, diminishing the likelihood of inter-group cooperation and enhancing that of conflict. For groups, societies or nations one solution to this problem is to stress the costs of conflict and the sometimes long-term common advantages of cooperation; this strategy is part of the process of negotiation leading towards 'conflict settlement' or 'conflict resolution' as discussed by Rubin (Chapter 15). Another possibility is to change the conception of 'we' versus 'they' and thus to create superordinate goals, as discussed by Feger (Chapter 16). If it is possible to shift the definition of the ingroup to a higher level, the conflict could give way to cooperation. (In this context it is worth noting that Feger, unlike Rabbie (Chapter 14), treats 'social group' and 'social category' as more or less equivalent concepts.) In addition to the social and emotional qualities of such a new definition of 'we', it is crucial to stress that in the long run it may increase the probability of reward for all.

However there are dangers on this road. If sub-national groups combine patriotically ('we as a nation'), they may come to perceive other nations as outgroups and augment possibilities for conflict at a higher level. As numerous historical events have shown, this is especially likely if other nations look or behave differently, or if economic interests clash. Yet history also provides a source for hope, for conceptions of humankind have changed over time. At one time peasants were considered to have no rights in many societies, and slavery was common until the last century. But with the growth of humanism and enlightenment what was once a merely philosophical idea of equality has created an understanding of other humans as individuals to be helped and protected. Even though societies differ in their norms and social structures, as discussed by Triandis, Stevenson and Goody (Chapters 4–6), there is hope that all humankind can be included in a 'we - definition' which still accepts the ethnic and cultural differences.

This, as emphasized especially by Feger (Chapter 16), requires communication between groups. Media of mass communication could play a central role here. Although often blamed for antisocial effects, the mass media nevertheless offer a global network which enhances worldwide communication, enabling the recipient to become acquainted with and even to participate in other groups and other cultures, thereby reducing their strangeness. They can even act as a 'moral witness', decreasing the incidence of political violence. Thus several analysts have argued that most Eastern European governments avoided the use of aggressive force against the population in 1989 primarily because they wanted to appear as peace-loving in front of the world media. On the other side of the coin, strict censorship during the Gulf War in 1991 made the war seem like an abstract game. Very few pictures of the suffering victims were shown, so that any empathy with them that the mass media might have created was held at a low level.

Voluntary communication and a definition of common goals may prove to be as effective in promoting international cooperation as the 'transcendental' norms provided by religions or the prosocial control systems of international law. The next two chapters provide case studies of how such processes can increase the likelihood of non-violent conflict resolution and international cooperation. In Chapter 17 the Director-General of UNESCO describes its role in creating a 'spirit of peace' between nations, and in the following chapter Beliaev and Marks show how, even before the end of the cold war, the common threat of terrorism could lead to cooperative action between the Soviet Union and the U.S.A. Finally, in Chapter 19, Czempiel's description of the course of Soviet–American relations of the last few decades exemplifies a theme that recurs throughout the chapters of this book, namely the need to cross and re-cross between the levels of social complexity: international relations between the two superpowers are seen as consequent upon and as affecting group interactions and individual personalities within each.

15

Changing assumptions about conflict and negotiation

JEFFREY Z. RUBIN

The field of conflict studies has grown dramtically over the last 25 years, mixing theory, research, and application, description and prescription, ranging from one discipline to another. It is an exciting time to be interested in conflict and negotiation, and the purpose of this paper is to summarize a few of the many conceptual shifts that have taken place in the field. To this end, the paper begins by focusing on some generalizations and assumptions about conflict and negotiation, and then points out some of the implications of this emerging 'ideology' for future research and practice.

Conflict settlement and resolution

For many years the attention of conflict researchers and theorists was directed to the laudable objective of conflict 'resolution.' This outcome denotes a state of attitude change that effectively brings an end to the conflict in question. In contrast, conflict 'settlement' denotes outcomes in which the overt conflict has been brought to an end, even though the underlying bases may or may not have been addressed. The difference here is akin to Herbert Kelman's (1958) useful distinction between the three consequences of social influence: compliance, identification, and internalization. If conflict settlement implies the consequence of compliance (a change in behavior), then conflict resolution instead implies internalization (a more profound change, of underlying attitudes as well as behavior); the third consequence, identification, denotes a change in behavior that is based on the target of influence valuing his or her relationship with the source, and serves as a bridge between behavior change and attitude change.

268

In keeping with the flourishing research in the 1950s on attitudes and attitude change, social psychological research on conflict in the 1950s and 1960s focused on conflict *resolution*. Only recently has there been a subtle shift in focus from attitude change to behavior change. Underlying this shift is the view that, while it is necessary that attitudes change if conflict is to be eliminated, such elimination is often simply not possible. Merely getting Iran and Iraq, Turkish and Greek Cypriot, Contra and Sandinista to lay down their weapons – even temporarily – is a great accomplishment in its own right, even if the parties continue to hate each other. And this simple act of cessation, when coupled with other such acts, may eventually generate the momentum necessary to move antagonists out of stalemate toward a settlement of their differences. Just as 'stateways' can change 'folkways' (Deutsch and Collins, 1951), so too can a string of behavioral changes produce the basis for subsequent attitude change.

The gradual shift over the last years, from a focus on resolution to settlement, has had an important implication for the conflict field: it has increased the importance of understanding *negotiation* – which, after all, is a method of settling conflict rather than resolving it; the focus of negotiation is not attitude change, *per se,* but an agreement to change behavior in ways that make settlement possible. Two people with underlying differences of belief or value (for example, over the issue of a woman's right to abortion or the existence of a higher deity) may come to change their views through discussion and an exchange of views, but it would be inappropriate and inaccurate to describe such an exchange as 'negotiation.'

Similarly, the shift from resolution to settlement of conflict has also increased the attention directed to the role of *third parties* in the conflict settlement process: individuals who are in some way external to a dispute and who, through identification of issues and judicious intervention, attempt to make it more likely that a conflict can be moved to settlement.

Finally, the shift in favor of techniques of conflict settlement has piqued the interest and attention of practitioners in a great many fields, ranging from divorce mediators and couples counsellors to negotiators operating in environmental, business, labor, community, or international disputes. Attitude change may not be possible in these settings, but behavior change – as the result of skilful negotiation or third party intervention – is something else entirely. Witness the effective mediation by the Algerians during the so-called Iranian Hostage Crisis in the late 1970s; as a result of Algerian intervention, the Iranian government came to dislike the American Satan no less than before, but the basis for a *quid pro quo* had been worked out.

Cooperation, competition, and enlightened self-interest

Required for effective conflict settlement is neither cooperation nor competition, but what may be referred to as 'enlightened self-interest.' Denoted here is a simple variation on what several conflict theorists have previously described as an 'individualistic orientation' (Deutsch, 1960): an outlook in which the disputant is interested simply in doing well for him or herself, without regard for anyone else, out neither to help nor hinder the other's efforts to obtain his or her goal. The word 'enlightened' refers to the acknowledgement by each side that the other is also likely to be pursuing a path of self-interest – and that it may be possible for *both* to do well in the exchange. If there are ways in which I can move toward my objective in negotiation, while at the same time making it possible for you to approach your goal as well, then why not behave in ways that make both possible?[1]

Trivial though this distinction may seem, it has facilitated recent work that, paradoxically, creates a pattern of *inter*dependence out of the assumption of *in*dependence. Earlier work, focusing as it did on the perils of competition and the virtues of cooperation, made an important contribution to the field of conflict studies; but in doing so, this effort also shifted attention away from the pathway of individualism, a pathway that is likely to provide a way out of stalemate and toward a settlement of differences. I don't have to like or trust you in order to negotiate wisely with you as a partner. Nor do I have to be driven by the passion of a competitive desire to beat you. All that is necessary is for me to find some way of getting what I want – perhaps even *more* than I considered possible – by leaving the door open for you too to do well. 'Trust' and 'trustworthiness,' concepts that are central to the development of cooperation, are no longer necessary, only the understanding of what the other person may want or need.

A number of anecdotes have emerged to make this point, perhaps the most popular being the tale of two sisters who argue over the division of an orange between them (Fisher & Ury, 1981; Follett, 1940). Each would

---

[1] Notice that what is being described here is neither pure individualism (where one side does not care at all about how the other is doing) nor pure cooperation (where each side cares deeply about helping the other to do well, likes and values the other side, etc.) – but an amalgam of the two.

like the entire orange, and only reluctantly do the sisters move from extreme demands to a 50/50 split. While such a solution is eminently fair, it is not necessarily wise: One sister proceeds to peel the orange, discard the peel, and eat her half of the fruit; the other peels the orange, discards the fruit, and uses her 50% of the peel to bake a cake! If only the two sisters had understood what each wanted the orange for – not each side's 'position,' but rather each's underlying 'interest' – an agreement would have been possible that would have allowed each sister to get everything that she wanted.

Similarly, Jack Sprat and his wife – one preferring lean, the other fat – can lick the platter clean if they understand their respective interests. The interesting thing about this conjugal pair is that, married though they may be, when it comes to dining preferences they are hardly interdependent at all. For Jack and his wife to 'lick the platter clean' requires neither that the two love each other or care about helping each other in every way possible; nor does it require that each be determined to get more of the platter's contents than the other. Instead, it is enlightened self-interest that makes possible an optimal solution to the problem of resource distribution.

The lesson for international relations is instructive. For the United States and the Soviet Union, Israel and its Arab neighbors, Iran and Iraq, the Soviet Union and Afghanistan, the United States and Nicaragua to do well, neither cooperation nor competition is required, but rather an arrangement that acknowledges the possibility of a more complex mixture of these two motivational states with enlightened individualism. While the United States and Soviet Union will continue to have many arenas of conflict in which their interests are clearly and directly opposed, and will also continue to find new opportunities for cooperation (as in the management of nuclear proliferation, hazardous waste disposal, or international political terrorism,), there are also arenas in which each side is not at all as dependent on the other for obtaining what it wants (e.g., the formulation of domestic economic or political policy). The world is a very big place; the pie is big enough for (almost) everyone to do well.[2]

A common process substrate

It has been fashionable for several years now to observe that conflicts are fundamentally alike, whether they take place between individuals, within or between groups, communities, or nations. Nevertheless,

---

[2] Two recent books (Lax & Sebenius, 1986; Susskind & Cruickshank, 1987) treat rather extensively the topic of enlightened self-interest, pointing out ways of

conflict analysts in each of these domains have tended not to listen closely to one another, and have proceeded largely as if international conflict, labor disputes, and family spats are distinct and unrelated phenomena.

Within the last decade or so, with the advent of conflict and negotiation programs around the United States, a different point of view has begun to emerge: one that argues for a common set of processes that underlie all forms of conflict and their settlement. Third party intervention – whether in divorce, international business and trade negotiations, a labor dispute, a conflict over nuclear siting or hazardous waste disposal, or an international border dispute – follows certain principles that dictate its likely effectiveness. Similarly, the principles of negotiation apply with equal vigor to conflicts at all levels of complexity, whether two or more than two parties are involved, negotiating one issue or many issues, varying in difficulty, etc.

Adherence to this bit of ideology has had an extremely important effect on the field of conflict studies, for it has made it possible for conversations to take place among theorists and practitioners, at work in an extraordinarily rich and varied set of fields. Anthropologists, sociologists, lawyers, psychologists, economists, business men and women, community activists, labor experts, to name but a few, have now started to come together to exchange ideas, to map areas of overlap and divergence. This, in turn, has made it possible for the development of conflict theory and practice to take shape under a larger umbrella than ever before. In fact, the symbolic location of these conversations is more like a circus tent than an umbrella, with beasts of different stripe, size, and coloring all finding a place under the Big Top.

Most recently, yet another twist has appeared. Having engaged in fruitful preliminary conversations about the nature of conflict and negotiation in their respective fields and disciplines, scholars and practitioners are now turning to the areas of *divergence* rather than *similarity*. Instead of homogenizing theory and practice in the different social sciences, analysts are now beginning to look beyond the areas of process similarity to the distinguishing features that characterize dispute management across the board.

At another but related level, conflict analysts are at last beginning to acknowledge the fact that our pet formulations have been devised by, and are directed to, a community that is predominantly White, Western, Male, and Upper Middle Class. Now that fruitful conversations have

expanding the resource pie, or finding uses for it that satisfy the interests of each side.

begun to take place among members of our own intellectual community, it is becoming clear that some of our most cherished ideas may be limited in their applicability and generalizability; other societies – indeed, other people within our own society – may not always 'play the conflict game' by the set of rules that scholars and researchers have deduced on the basis of American paradigms.

To give but one example, 'face-saving' has been an extremely important element of most conflict/negotiation formulations: the idea that people in conflict will go out of their way to avoid being made to look weak or foolish in the eyes of others and themselves. While face-saving seems important in the United States and in countries such as Japan or Korea, less obvious is the extent to which this issue is of *universal* significance. Do Pacific Islanders, Native Americans, or South Asians experience 'face,' and therefore the possibility of 'loss of face?' It's not clear. Do women experience face-saving and face loss, or is this a phenomenon that is largely restricted to the XY portion of the population?

What does it mean to set a 'time-limit' in negotiations in different cultures? Do other cultures measure a successful negotiation outcome the same as we tend to in the United States? Are coalitions considered equally acceptable, and are they likely to form in much the same way, from one country to the next? Do different countries structure the negotiating environment – everything from the shape of the negotiating table to the presence of observing audiences and various constituencies – in the same way? The answers to questions such as these are not yet available, and we must therefore learn to be cautious in our propensity to advance a set of 'universal' principles.

The importance of 'relationship' in negotiation

Much of the negotiation analysis that has taken place over the last 25 years has focused on the 'bottom line:' who gets how much once an agreement has been reached. The emphasis has thus largely been an *economic* one, and this emphasis has been strengthened by the significant role of game theory and other mathematical or economic formulations.

This economic focus is being supplanted by a richer, and more accurate, portrayal of negotiation in terms not only of economic, but also of relational, considerations. As any visitor to the Turkish Bazaar in Istanbul will tell you, the purchase of an oriental carpet involves a great deal more than the exchange of money for an old rug. The emerging relationship between shopkeeper and customer is far more significant, weaving ever so naturally into the economic aspects of the transaction. An initial conversation about the selling price of some item is quickly transformed into an exchange of a more personal nature: Who one is, where one is from, stories about one's family and friends, impressions of the host country, and lots more. When my wife and I purchased several rugs in Turkey some years ago, we spent three days in conversation with the merchant – not because that's how long it took to 'cut the best deal,' but because we were clearly having a fine time getting to know one another over Turkish coffee, Turkish delight, and Turkish taffy. When, at the end of our three-day marathon transaction, the shopkeeper invited us to consider opening a carpet store in Boston that could be used to distribute his wares, I was convinced that this invitation was extended primarily to sustain an emerging relationship – rather than make a financial 'killing' in the United States.

Psychologists, sociologists, and anthropologists have long understood the importance of 'relationship' in any interpersonal transaction, but only recently have conflict analysts begun to take all this as seriously as it deserves. Although it seems convenient to distinguish negotiation in one-time-only exchanges (where you have no history with the other party, come together for a 'quickie,' and then expect never to see the other again) from negotiation in ongoing relationships, this distinction is more illusory than real. Rarely does one negotiate in the absence of future consequences. Even if you and I meet once and once only, our reputations have a way of surviving the exchange, coloring the expectations that others will have of us in the future.

Negotiation in a temporal context

For too long, analysts have considered only the negotiations proper, rather than the sequence of events preceding negotiation, and the events that must transpire if a concluded agreement is to be implemented successfully. Only recently, as analysts have become more confident in their appraisal of the factors that influence effective negotiation, has attention been directed to the past and future, as anchors of the negotiating present.

Analysts of international negotiation (e.g., Saunders, 1985) have observed that some of the most important work takes place before the parties *ever come to the table*. Indeed, once they get to the table, all that typically remains is a matter of crossing the t's and dotting the i's in an agreement that was hammered out beforehand. It is during *pre-*

negotiation that the pertinent parties to the conflict are identified and invited to participate, that a listing of issues is developed and prioritized as an agenda, and the formula by which a general agreement is to be reached is first outlined. Without such a set of preliminary understanding, international negotiators may well refuse to sit down at the same table with one another.

Pre-negotiation is important in other contexts as well, something I discovered in conversation with a successful Thai businessman. He observed that Thais are extremely reluctant to confront an adversary in negotiation, to show any sign whatsoever of disagreement, let alone conflict. Yet many Thais have succeeded admirably in negotiating agreements that are to their advantage. The key to their success is pre-negotiation, making sure beforehand that there really *is* an agreement before labeling the process 'negotiation', before ever sitting down with that other person; in effect, they use pre-negotiation to arrange matters to their advantage, and they do so without ever identifying the relationship with the other party as conflictful, or signalling in any way that concessions or demands are being made.

At the other end of the spectrum lies the matter of follow-up and implementation. To reach an agreement through negotiation is not enough. Those parties who are in a position to sabotage this agreement, unless their advice is solicited and incorporated, must be taken into account if a negotiated agreement is to succeed. (Witness the failure of the Dukakis presidential campaign to consult sufficiently with Jesse Jackson and his supporters, prior to the Democratic Party Convention in Atlanta.) Note the trade-off here: the greater the number of parties to a negotiation, the more difficult it will be to reach any agreement at all. But only if the relevant parties and interests are included in the negotiations is the agreement reached likely to 'stick.'

As negotiation analysts have broadened the temporal spectrum to include pre- and post-negotiation processes, more work has been given over to devising creative options for improving upon the proceedings. To cite but one example, Howard Raiffa (1985) has proposed a procedure known as 'Post-Settlement Settlement,' by which parties who have already concluded an agreement are given an opportunity – with the assistance of a third party – to improve upon their agreement. The third party examines the facts and figures that each side has used in reaching a settlement; based on this information, which is kept in strict confidence, the third party proposes a settlement that improves upon the agreement reached. Either side can veto this post-settlement settlement, in which case the *status quo ante* is in effect. However, if both sides endorse the proposed improvement on the existing contract, then each stands to benefit from this proposal – and the third party, in turn, is guaranteed a percentage of the 'added value' of the contract.

Negotiating from the inside out

Conventional wisdom for effective negotiation calls for the parties to start by making extreme opening offers, then conceding stepwise until an agreement is reached. If you want to sell a used car, purchase a rug, secure a new wage package, or settle a territorial dispute with a neighboring country, begin by asking for more than you expect to settle for, then gradually move inward until you and the other side overlap; at that point you've got a negotiated settlement.

A large body of negotiation analysis has proceeded in accordance with this conventional wisdom. Moreover, this way of negotiating 'from the outside in' makes good sense for several reasons: it allows each negotiator to explore various possible agreements before settling, to obtain as much information as possible about the other negotiator and his or her preferences, before closing off discussion (Kelley, 1966); it also allows each party to give its respective constituency some sense of the degree to which the other side has already been 'moved,' thereby maintaining constituency support for the positions taken in negotiation.

On the other hand, this 'traditional' way of conducting the business of negotiation ignores an important and creative alternative: working from the 'inside out.' Instead of beginning with extreme opening offers, then moving slowly and inexorably from this stance until agreement is reached, it often makes sense to start with an exchange of views about underlying needs and interests – and, on the basis of such an exchange, building an agreement that both parties find acceptable. The key to such an approach is, as negotiation analysts have observed (e.g., Fisher and Ury, 1981), to work at the level of interests rather than positions: what one really needs and wants (and why), rather than what one states that one would like to have.

It was precisely this that happened in October of 1978 at Camp David where, with the mediation assistance of President Carter and his subordinates, President Sadat of Egypt and Prime Minister Begin of Israel were able to settle the disposition of the Sinai Peninsula. The Sinai had been taken by the Israelis in 1967, and its complete and immediate return had been demanded by the Egyptians ever since. Had the discussions about the fate of the Sinai been conducted solely at the level of positions – with each side demanding total control of the land in question, then making step-wise concessions from these extreme opening offers – *no* agreement would have been possible. Instead, with assistance from President Carter, the Egyptians and Israelis identified their own respective underlying interests – security in the case of Israel, hegemony in the case of Egypt – and were able to move to an agreement that allowed the Israelis to obtain the security they required, while the Egyptians obtained the territory they required. 'Security in exchange for territory' was the formula used here, and it was a formula devised not by moving from the outside in, but by building up an agreement from the inside out.

A useful variation on this inside out idea is the 'one-text' negotiation procedure (Fisher, 1981), whereby a mediator develops a single negotiating text that is critiqued and improved by each side, until a final draft is developed for approval by the interested parties. Instead of starting with demands that are gradually abandoned, the negotiators criticize a single document that is rewritten to take these criticisms into account, and eventually – through this sort of inside out procedure – a proposal is developed in which both sides have some sense of ownership.

The role of 'Ripeness'

Although it is comforting to assume that people can start negotiating any time they want, such is not the case. First of all, just as it takes two hands to clap, it takes two to negotiate. *You* may be ready to come to the table for serious discussion, but your counterpart may not. Unless you're both at the table (even at the distance of a telephone line or cable link), no agreement is possible.

Second, even if both of you are present at the same place, at the same time, one or both of you may not be sufficiently motivated to take the conflict seriously. It is tempting to sit back, do nothing, and hope that the mere passage of time will turn events to your advantage. People typically do not sit down to negotiate unless and until they have reached a point of 'stalemate,' where each no longer believes it possible to obtain what he or she wants through efforts at domination or coercion (Kriesberg, 1987). It is only at this point, when the two sides grudgingly acknowledge the need for joint work if any agreement is to be reached, that negotiation can take place.

'Ripeness,' then, denotes a state of conflict in which all parties are ready to take their conflict seriously, and are willing to do whatever may be necessary to bring the conflict to a close. To pluck fruit from a tree before it is ripe is as problematic as waiting too long. There is a *right* time to negotiate, and the wise negotiator will attempt to seek out this point.

It is also possible, of course, to help 'create' such a right time. One way of doing so entails the use of threat and coercion, as the two sides (either with or without the assistance of an outside intervenor) walk (or are led) to the edge of 'Lover's Leap,' stare down into the abyss below, and contemplate the consequences of failing to reach agreement. The longer the drop – that is, the more terrible the consequences of failing to settle – the greater the pressure on each side to take the conflict seriously. There are at least two serious problems with such 'coercive' means of creating a ripe conflict: first, as can be seen in the history of the arms race between the United States and the Soviet Union, it encourages further conflict escalation, as each side tries to 'motivate' the other to settle by upping the ante a little bit at a time; second, such escalatory moves invite a game of 'chicken,' in which each hopes that the other will be the first to succumb to coercion.

There is a second – and far *better* – way to create a situation that is ripe for settlement: namely, through the introduction of new opportunities for joint gain. If each side can be persuaded that there is more to gain than to lose through collaboration – that by working jointly, rewards can be harvested that stand to advance each side's respective agenda – then a basis for agreement can be established. In the era of *glasnost*, the United States and Soviet Union are currently learning this lesson: namely, that by working together they can better address problems of joint interest, the solution of which advances their respective self-interest. Arms control stands to save billions of dollars and rubles in the strained budgets of both nations, while advancing the credibility of each country in the eyes of the larger world community. The same is true of joint efforts to slow the consequences of the 'greenhouse effect' on the atmosphere, to explore outer space, to preserve and protect our precious natural resources in the seas.

A 'residue' that changes things

It is tempting for parties to a conflict to begin by experimenting with a set of adversarial, confrontational moves – in the hope that these will work. Why not give hard bargaining a try at first since, if moves such as threat, bluff, or intimidation work as intended, the other side may give up without much of a fight? Moreover, even if such tactics fail, one can always shift to a more benign stance. The problem with such a sticks-tocarrots approach is that once one has left the pathway of joint problem solving, it may be very, very difficult to return again. It takes two people to cooperate, but only one person is usually required to make a mess of a relationship. The two extremes of cooperation and competition, collaboration and confrontation, are thus *not* equally valenced; far easier to move from cooperation to competition than the other way around.

In the course of hard bargaining, things are often said and done that change the climate of relations in ways that do not easily allow for a return to a less confrontational stance. A 'residue' is left behind (Pruitt & Rubin, 1986), in the form of words spoken or acts committed, that cannot be denied and which may well change the relationship. The words, 'I've never really liked or respected you,' spoken in the throes of an angry exchange, may linger like a bad taste in the mouth, even when the conflict has apparently been settled. Similarly, a brandished fist or some other threatening gesture may leave scars that long outlive the heat of the moment. Thus, the escalation of conflict often carries with it the presence of moves and maneuvers that alter a relationship in ways that the parties do not readily anticipate.

The implication of all this for conflict and negotiation studies is clear: insufficient attention has been directed to the lasting consequences of confrontational tactics. Too often scholars, researchers, and practitioners have assumed that cooperation and competition are equally weighted, when in fact cooperation is a slippery slope; once left, the pathway leading to return is difficult indeed. Required for such a return journey is a combination of cooperation and persistence: the willingness to make a unilateral collaborative overture, then to couple this with the tenacity necessary to persuade the other side that this collaborative overture is to be taken seriously (Axelrod, 1984; Fisher & Brown, 1988).

Conclusions

This paper has attempted to outline several of the changes in thinking about conflict and negotiation that have occurred in recent years. None of these changes, taken by itself, is surprising or even of clear importance. Moreover, despite the title of this chapter, it is less a matter of wise vs. mistaken assumptions than a shift in outlook that has begun to emerge, and that has helped to create a new field of conflict and negotiation studies.

Interestingly, this new 'field of study' is housed in no single academic department; instead, nooks and crannies have sprung up in all sorts of strange places (law schools, programs in urban and environmental studies, schools of industrial and labor relations, business schools, schools of international diplomacy) where multi-disciplinary conversations can take place under one roof. It is no longer only sociologists and experimental social psychologists, but a diverse group of other professionals, who study conflict and negotiation, and persist in efforts to test their formulations in the crucible of practice.

Acknowledgements

This chapter is a revised version of the author's Presidential address to the Society for the Psychological Study of Social Issues in Atlanta, Georgia on August 12, 1988: that address was published in the *Journal of Social Issues*, 1989, **45**, 3. Thanks to Walter Swap & J. William Breslin for helpful comments on an earlier draft of the manuscript.

References and further reading

Axelrod, R. (1984). *The Evolution of Cooperation*. New York: Basic Books.

Deutsch, M. (1960). The effect of motivational orientation upon trust and suspicion. *Human Relations,* 13, 123–39.

Deutsch, M., & Collins, M. E. (1951). *Interracial Housing: A Psychological Evaluation of a Social Experiment*. Minneapolis, MN: University of Minnesota Press.

Fisher, R. (1981). Playing the wrong game? In J. Z. Rubin (ed.). *Dynamics of Third Party Intervention: Kissinger in the Middle East*. New York: Praeger.

Fisher, R., & Brown, S. (1988). *Getting Together*. Boston: Houghton Mifflin.

Fisher, R., & Ury, W. L. (1981). *Getting to YES: Negotiating Agreement Without Giving In*. Boston: Houghton Mifflin.

Follett, M. P. (1940). Constructive conflict. In H. C. Metcalf & L. Urwick (eds.). *Dynamic Administration: The Collected Papers of Mary Parker Follett*. New York: Harper.

Kelley, H. H. (1966). A classroom study of the dilemmas in interpersonal negotiations. In K. Archibald (ed). *Strategic Interaction and Conflict: Original Papers and Discussion*. Berkeley, CA: Institute of International Studies.

Kelman, H. C. (1958). Compliance, identification, and internalization: Three processes of attitude change. *Journal of Conflict Resolution,* 2, 51–60.

Kriesberg, L. (1987). Timing and the initiation of de-escalation moves. *Negotiation Journal,* 3, 375–84.

Lax, D. A., & Sebenius, J. (1986). *The Manager as Negotiator*. New York: The Free Press.

Pruitt, D. G., & Rubin, J. Z. (1986). *Social Conflict: Escalation, Stalemate, and Settlement*. New York: Random House.

Raiffa, H. (1985). Post-settlement settlements. *Negotiation Journal,* 1, 9–12.

Saunders, H. H. (1985). We need a larger theory of negotiation: The importance of prenegotiating phases. *Negotiation Journal,* 2, 249–62.

Susskind, L., & Cruickshank, J. (1987). *Breaking the Impasse*. New York: Basic Books.

16

Cooperation between groups

HUBERT FEGER

Introduction

This chapter is an attempt to integrate research in social psychology on cooperation between groups. To begin with, the concept of cooperation deserves some discussion, and the conditions favorable or detrimental to cooperation are treated. The experience and behavior of members of cooperating groups are described as well as ways of inducing cooperation. The chapter provides a general overview of the main ideas and results of empirical research without quoting the respective evidence in detail. References can be found in the publications listed at the end of the chapter.

Several topics which could be subsumed under the title of this chapter will be found in other parts of this book: cooperation within groups, which has many parallels with cooperation between groups, is treated in the chapter by Rabbie. Good discusses the evidence coming from experimental games, while Boon & Holmes, and Lund, are concerned with trust – which can be seen as a prerequisite for a phenomenon of cooperation. Several papers analyze the international aspects of this topic.

Interacting persons may be oriented toward behavior and features shown by the partner of the interaction *as an individual*. Research using this perspective is called 'interpersonal'. But interacting persons may relate their behavior to properties which the other person shows *as a member of a social group*, and the acting person may base his or her behavior on his or her membership in a group. Research taking this perspective analyzes 'intergroup behavior.' Of course, this distinction can not always be maintained, and 'individual oriented' and 'membership related' should not be considered as a strict dichotomy.

Only if persons *accept some criteria* or features as defining themselves as a social group, and as defining some other persons as a social group, can intergroup behavior occur. Quite often we observe that groups mutually accept – maybe reluctantly – the same defining criteria, or just one complex and dichotomous criterion, to create contrasting groups. Examples are skin color (race?) with black and white, political orientation with conservative and liberal, nationality and culture with Dutch and Indonesian. Then the membership in one group *excludes the membership* in one of the contrast groups, and the very selection of this kind of criteria suggests conflict between groups rather than cooperation.

Indeed, social psychologists have preferred to study conflict and competition between groups; only a few experiments are concerned with cooperation that is clearly not within but between groups. Attitudes as prejudices and stereotypes, as well as behavior in the form of discrimination and hostility have been intensively analyzed empirically since the beginning of this century. Cooperation, or just friendly relations between groups, remained something like a background for comparison.

Throughout this chapter 'group' means any aggregate of persons, varying in size from the pair or dyad to whole nations. An equivalent term to the group in this context is 'social category'. This term emphasizes the fact that persons constituting a group have some features in common: those features which define the social category. Nothing is implied about a specific kind or amount of interaction between the members of a group; only rarely will the members know each other from face-to-face contact. It will make the following discussion easier if we define the social category of which a person is a member as the 'ingroup', and correspondingly call the group in which he or she is not a member but which is the target of the ingroup's interaction the 'outgroup'.

We have to realize that groups are socially created. While some of the features which are used in defining social categories seem to be provided by nature it is nevertheless a process of selection to use just these features and not others. While in most Western civilizations the color of the skin is used to define social categories, the color of the eyes is not. The features most often used to constitute social categories are race, religion, sex, age, socioeconomic status, nationality, culture, and language.

The selection of features to construct groups is not a chance process. If features are visible, cannot be hidden, and are of high discriminatory power, they are good candidates. Quite often, some of the features correlate highly or even perfectly with one another. Sometimes, features have to be created because discrimination is intended; e.g., the Jews in

Nazi Germany were required to carry a 'David-Stern'. Sometimes, there exists ingroup training of feature recognition, e.g., by pointing out differences in clothing or hair style. It is a good indicator of harmony or tension between groups how discriminative features are reacted to by ingroup and outgroup.

As mentioned, social categories are societal constructions. While the media and conversations with other group members will contain statements about an outgroup in general, the everyday encounter is a contact with one or a few persons as members of one or several outgroups. Only rarely, e.g., in race riots or in the performance of a religious cult, a larger number of persons may be directly experienced as a representation, or even as the representation, of the outgroup. This is quite often associated with the vivid actualization of one's own attitudes toward this outgroup. But for most of us these experiences are only a few in a lifetime.

An everyday encounter allows the placement of the same person into several social categories simultaneously, e.g., into the categories of black, female, young, medical doctor. Personal and situational circumstances decide whether a placement favoring one category above the other is made, and which category is selected. This *cross-secting of categories* within any individual is one of the reasons why interpersonal and intergroup relations are separate topics in principle, even if the boundaries of this distinction cannot always be drawn sharply when reporting the results of empirical research.

Among groups, we may distinguish social categories from *ad hoc organizations*. These organizations may vary in size but their members have (or had at the time when the organization came into life) goals in common which determine the structure of roles, positions, and norms of this kind of social group. They are much more flexible in the responses to a situation allowing for cooperation or competition. For example, firms or other units of production can be changed in size, structure, and composition as a response to or in anticipation of cooperation and competition. A firm may buy up a supplier to avoid dependence. *Ad hoc* organizations, even those with a long tradition, lack the fixation of their boundaries created by unchangeable discriminatory features. They usually allow membership in other groups as well; for example, being a member of a soccer club does not interfere with being a member of the Red Cross organization.

In the process of *socialization*, the child has to learn which groups are defined by which features, and which attitudes and behaviors are appropriate toward those groups. He also learns whether the goals of those groups are compatible with each other or not. In this process, the child simultaneously learns to which group he belongs, and the child learns the position of his group relative to other groups. This implies the acquisition of expectations of how members of other groups are likely to behave toward the child. And all this, of course, is learned from the perspective of the ingroup. Socialization also extends to the learning of behavior towards members of the ingroup and the various outgroups. It has been demonstrated that children can learn cooperative behavior by techniques of operant conditioning as well as by imitating a model.

The concept of cooperation

Cooperation refers to interactional behavior or a relationship between at least two parties, be they persons, groups or institutions. Their behavior is coordinated in such a way that some actions of one side facilitate the goal attainment of the other side. Usually, this behavior is conceived to be voluntary and not the result of yielding to power. It is the rule that both sides support each other in a balanced or symmetrical fashion, at least in the long run. The cooperating partners work toward the same or toward different but mutually compatible goals.

Cooperation may vary in degree, i.e., the number and importance of cooperative actions can be large or small. It may differ in quality, duration, and pervasiveness; there may even be cooperation in some aspects and competition in others. Cooperation with one group could automatically imply conflict with another group, depending on the inter-relations between goals, resources, and the history of group interactions. The characterization of a group as cooperative thus may reveal at least as much about its social environment as about its members.

'Help' or 'support' may require or induce cooperation but both are usually conceived as being much less symmetrical; there is, at least for some time, one party defined as the helper, and the other party is correspondingly the one receiving the benefit. This does not rule out that the supporter also obtains positive reinforcement from the act of helping, e.g., an increased respect in the eyes of others or in his own eyes.

We differentiate between attitudes, effects on the internal states of the groups, and the behavioral relations between groups. Attitudes include, as conceived within social psychology, stereotypes and prejudice, positive and negative affect, intentions, hostility and discrimination. In principle, and as usually measured, attitudes could vary independently from the relationship between groups, but they do not, as will be detailed later.

More than one alternative to cooperation exists: groups may be in competition, in conflict, or working alone with or without contact with outgroups. Experiments manipulate further variables, e.g., whether feedback about the results of cooperation is provided or not. Two groups are *competing* if both try to reach the same goal but only one group can obtain it, or get the lion's share. For example, each team in a football competition seeks victory but only one can be the winner. Some authors specify that *rivalry* is competition in front of some audience. Groups are in *conflict* if they want different things but only one state of affairs is possible. Examples are labor and management negotiating an agreement, or countries disputing a boundary. Thus, inflicting damage upon the opponent is not a part of the definition, but it is an empirical observation that hostile intentions and actions often accompany conflict.

One should realize specific problems for the experimental analysis of cooperation. It is not difficult to create certain conditions and then observe, as a dependent variable, whether cooperation occurs, how much, or which other patterns of interaction appear. Because cooperation is interactive behavior, one may instigate it and hope that it results from the conditions produced by the experimenter. But there exists no guarantee that one can definitely create cooperation as an independent variable.

The *operationalizations* of a situation that should induce cooperation quite often use different schedules of allocating gains and losses (of money, prestige, etc.). The amount of gains and losses follows from the kind of previous interaction between the groups. These schedules are either communicated to the groups in an 'instruction' by the experimenter, or the groups have to find out the arrangement by themselves. Both procedures run the risk of confounding the learning and understanding of the situational frame with intentions to behave cooperatively. Of course, the experimenter may use a simple situation and just ask the (small) groups to cooperate, or only to expect either competition or cooperation. As so often in social psychology, it is then up to the participants of the experiment to interpret what is meant by the phenomenon in question, and act accordingly.

A well-known study in this field (Brown and Abrams) treats this problem with an unusual explicitness. Manipulation checks are used, implying that the experimenter tries to produce a specific orientation (cooperativeness) and a certain behavior (cooperation) but that the researcher has to make sure that the desired state or process within the participants of the experiment has indeed been achieved. The manipulation checks involved asking the participants after the experiment whether they agreed with the description of the intergroup relation as 'two teams working together' – to which persons of the cooperation condition agreed more, or as 'being on opposite sides,' to which children in the competitive group agreed more. Further, persons in the competitive groups reported themselves to feel more competitive and judged that the two groups would work less well together.

The cooperative condition was experimentally produced by telling the 12-year-old boys and girls that the experimenters were interested in how working *with* another school affects people's performance. A prize for the joint performance of the two schools was announced. For competition, it was announced that the relative performance of the schools would be compared, and also that the prize would depend on the relative performance. Thus, the groups were described to be positively interdependent in the cooperative, and negatively interdependent in the competitive condition.

The actual task, given for five minutes, consisted of answering IQ-type test items, to which two answers had to be found. But the participants had to give only one answer because the other reportedly had already been given by a member of the other school. Any participant was matched at random with a pupil of the other school – a procedure which was explained to the participants. The scoring ensured that in both conditions less effort would result in a poorer outcome for the ingroup.

The random matching was used, as the researchers argue, to maintain the encounter at an intergroup level. But the 'partner' of the cooperation or competition was at no time physically present, and correspondingly, no co-acting or co-orientation took place. It is a kind of degenerated cooperation without interaction. Maybe the fictitious partner was imagined. The participants in the respective experimental conditions only had the choice of being more or less active in problem solving. This activity would be interpreted as cooperation in the cooperative condition, and in the other groups as competition.

The *compatibility of goal attainment* and the relations of the means to the various goals constitute the structural frame for the relationships between groups. This frame has – as every situation – an objective and a subjective aspect. The objective side is what can be described with perfect or high interindividual agreement because every perceiver has or could have in principle the same access to this information. The subjective side is the perception of the structure by members of the groups.

When stressing the importance of the perceived goal interdependence, one should do this with some reservations. Great Britain and France can be seen as cooperating for quite some time. But it is difficult to spell out in detail and precisely the goal structure responsible for this cooperation. It seems more natural to assume a very general attitude of cooperativeness on both sides that leads to a repeated and continuous interpretation and adjustment of mutual goals. So we may differentiate between situations in the laboratory in which the participants have no specific common tradition nor expect any further interaction, and on the other side the cooperation being a part of a political tradition and common culture. Even studies using experimental games show the importance of previous relations between the participants and the expectation of future interaction for the amount of cooperation between parties.

Groups forming *coalitions* may or may not cooperate. Research on coalitions has focused on questions of coalition formation: which parties will go together? The differences between parties in these studies were provided by the distribution of resources arranged by the experimenter. Furthermore, the question of how to divide the joint result of a coalition agreement has found much interest in coalition research. What is typically lacking in laboratory experiments on coalition formation but what is essential for cooperation is the necessity to establish co-orientation and co-action between groups to begin or to continue cooperation.

Co-orientation is the cognitive activity of at least two persons or groups directed to a part of their worlds deemed as commonly accessible. Pilot and co-pilot steering a plane, or the UN assembly trying to agree on the state of affairs in Cambodia are examples. The common part of the world, i.e., the target of the co-orientation, will have to be described at least partially in terms which are mutually understood; some agreement on the features and relationships of the target in question must be reached. Therefore, some prerequisites of communication have to exist for cooperation to be achieved and maintained. *Co-acting* requires learning of synchronization and timing, the common planning of the use of resources, and sometimes supervised training of the interlocking of specific actions.

Summarizing this section, we emphasize the following points: (1) cooperation is the *behavior* of at least two parties (persons, groups, institutions) pursuing compatible or identical goals by coordinating their actions and establishing co-orientation towards their common targets; (2) a *relationship* between two or more parties may be characterized as cooperative if cooperation (and not conflict, competition, or isolation) dominates their interaction; (3) some structural or situation *conditions* have been identified which foster cooperative behavior and relations, e.g., a goal structure perceived as compatible, and the assumption of a common future.

Strategies of inducing cooperation

The research on creating cooperation between groups has centered on basically two approaches: *establishing common goals* for all groups involved, and *providing contact* between members of the respective groups. The first approach originates from Sherif's famous field experiment, 'the robbers cave' studies. The second approach is known as the 'contact hypothesis'.

Creating superordinate goals

Sherif's original experiment, which has been repeated with basically identical results by himself and by others, had 20–24 boys as participants of a summer camp. The boys were carefully selected; they were 11–12 years old, all were white, did not know each other before the start of the summer camp, and were without any known behavioral difficulties or deviancies. The camp lasted several weeks. At the beginning, the boys were divided into two groups, and the criterion of the allocation to these groups was selection at random. In the first phase of the study, the experimenters tried successfully to create hostility and conflict between the two groups by arranging competitive games in which only one group could win. Results of the competition were announced in the camp, and their importance was stressed by the experimenters. Soon the conflicts became very strong. The groups identified themselves by names and symbols, avoided contact with members of the other group, and showed hostility also in situations without a competitive nature. As Sherif points out, those conflicts cannot be derived from psychopathological traits or habits of the boys.

In the second phase, the experimenters attempted to reduce the conflict. Several measures were tried, e.g., moral appeals to both parties, but without success. Providing pleasant situations simultaneously to both groups – e.g., showing a movie – also did not lead to positive results. It was only the repeated and simultaneous work of both groups for 'common goals' which reduced the conflict and lead to cooperation. Examples of the activities that were successful are repairing the water supply system or pulling a car which contained the camp's food supply out of trouble using combined strength of both groups.

Common goals are goals which strongly motivate the members of both groups but cannot be reached by the members of one group alone without the help of the other group. Functional interdependence probably does not constitute a sufficient condition for cooperation. We mention some qualifications. One requirement is a *minimum of identification* of the members with their respective groups. Experiments in this field thus have the problem of identifying identification, which was certainly given in Sherif's studies: the members provided their groups with names and symbols and called themselves by their group's name.

The interdependence has effects *perceived* by the members of both groups. Collective behavior will occur to the extent that a number of persons interpret the situation in the same way. This implies that the persons consider themselves as members of the social categories, and that category membership is behaviorally relevant in this situation. In this context one may interpret Sherif's results as a shift in saliency of group membership: in a situation interpreted as requiring common action to reach common goals, subgroup membership lost relevance. Certainly the boys did not simply forget their hostilities of some minutes ago.

The measures taken in Sherif's experiments were, compared with some other studies, relatively severe. It has been one of the astonishing results of recent laboratory studies that procedures like a *random assignment* of persons to groups creates intergroup bias. Just imagine that you are sorted together with some other persons you do not know by a procedure like drawing lottery tickets, and then you are a member of 'your group', and if it comes to competition with another group created in the same manner there will be, for example, overevaluation of the products of your group compared with those of the outgroup (Ed.: but see Rabbie Chapter 14). Furthermore, it has been shown that some of the effects of cooperation can be produced by just instigating the *expectation* that one will be cooperating (versus that one will be in competition) with another group. The ingroup members then expect they would like the members of the outgroup and that they would like to cooperate with them.

On the other hand, the effect of establishing positive interaction between groups may depend on the *outcome of cooperation*; it should be a success, not a failure. Some experiments by Worchel and his coworkers used a design with two phases. In the first part, the interaction between the groups was either cooperative, or competitive, or the groups worked independently without being informed about the result of their work. After this first phase, the attitude toward the outgroup was most favorable after cooperation, and least positive after competition. The orientation toward the ingroup showed the opposite pattern: the attraction toward the ingroup was largest after competition with an outgroup. Thus, even after such a short past, history of previous interaction demonstrated its effect.

In the second phase of the experiment, the cooperating groups learned whether they had succeeded or failed. But irrespective of success or failure, groups working independently or cooperatively became friendlier to the outgroup. The exception is the groups which were in competition in the first part of the experiment. If they failed in the second phase, they became less positively oriented toward the outgroup. Under these circumstances, the outgroup may be blamed for the joint failure to reach their superordinate goals – except if there was a good excuse for failure, like poor working conditions. If they succeeded they became more favorable toward the outgroup even after competition.

In Sherif's experiments, the interaction between the groups did not last long enough to lead to the establishment of *institutions* within each group. Consider the ANC and other organizations on the side of the black population in South Africa, and the parties and the government on the side of the white population. Both sides have created structures which will have to be changed if integration is to make progress. Usually, forces resisting those changes appear within the institutions because they are interested in prolonging their own existence.

A final comment on Sherif's technique to create cooperation. The experimenters almost *completely controlled* the situation. Such a superpower is not always available in conflict situations. Furthermore, the tasks which the experimenters provided were of a relatively simple nature. To be solved, they did not need detailed planning or deliberate organization of co-acting. Outside a summer camp, the ways to achieve common goals may be more complicated, may require an accepted division of labor and the development of distinctive roles.

A slightly different approach to the question of how to motivate cooperation can be gained from *social identity theory* (Tajfel). While the theory has mainly been used to explain intergroup bias it also sheds light on possible conditions for cooperation. The theory assumes that people derive their social identity partially from their membership in groups. Because, as this theory assumes, people want to maintain positive self-regard, they want to belong to groups which are highly evaluated, and also wish that those groups to which they (inevitably) belong appear in a positive light. The evaluation of a group is revealed and appears in comparisons with other groups. Then self-esteem can be enchanced if the behavior or products of one's ingroup are more positively evaluated – even if this is not justified – than performances of the outgroup. Intergroup bias is the result of such judgemental tendencies with the function of enhancing self-regard.

Social identity theory postulates what has been called a 'need for positive distinctiveness in intergroup relations.' This need would explain why high attitudinal *similarity* between groups when experimentally induced (and maybe not identical in its effects with intercultural similarity as perceived between groups outside the laboratory) leads to increased liking and cooperativeness, while at the same time – even under conditions of cooperation – increased intergroup bias is observed in ratings of group performances. If persons are free to choose whether to cooperate or to be competitive, one should consider a curvilinear relationship between similarity and closeness on the one hand, and the kind of behavior on the other hand. If an outgroup is very similar and close, competition is possible, sometimes even a strong one, because it would not endanger the good relations already existing between the partners. The same amount of competition against a group of medium similarity would risk the chance of positive future interaction.

From the viewpoint taken by social identity theory, measures chosen to make the group identity more salient should lead to competition, while measures de-emphasizing the distinctiveness of the groups involved, and measures playing down the need for positive self-evaluation, should increase the willingness to cooperate. Persons with high self-esteem should cooperate more often than persons with low self-regard. And persons which tend to categorize other human beings into dichotomously conceived social categories and do not react to the criss-crossing of these categories should show little desire to cooperate.

The contact hypothesis

The initial, optimistic version of the *contact hypothesis* is: in order to create positive attitudes and friendly behavior (including cooperation) it is sufficient to bring people from various social categories together. If they meet each other they will learn that people are the same in all important aspects, and this will lead to a vanishing of prejudice. Intensive research on this assumption has shown that some qualifications are necessary. Some pertain to the contact situation. The contact should not be superficial but the situation should allow for *intensive and repeated contact*. The situation itself should be *pleasant* for members of both

parties. The social *environment*, especially the groups in which the participants are members, should support the contact activities. Other qualifications pertain to the participant. Members of the different groups
should be, if possible, of *equal status*. They should be *prepared* for the
contact situation; e.g., the participants of student exchange programs
should be made familiar with the foreign culture before they are sent to a
foreign country.

There exist several problems with this approach. Imagine that the two
groups involved are a majority and a minority group. Then it will be
difficult to have a large number of members with the same status because
typically members of minority groups are not represented equally in
higher income groups, for example. If contacts are voluntary they tend to
be more successful. On the other hand, it is known that persons with
strong prejudices try to avoid contact with the target of their prejudice.
Thus, contact programs may fail to reach those who need them most.

If contact is not voluntary as in such large projects as mixed racial
housing, school desegregation, integrated units in the army, or mixed
teams at working places, the result depends on several factors. One of
them is whether the members of the outgroup show – as judged by the
ingroup observers – just those patterns of behavior which were expected
anyhow as a part of the existing stereotypes and prejudice. In such a case,
the negative attitude may even increase. Furthermore, it has frequently
been observed that while prejudice and open discrimination were abandoned for example at the working place, no *generalization* occurred, that
is, while black and white cooperated at work, there was no contact during
leisure time. Furthermore, as mentioned above, the social surroundings
should be favorable to the contacts. This is quite often not the case,
especially if the conflict situation is embedded in racial, economic, and
political antagonisms. Even if an excursion of black and white children in
South Africa should receive the consent of the authorities, the political
context would limit the effects on the attitudes of the participants, and the
value of such an action might be found rather in its symbolic nature and
its properties as a signal.

Usually, the contact situation provides opportunities to engage in
interaction for some members or even representatives of groups. The
hope is that positive experiences spread to those who could not or did not
want to participate. Quite often, the creation of a contact situation (and
not necessarily the interaction of the participants) is a *signal* to both
groups that expectations or even pressure exists to make the interaction
between the groups more favorable. From this point of view it may be

meaningful also from a social-psychological point of view to arrange a folklore festival with Turkish performers for a mixed audience even if the Turkish and the German visitors form neatly separated clusters in the festival hall.

To obtain truly intergroup effects the contacts should be repeated and continued for long time intervals. This might avoid the possibility that only certain segments of the different groups participate. This also makes it difficult for the participants to attribute their experience to 'exceptional cases', like 'Indians in general are ... but this one is (not) ...' Furthermore, intensive contacts increase the likelihood that the positive experiences are associated with the feature defining the social categories, and not with other features which also (but in fact accidentally) defined those members of the outgroup with whom contact was actually established.

There exists some research on the fine details of the contact situation, especially on educational programs for racially mixed school classes. Small mixed teams of children are organized into *cooperative learning groups*.' Part of the discussion in this context is related to the following questions. If one (the teacher or the training program) de-emphasizes the membership to different groups, a positive development of the relations between members of different groups seems to be more likely. But then the generalization from these individual contacts to a positive orientation toward the other social category as a whole seems to be less likely. And further, the attitude to a black medical doctor tends to be more positive than the attitude to black people in general. So, having a black doctor acting in the media or in textbooks may lead to more positive attitudes toward black people – but how can one be sure that the positive effect is not almost exclusively integrated into the stereotype of medical doctors in general? Again, the unwanted attribution may occur mainly with those participants who are strongly prejudiced.

Research on cooperative learning groups measured not only attitudes but also registered sociometric friendship choices and patterns of interaction. Choices crossing race boundaries are of special interest. In longitudinal studies, however, racial segregation in friendship choices was observed actually to increase despite the educational engagement, so that the term 'resegregation' came into use. Under which conditions could this longitudinal effect be altered?

Is there an effect of *group composition*, i.e., of the proportion with which one of the parties, say the minority, is represented in the desegregation setting? The answer is far from definite. Any deviation from an equal representation of the social categories seems to draw more (unwanted) attention of the majority to the minority, and tends to increase the self-attention of the minority. But even if equal representation is provided, this proportion is compared with what the participants have experienced in the past, and to what they are used to in other settings at present. Furthermore, the number of social categories involved, and the absolute size of the groups have to be taken into account.

The *classroom climate and structure* influences the expression of hostility and cross-racial friendship choices. Cooperative learning programs in which the outcomes are positively interdependent are more successful in inducing positive intergroup relations than the traditional learning programs; the positive effects even tend to generalize to situations outside the schools.

The effect of *academic tracking* (forming/evaluating teams by their achievement in school) seems detrimental to intergroup relations. Apparently, differences in achievement lead to status differences which are known to diminish contact effects. Of course, status differences have their effect in the way they are perceived, which depends on several aspects of the immediate situation as well as the global societal and historical positions of the social categories. If status differences are experienced as threatening, both sides are motivated to increase intergroup differentiation.

We have to mention that most of this research is mainly concerned with attitudes and, if it is related to behavior, cooperation is just one of several possibilities. Thus the evidence is not very specific. Nevertheless, in summary of this section it can be said that the motivation of a group to seek and continue cooperation as well as the group's behavior depends on several factors:

(1) The group *expects to gain* by cooperation relative to working alone or to working in competition, and relative to any additional costs of working together, and despite the risk of being exploited.

(2) The group has *positive experiences* with cooperation in general as a method to reach goals, and the previous experiences with the target group in particular were positive; the group assumes that the target group is able and willing to cooperate.

(3) The target group is perceived as possessing sufficient means and *resources* to reach the aims of the intended cooperation; the group further assumes the use of these means by the outgroup will be compatible with the goals and also with the use of the means made by the ingroup.

(4) The structure of the *aims* and purposes of both groups is seen as *compatible*. Perceiving this structure as in zero sum games – the gain of the one is the loss of the other – is almost certain to destroy any intent to cooperate.

(5) Rules and *traditions* on how to initiate and maintain cooperation are available, also covering the case of defection.

Experiencing cooperation

Changing to or being in cooperation – as contrasted to other styles of social relations – creates specific experiences for group members. These experiences pertain to one's own position and identity, to the perception of the features and situation of the partner group, and to the perceived relationship between the groups.

The members of the groups have, to begin with, to realize that various ways to interact exist, and that the relationships formed depend at least partially upon the group itself. Whether cooperation is among the options depends upon the trust of the ingroup toward the outgroup, on the perceived trustworthiness, and on the *'interpersonal risk'*. This risk includes the judged possibility of being deceived and exploited, and the increased possibilities of being disadvantaged in any way by someone who comes into close interaction with oneself. Cooperation also moves the others closer to oneself in another meaning: they appear to be more similar to oneself, maybe even more human. The partners have repeatedly to take the perspective of the other side. They do it by using their own experiences, and if it is successfully done, this provides positive reinforcement.

One should not forget that the information about the outgroup, and even about the ingroup, is usually scanty and not acquired by direct interaction. Therefore, a little (new) information communicated by credible and consistent sources can easily change the attitude toward an outgroup. And it is not unrealistic to imagine that group members are at least dimly aware of all this, and keep in touch with their fellow members so as usually to swim with the main stream, especially if some currents seem to change. The cues provided within the ingroup might be subtle and indirect. For example, an exchange program may show effects even before any exchange took place simply because it was announced. The announcement provides the cue that now – perhaps in contrast to earlier times – the relation toward an outgroup is intended to be changed.

Competition is quite often experienced as more stressful and demanding the use of all, even the ultimate resources, especially if the 'honor' of the group is at stake. It may lead to an increased sensitivity to failures by members of the ingroup. Cooperation seems to be *more satisfying*, to make work easier, to provide – by synchronization and planning – a fixed temporal and spatial frame for actions so that one does not have to decide again and again.

Cooperation is a process taken to have a certain *duration*. As such it implies expectations about involvement with the outgroup and how the group's goal attainment will be affected. These expectations have to be mutually adjusted to create *complementary* expectations and intentions. Violations of expectations have to be reacted to in a way seen as fair by both sides; this is easier if arbitration is institutionalized in advance, and if mutually accepted traditions and norms on how to restitute cooperation exist. Especially if cooperation is regulated by contracts, *sanctions for defection* are included.

Sometimes it is possible that one or both parties *use threat* against the partner, i.e., are able to inhibit the goal attainment of the other party by specific actions, which are then usually seen as hostile. Although the famous studies by Deutsch and Krauss analyzed the behavior of individuals, some results might be generalized to relations between groups. The existence of the possibility of using threat seems to increase the competitiveness of a situation, seems to increase the complexity of the decision situation of the partners, and makes it more difficult to anticipate correctly the actions of the other person or group. When large amounts of money were at stake, high cooperation appeared despite the possibility of using threats, but if the status and self-esteem of the participants was involved, cooperation decreased even without threat potential.

Cooperation itself may be threatened from the outside, for example, by a third party. If the threat is experienced as common, it will make the attitudes between ingroup and outgroup more positive, intensify cooperation if that is seen as meaningful, and could lead to a common fight against a common enemy or fate.

Cooperation often leads to products, to gains and losses, and for a long lasting cooperation it should be made clear in advance how the benefits and loads will be shared. The ethics of cooperation should include the distribution of the results of cooperation.

Effects of cooperation on the parties involved

Because most studies are correlational in nature we will not state that cooperation between groups causes certain effects but will rather report which behaviors or experiences are often observed in conjunction with cooperation. The observations are concerned with processes within a group cooperating with another group, and to orientations toward the other group.

Members of the ingroup and members of an outgroup are *perceived differently*. Judgements on outgroup members show *less variability* compared with judgements about persons in the ingroup. The higher the homogeneity of the outgroup related judgements, the higher the confidence in these judgements tends to be – which themselves usually have a negative tendency. One reason for the increased variability of ingroup judgements is the higher amount of contact with ingroup members. The increased contact may provide experience with more differing 'exemplars' as well as a larger number of ways in which they differ. Since information about outgroup members is indirectly communicated, this also may contribute to lower variability. Furthermore, contacts with ingroup members probably differ qualitatively from those with outgroup members; if the former are more intense they could lead to more differentiation. If the probability of future interaction with outgroup members is high, then the cognitive representation may be more differentiated.

If a group cooperates with another, its members as well as independent judges report a higher 'we-feeling', a stronger experience of belonging to one's own group. Competing groups judged themselves to be more ego-centered, to have more difficulties in communicating with the members of their own group, and these judgements were in agreement with the ratings of independent observers. Cooperative groups exchanged more communications which were relevant for the solution of the task given to the group either by an experimenter or chosen by the group itself.

Cooperation generally leads to a more friendly climate within the group, to a stronger *cohesion* of the group – that is a tendency to stay together for a longer period of time, and despite some obstacles. The pressure towards uniformity is stronger in cooperating groups, and more yielding to this *conformity* pressure is observed. At the same time, the members of a cooperating group tend to like their tasks more and are more satisfied with their handling of these tasks. Quite often, tasks are solved in a shorter time, while the question whether cooperative or competitive groups are more efficient depends on a larger set of special conditions. Within cooperating groups, team members are evaluated more positively. These groups are more likely to elect a leader.

We can offer only some speculations why people in cooperating groups are more satisfied with their tasks and are friendlier to their team members and also to members of interacting outgroups. In Western cultures, to be cooperative is a general positive *norm*. To this norm, only a few exceptions exist, like competition among different teams in sports. But even for these exceptions, special norms are provided, like fairness and adherence to the rules regulating the competition. Thus, behavior according to a norm may well be accompanied by a pleasant feeling out of which a positive orientation toward the social environment may be derived. If some of the reasoning underlying the contact hypothesis is correct, the intensified interaction with members of the outgroup – and this contact takes place under more favorable conditions when cooperating – should also lead to a higher evaluation of the outgroup and therefore to a more positive judgement of one's own situation.

Explanations of cooperation

Several theories exist to explain why groups enter into conflict, even war. The fact that groups cooperate, some very successfully and over a long period of time, does not seem to demand an explanation. Maybe the very success of this behavior pattern, experienced by members of cooperating groups as well as by observers, is reason enough to continue cooperation and to consider it to provide reinforcement by itself. But such an assumption does not cover the cases with mixed success, does not explain the breaking up of cooperation, and is insufficient to account for those instances where the sheer expectation of cooperation leads to similar effects as cooperation itself.

In contrast to conflict theories (cf. Adorno's concept of 'the authoritarian personality'), the idea that specific personality traits of members in interacting groups are responsible for cooperation has not been favored in research, perhaps with the exception of a few explanations for results of experimental games. While there may well exist a social norm or a learned habit to cooperate, an 'inborn drive', 'primary motive' or 'fundamental personality trait' has not been postulated. And while certain personality structures, including those with generalized tendencies to cooperate, may be fostered in particular social climates, one still has to postulate that (under certain conditions to be specified) these persons attain control in their groups and steer them into cooperation.

Campbell's 'realistic conflict theory' may as well be interpreted as a theory deriving cooperation from the material interests of the interacting groups. Politicians, opinion leaders, or experimenters in their instructions to the participants of an experiment, provide information to the group members. From this information they may derive the appropriateness of cooperation, conditional to their goals, and given the distribution of resources in a specific situation. Cooperation, in the extreme case, would be the result of a higher *expected utility* of cooperation relative to other perceived options. The processes by which a group arrives collectively at such an evaluation would be a research topic in itself. Looking at the way politicians try to instigate cooperation reveals that the appeal to ethical principles is among the tools, as well as demonstrating and promising higher material utility.

Concluding comments

There exists no theory dominating the field of cooperation research, only a few hypotheses have been tested more than once, and even a commonly accepted terminology is in its first stage of development. And there does not seem to exist a consensus of what constitutes the phenomenon of cooperation, what is essential and what is just one aspect. Difficulties of integrating results increase if in some studies cooperation means just its anticipation, in others the intent to cooperate or an attitude of cooperativeness, and even if actual behavior is analyzed, it may be acting with or without a partner.

Cooperation between social categories and between *ad hoc* groups, both in the field and in the laboratory, was not treated separately in this chapter because there is not enough evidence available to do so. But we expect that future research will reveal distinctions between the two kinds of group, the field and the laboratory, and different aspects of cooperation as major division lines limiting generalization of results.

Long-term studies of cooperation, including its planned or unintentional results, are missing. Projects like the launching of the Apollo missions to fly to the moon are also a triumph of human cooperation, planned with the aid of elaborate computerized techniques developed in *operations research*. Every house built requires the cooperation of several teams of craftsmen, and it is well known that the degree and kind of cooperation vary widely, with extreme results.

We expect further progress if research on *social networks* includes a comparative perspective because helping and support, and maybe cooperation, are among the constituting factors of most social networks. It is reassuring from a theoretical as well as from an applied viewpoint that social psychology has begun in the past two decades to study with more intensity the phenomena of positive relations between persons and groups.

Further reading

Austin, W. G. & Worchel, S. (eds). (1979). *The Social Psychology of Intergroup Relations*. Monterey, CA: Brooks & Cole.

Brown, R. (1988). *Group Processes. Dynamics within and between Groups*. Oxford: Basil Blackwell.

Feger, H. (1972). Gruppensolidarität und Konflikt. In C. F. Graumann (Hrsg.) *Handbuch der Psychologie, Bd. 7, 2. Halbband: Sozialpsychologie* (S. 1594–653). Göttingen: Hogrefe.

Feger, H. (1979). Kooperation und Wettbewerb. In A. Heigl-Evers (Hrsg.). *Lewin und die Folgen. Die Psychologie des 20. Jahrhunderts, Bd. VIII* (S. 290–303). Zürich: Kindler.

Hendrick, C. (ed). (1987). *Group Processes and Intergroup Relations*. Newbury Park: Sage.

Marwell, G. & Schmitt, D. R. (1975). *Cooperation: An Experimental Analysis*. New York: Academic Press.

Messick, D. M. & Mackil, D. M. (1989). Intergroup relations. *Annual Review of Psychology*, **40**, 45–81.

Miller, N. & Brewer, M. (eds). (1984). *Groups in Contact: The Psychology of Desegregation*. New York: Academic Press.

Stephan, W. G. (1985). Intergroup relations. In G. Lindzey & E. Aronson (eds.). *Handbook of Social Psychology, (3rd edn)*, Vol. II, pp. 599–658. New York: Random House.

Turner, J. C. & Giles, H. (eds). (1981). *Intergroup Behavior*. Oxford: Basil Blackwell.

17

The role of UNESCO in the development of international cooperation

FEDERICO MAYOR, DIRECTOR-GENERAL OF UNESCO

Introduction

As is well known, peace-building is the fundamental purpose of Unesco: all the varied forms of cooperation it undertakes in the fields of education, science, culture and communication are intended to contribute to the supreme objective of banishing – in the words of its Constitution – 'that suspicion and mistrust between the peoples of the world through which their differences have all too often broken into war'. This approach rests on the postulate that – in the phrase of Clement Attlee, spoken at the 1945 London Conference that brought Unesco into being[1] and subsequently incorporated in its Constitution – 'wars begin in the minds of men', from which it follows that 'it is in the minds of men that the defences of peace must be constructed.' Léon Blum, another of the founding fathers of Unesco, gave more positive expression to this same thought when he spoke at the London Conference of Unesco's function as that of contributing to a 'world in which the *spirit of peace* shall become one of the guarantees, and perhaps the surest guarantee, of Peace.'[2] Arguably, his formula encapsulates better than any other the essential task of Unesco.

Unesco's most visible contribution to the task of fostering a 'spirit of peace' – or, as we might today prefer to say, a 'culture of peace' – has undoubtedly been made through its direct peace-promoting activities. Foremost among these has been its work in the sphere of international

---

[1] Conference for the establishment of Unesco, Second Plenary meeting, 1 November 1945.

[2] *Idem.*

education: clarification of the concept through conferences and seminars, promotion of standard-setting instruments[3], preparation of teaching materials, awarding of prizes[4] and – most important – the practical application of the philosophy through the Associated Schools Project, a worldwide network of some 2300 institutions where innovative approaches to the teaching of essential international themes and problems can both be developed and implemented. Unesco has also been active in the promotion of multidisciplinary research into the causes of war, the roots of racial prejudice and ways of building peace. Since 1980, the results of research in such domains have been circulated through the *Unesco Yearbook on Peace and Conflict Studies*. Following the 1989 Yamoussoukro International Congress on Peace in the Minds of Men, the Organization plans to give increased emphasis to such research, notably by furthering the process of reflection inaugurated by the Seville Statement on Violence, encouraging studies of factors conducive to peace and initiating interdisciplinary research that explores the interconnections between peace, human rights, disarmament, development and the environment.

In stressing the links between peace, development and the protection of the environment, the Yamoussoukra Declaration highlighted a central preoccupation in Unesco's Constitution and one of the main lines of emphasis of its recently adopted Third Medium-Term Plan for 1990–1995. The challenge of building peace is indissociable from that of promoting equal development, just as viable development is inconceivable without proper regard for the environment. These challenges call for a comprehensive response based on solidarity among nations and peoples. In other words, they call for a partnership: a partnership between human beings in the search for mutual understanding and well-being; and a harmonious partnership of humanity with nature. In a world in which the interconnectedness of phenomena and interdependence of peoples is becoming ever more apparent, the whole gamut of Unesco's activities in education, science, culture and communication will in this way be directed to the overarching goal of laying the foundations for the development of a culture of peace.

Unesco's programmes – whether directly or indirectly related to peace –

---

[3] In particular, the seminal 1974 Recommendation concerning Education for International Understanding, Cooperation and Peace, and Education relating to Human Rights and Fundamental Freedoms.

[4] Notably, the Unesco Prize for Peace Education, awarded annually, and the biennial Unesco Prize for the Teaching of Human Rights.

are, as it were, the manifest aspect of its promotion of international cooperation. There is, however, another less obvious – and certainly more neglected – aspect of the question on which I propose to focus in the pages that follow. I refer to the actual mechanisms whereby international cooperation works, the process whereby the myriad intellectual, institutional, human and financial components of the Organization come together and find expression in its programme of activities. International cooperation, thus understood, does not exist once and for all nor can it be fully created *ex nihilo*. Rather, it is a complex and evolving process whereby a network of interrelations is built up in the pursuit of common goals. These institutional structures and processes, on which the implementation of specific activities depends, are inseparable from the notion of international cooperation and constitute, in a sense, its physical embodiment.

Co-operation through intergovernmental organizations (IGOs) involves a continuous process of learning and adaptation. Since the mid-1940s when the United Nations family was set up, the world has obviously changed radically and so have issues of foremost public concern. Agendas that seemed urgent in the wake of the Second World War have given way to different priorities, if not always of substance then certainly of procedure and emphasis. Thus, for example, appeals to world solidarity expressed in terms of the threat of nuclear destruction, which has apparently receded somewhat of late, can now be most cogently couched in terms of facing a common enemy: environmental catastrophe through the degeneration of the biospheric habitat of all living species. Furthermore, participation in the universalistic IGOs by formerly voiceless communities as these emerged into statehood from their colonial condition has revealed clearly that, even if 'world problems' exist in the sense of affecting all communities and their welfare around the globe in some fashion or another, the manifestations and perceptions of these same problems vary widely from location to location. This in part explains the so-called 'politicization' of many issues in Unesco's domains – a phenomenon that is hardly surprising given that education, science, communication and culture have been the very stuff of political controversy at national levels ever since they entered prominently into the public arena. The learning process has thus been put through a number of painful tests as familiar assumptions have had to be adjusted or abandoned to accommodate a widening range of viewpoints and interests.

It is perhaps appropriate to imagine international cooperation as a flow. The mainstream, originating in a lake of pooled resources, is fed by numerous tributaries and the resulting volume is tapped to irrigate different areas. Over time, everything changes: the volume of the flow, the meanders of the beds of the main stream and of its tributaries, the needs in the irrigated areas, and the speed of currents, so that the entire configuration becomes transformed. IGOs can, in this metaphor, be likened to major pumping stations dotted along the banks, of interest chiefly for the efficiency with which they are able to focus and accelerate currents. They must moreover constantly adjust or risk being left high and dry.

In keeping with this image, I propose to discuss international cooperation through Unesco under three headings, emphasizing the adaptive or dynamic processes under each. The first section concerns the search for partners – the 'tributaries' – which an Agency without real forerunners had to undertake gradually after its establishment. The second section deals briefly with what goes on in one 'pumping station' as the inflow is handled with increasing effectiveness and flexibility. The third section provides certain details about the outflow. Throughout, selected examples of Unesco actions are cited.

Functionalism and the search for partners

The spirit attending the creation of the Specialized Agencies of the United Nations was that of functionalism. As persuasively put forward by David Mitrany in his *A Working Peace System: An Argument for the Functional Development of International Organizations* (1943) the functionalist case rests essentially on the assumption that a largely value-free, technocratic approach to matters is the optimal way of avoiding or overcoming inevitably value-loaded political conflicts and confrontations. Two further assumptions flow therefrom: firstly, that the technical aspects of matters can in practice be clearly identified and separated out from other parameters, and secondly that suitable partners who possess appropriate competences for operational purposes can be found wherever action is called for. Unfortunately, none of these assumptions fit Unesco's domains at all closely. Nor were there earlier organizational models on which to build. For, following the establishment of the first modern IGO by the Congress of Vienna in 1815 (the Central Commission for Navigation on the Rhine) only about 30 such bodies survived into the 1930s. Some, like the International Telegraph Union (1865) and the Universal Postal Union (1874), were indeed narrowly technical – outgrowths of existing national structures which became their natural partners. The few surrounding the League of Nations were not primarily functional in design, this being certainly true of the International Institute of Intellectual Cooperation, sometimes said to be Unesco's immediate predecessor, but in fact a non-governmental organization (NGO) with diffuse aims. The International Bureau of Education, founded in 1925 also as an NGO, became an IGO in 1929 and might have been a more relevant model (formally integrated with Unesco in 1969), but its main activity before the Second World War consisted in convening periodic conferences on public education and its membership was limited, even by the standards of the time.

Unesco's mandate is the broadest and least technical amongst the Specialized Agencies; and it was realized from the start that finding suitable partners for an IGO which transcends narrow functionalism and hence cannot rely on a spontaneously self-selective network needed to be faced. The very novelty of the Unesco experiment meant that such partners had, in some cases, to be traced, in others transformed the better to serve Unesco's interests, in others still actually created. Other IGOs, as incipient world ministries, might rely – in the first instance at least – on national ministries and administrative structures as natural partners. That solution was not always open to Unesco. True, ministries of education or their equivalent existed in all the founding Member States as in those that joined later. But these hardly ever covered all Unesco domains; ministries of science, culture or communication were rare enough in the 1940s and are by no means universal today. Apart from that, it was deemed essential to delve below the governmental level to reach professional communities, institutions and individuals directly. To gain access to such partners the establishment of Unesco National Commissions was encouraged. They were to be committees of specialists meeting occasionally to act as two-way transmitters and amplifiers for the programme priorities adopted by General Conference, advising delegates to the latter, members of the Executive Board and the Secretariat on the one hand while seeking national counterparts, contractors or contact points and spreading the Unesco message on the other. As it would turn out, some functioned creatively while others acted simply as administrative screens or switchpoints.

Another important set of partners are the many non-governmental organizations (NGOs) with consultative status at Unesco. They are the principal links with professional communities and institutions, such as museums, libraries or scientific bodies, invaluable as channeling and selection mechanisms. Some, like the International Council of Scientific Unions (ICSU) have a longer history than Unesco, but many others, like the International Council of Philosophy and Humanistic Studies, the International Theatre Institute or the International Social Science Council, were set up under Unesco auspices often as federations of disciplinary associations also on many occasions created with Unesco assistance. Support for NGOs absorbs a sizeable proportion of Unesco's budget and is sometimes criticized because the priorities of these bodies do not strictly coincide with Unesco's own. This may however be seen as a valuable corrective to governmental concerns, entirely consistent with the role of NGOs in civic society; all the more so now that the trend is towards restricting governmental involvement. In any case, it is hard to imagine how Unesco could operate without NGOs as intermediaries. If they still require Unesco support, it is also true that many have acquired status and resources over the years which often make them less dependent on this source. Unesco's midwifery has thus allowed a vigorous brood to contribute to international cooperation in its own way.

An early example of programme dynamics illustrating partnership problems is the so-called Tensions Project pursued in the Social Science Department from 1947. In the immediate post-war climate, and given Unesco's then still limited range of Member States, a natural impulse arose to pursue the logic implicit in the reference in Unesco's Constitution to 'the great and terrible war ... made possible by the denial of the democratic principles of the dignity, equality and mutual respect of men.' To this end, it was decided to study tensions which cause wars. Scientifically timely (following a good deal of research conducted largely under military auspices, chiefly in the United States) and politically attractive, the project was executed essentially as a collaborative venture between top specialists recruited by Unesco precisely for this purpose and recognized outside experts in socio-psychological and political fields. A classically academic exercise in research and reasoned persuasion, it resulted in a number of books brought out by various publishers and in a series of popular pamphlets on race – designed to counter the poisonous myths propagated by Nazi and fascist regimes – published by Unesco and widely translated. Within a decade, the Tensions Project fed into broader themes associated with tensions: industrialization, migration, ethnic conflicts, urbanization, demographic structures, knowledge transfer and others, finally to embrace socioeconomic development issues on the one hand and human rights on the other.

While it illustrates an intellectual progression in keeping with more inclusive world-views and improved monitoring capacities the trajectory of the Tensions Project also points to the limitations inherent in initiatives of this sort. The partnership was mainly one of scholars along with some NGO support; the output had the limited appeal and impact to be expected at their level of discourse. No serious effort was made to translate findings into operational terms, for instance by influencing policy, legislation or teaching, as attempted much later when a wider network of partners could be found, particularly in the area of human rights. Unesco's early endeavours suffered the fate common to pioneering efforts in being somewhat isolated and tentative. Wider articulation through extended partnership came later as the Unesco network was perfected.

That this network now exists as a community of shared aspirations and commitments indeed demonstrates that the Unesco idea – no doubt interpreted in many different ways, in different locations and for a variety of purposes, to an extent independent of the central structure from which it partly derives and upon which it partly converges – is a living reality. It was decisively strengthened by the knowledge transfer and institution-building tasks which Unesco shouldered from the 1960s, drawing a large number of bodies into the Unesco orbit at operational levels, especially in the Third World. The mix of functionalism with normative concerns has thus achieved a degree of maturity. Its most concrete expression, along with respect for human rights, is perhaps the governmental policies to promote education, science and culture increasingly crystallizing around the world. No doubt impelled by objective necessities, these have been accelerated and clarified in the Unesco framework through inter-ministerial conferences, meetings, documentation and the continuing services offered by the Secretariat, its associated institutes, regional offices, field missions and regular national and international partners.

Inflow processing and administrative innovation

Once active partnership linkages were consolidated, a great inflow of information, claims, desires, ideas, plans, data sets and other impulses was set in motion. Unesco is peculiarly attractive to such inflows since it covers areas in which imagination is at a premium and which have undergone rapid transformation almost everywhere over the past four decades. Yet the components of this inflow are not always fully concep-tualized or ripe, nor do they necessarily take due account of the capacities and constraints of an IGO. A major task of the Secretariat is therefore to articulate such raw material into a plausible programme consistent with the Agency's human, financial and organizational resources. It requires a constant effort to sift, integrate, balance, evaluate and allocate disparate elements, and this absorbs much energy. The results are fashioned into the biennial Programmes and Budgets submitted every second year as drafts to the General Conference as well as the six-year policy projections known as the Medium Term Plans which sketch out in broad strokes the general orientations of three programming and budgeting exercises. Once these formal documents have been approved individual projects can be tackled.

Since an IGO like Unesco does not actually manage enterprises on a day-to-day basis over the long run, as a national ministry may be responsible for the overall running of an education or health system, it must determine its strategy according to the limited operational modes at its disposal. These include the granting of subsidies, the award of fellowships or travel grants, the convening of conferences, expert meetings or discussion groups, the collection of data and information (directly by questionnaire, or indirectly through agents), acting as a clearing house for information exchange, the commissioning of research, documentary instruments, listings and the like, expert missions of longer or shorter duration, and the preparation and production of written and recorded materials. Problems of continuity must constantly be faced, a task made the more difficult by the tendency of Member States to press for innovations in keeping with their own shifting domestic priorities and approaches. While the argument that worthwhile endeavours should, after a time, find their own support is perfectly sound, it must also be remembered that international initiatives often require continuing help from IGOs precisely because their international scope cannot be guaranteed by any other sponsors. While Unesco has been able to phase out its role in, say, the restoration of monuments, institution-building or research efforts as the tasks set were completed or alternative arrangements emerged at the national, regional or international levels, there are other elements which have had to be carried by it for decades precisely because they would otherwise not prove viable. Unesco thus acts, now as a catalyst or broker, now as the base for unique, ongoing activities which it is best fitted to conduct.

Aside from assemblage, however, there are tasks of a more challenging intellectual and analytical nature to perform, especially to assess situations and monitor their dynamics. Education, science, culture and communication are in constant flux. Technological advances since the Second World War, together with changes in economic structures which have made knowledge-intensivity a leading production factor, and progressive globalization have accelerated developments everywhere. As a result, the gap between leaders and laggards has not only become greater as the latter fall ever further behind but also imposes cumulative handicaps. Unesco has been central in raising awareness on such issues in world-wide perspective, relying on a great range of sources to do so.

One primary source over which it exercises practically no control but must take largely as supplied are the relevant national statistical series. Unesco collates these and publishes an annual compilation widely referred to and often quoted. It is instructive to compare the earliest with the latest output in this series. *Basic Facts and Figures* (1952) is a slim booklet covering principally a few categories in the educational and cultural domains for a limited number of countries and territories: it contains nothing at all about science. The massive *Unesco Statistical Yearbook* (1989) is infinitely more elaborate, covers all Unesco domains and contains many fine breakdowns, such as book production by subjects, education by levels or expenditures on scientific research and experimental development. This, of course, reflects the emphasis on monitoring placed by Member States on domains previously but roughly surveyed. Greater sophistication since the 1950s of statistical series at least in the more advanced countries has revealed numerous trends formerly hidden or considered inconsequential, adding decisively to analytical capacities and the observation of dynamics over time and space. For all that, the available figures pose great problems of interpretation and comparability since statistics are notoriously poor indicators of quality. What does it require to be classified as 'literate'? What exactly constitutes a 'book'? What significance is to be attached to cinema attendance, newspaper readership or the output of science graduates?

Making sense of data sets, acquiring a balanced critical input and analysing situations in global perspective with the aid of the best instrumentation available has become an acknowledged specialism of major IGOs resulting from, while equally influencing, international cooperation. Matters but dimly apprehended at national levels a couple of decades ago: the mismatch between curriculae content and the employment market, the cultural implications of satellite broadcasting or the confluence of biospheric pollution agents, for example, are now quite rapidly conceivable at the global level thanks in large part to the transmission capacities of IGOs like Unesco. Admittedly, practical abilities to solve what are correctly perceived as increasingly interlinked problems have not kept pace with the accumulation of evidence, but at least an overall vision exists. In the coming years, systemic characteristics will

surely be clarified with the help of computerized link-ups and the conco-
mitant analytical powers which should make possible a quantum jump
from information acquisition and archiving to the marshalling of
elements tailored to policy and operational needs in all locations. The
often somewhat underrated pioneering and preparatory work of IGOs
like Unesco should then come fully into its own.

At the outset, international cooperation through Unesco occurred
exclusively through the framework of the regular programme and budget.
This meant rather long lead-times and the crossing of many hurdles
before a given idea actually became an approved project, besides running
risks of failure or discontinuity through the imponderables of a cumber-
some mechanism. Different administrative arrangements were therefore
invented to diversify possibilities and to accommodate real needs opti-
mally. At the international level, one of these consisted in direct inter-
IGO meshing through complementarity and a functional division of
labour. Thus, for instance, the World Bank or the UN Fund for Population Activities (UNFPA) might finance a number of posts and activities
within Unesco, which thus became their executive agent in a common
cause. This device also resolves possible borderline frictions concerning
mandates that are inevitably somewhat overlapping. Educational dimensions of given problems, currently for example the AIDS epidemic, were
thus often defined as falling within Unesco's province.

A quite different device consists of basing a fund or the administration
of funds-in-trust at Unesco. This offers escape from the rules binding an
IGO in the acceptance of gifts or the observance of certain criteria applied
to its own projects. An example is the International Fund for the Promotion of Culture, created in 1974, dedicated to creativity, directly
accessible to private and public institutions, groups and individuals with
voluntary contributions originating from governments, institutions and
individuals. It is steered by a Council of 14 eminent persons from as many
countries. Only the director and his office form part of the Unesco
Secretariat. A comparable arrangement is that of intergovernmental entities like the General Information Programme (PGI), the Man and Biosphere Programme (MAB) and the Intergovernmental Oceanographic
Commission (IOC), which enjoy substantial functional autonomy within
Unesco. These programmes reproduce the structure of Unesco itself, in
the shape of steering bodies whose members are governmental representatives at the direct programme level. Their advantage consists in pre-
determining substantive orientations without going through the complexities of General Conferences to which, however, the steering bodies must

report. Funding is provided by Unesco and programme execution is the responsibility of its Secretariat. Intergovernmental programmes are thus by way of being miniature IGOs integrated into a larger whole.

A final example of administrative innovation is the Participation Programme financed by setting aside a fixed percentage under principal programme headings for spot assistance to member states. As the name implies, the object here is to supplement an ongoing activity or assist a national institution, perhaps by the offer of books or equipment, a fellowship for study abroad, a short expert mission, the organization of an exhibition or defraying the translation and production expenses of a publication. Such limited interventions are sufficiently flexible to respond to specific demands while often, if not necessarily, allowing a given domestic effort to attain a broader and internationally more visible profile.

How to adapt administrative structures to the inflow reaching Unesco has been a process as yet surely far from complete. On the debit side it has resulted in degrees of compartmentalization which make the overall governance of the Organization more complex, introducing certain constraints and encouraging special lobbies. On the credit side, however, it has enabled a bureaucracy originally designed as a monolith to respond more flexibly to the increasingly varied and to an extent contradictory demands placed upon it, thus responding to the enhanced intricacies and interdependencies of the international environment. Had such innovations not been accepted it is certain that Unesco's universalistic mission could not have extended as it did, particularly to the Third World where it is most gratefully acknowledged. The initial aim of intellectual cooperation has not been lost but rather refined and channeled in ways consistent with the potential of the partnership network.

Spreading the messages

Unesco does not convey a single, overwhelming message but rather a set of general values to influence public decision-making, group and individual attitudes plus specific information. If it is sometimes perceived as blandly on the side of the angels, at others as propagating the controversial causes of only one set of its constituents, such differences in fact attest to its vigour. For what is acceptibly orthodox in certain quarters is provocative, even outrageous in others, schisms and oppositions being what they are in our contemporary world. In no sphere is it then more difficult to walk the tightrope upon which an IGO must constantly balance than in that of its communicational standards and practices. If too little is said nobody will be offended but few will be interested; if excessively categorical positions are endorsed, protests may be loud and immediate. Governmental views may conflict with Unesco's commitment to citizens at large, or the latter may deplore Unesco's allegedly excessive respect for the former. Communication is consequently a constant dilemma. The adaptive process incumbent on Unesco must be accompanied by a parallel effort in Member States which must face up to possible incompatibilities between respect for the Constitution of an IGO to which they have formally subscribed and cherishing contradicting official ideologies or aspirations. The easy way out could be for Unesco to confine itself to strictly technical communications, in the spirit of narrow functionalism, or to documentation directly linked to assistance formally requested by Member States. There are few reservations either about guides, repertoires, bibliographies, trend reports, statistical compilations and the like, of which Unesco indeed brings out numerous series, nor is there much ado about highly specialized material addressed to limited audiences or specifically concerned with local matters (like the recommendations of field missions) and unlikely to get beyond them. Where difficulties arise is when Unesco attempts to fulfil its more normative mandate, either by direct communication at a fairly popular level or by the diffusion of general policy recommendations or criteria of desirability. Inevitably, such messages convey value assumptions; equally inevitably, these do not meet with universal approval and objections arise at many levels. All media are well acquainted with such feedback. Unesco's row, however, is particularly difficult to hoe.

A medium-sized publisher or co-publisher by the number of titles issued annually, Unesco has laid special emphasis on the linguistic diversification and low pricing of its production. To the three original official languages (English, French and Spanish) in which it continues to be most active, three more (Chinese, Russian and Arabic) were added subsequently, in which production (also by arrangements within the countries where they are spoken) remains more restricted. Furthermore, efforts have been made to supply material in languages such as Swahili, Hindi, Bengali, Farsi, Indonesian or Korean, widely spoken but not always well endowed with quality publications for various audiences. In addition, there have been programmes to support the translation and publication of literary masterpieces as well as initiatives to diffuse Unesco material in non-official languages with strong local publishers and markets, like Portuguese, German or Italian. The Unesco-sponsored

History of Mankind has thus appeared in many languages as a series of books and even in instalments as a richly illustrated magazine. The record, however, is held by the monthly *Unesco Courier*, an illustrated general periodical aimed at the high school level and above, now appearing in 32 language editions with by far the largest circulation of any of the countless serials published out of the U.N. family. Unesco's major quarterlies, which maintain continuous links and dialogues with the professional communities to which they cater, equally have multiple, identical language editions: seven (with two more to come soon) in the case of *Prospects*, the education quarterly, six in the case of the *International Social Science Journal* and of *Impact of Science on Society* and five in the case of *Museum*. A new venture is the monthly news bulletin *Sources*, published in English and French, which provides information on Unesco's programme and activities for a wide general public. The *Bibliography of Publications Issued by Unesco or under its Auspices* (1973) in the first 25 years records 5475 items; there may be as many again since. In some domains, Unesco holds a virtual monopoly, for instance in that of scientific maps and atlases arising out of such important undertakings as the Man and the Biosphere Programme, the International Geological Correlation Programme, the International Hydrological Programme and the Intergovernmental Oceanographic Commission, published jointly or alone.

Investment in audio-visual supports has been comparatively more modest than in paper ones, for audience-specificity and distributional barriers are acute for such material. Nevertheless, Unesco-sponsored films, recordings, sets of slides, posters and film-strips exist, some designed for special uses, others for wider audiences. It would no doubt be desirable for Unesco's presence on television to be developed to enhance visibility and impact. Beginnings have been made but they are yet tentative, as are those of the U.N. family of organizations as a whole – for reasons of cost, to begin with.

Finally, it is worth mentioning that Unesco's headquarters in Paris have become a familiar attraction for the city's residents and visitors. Guided tours of the buildings and its art are a popular feature for tourists while exhibitions, film and theatrical shows, concerts, lectures and other manifestations enliven the facilities they contain. This form of communication, even if it is locationally restricted, encourages the view of Unesco as an open undertaking, accessible to all, which was surely a symbolic intention of its Member States in opting for the memorable architecture of the much-photographed, Y-shaped main structure inaugurated in 1958 and generously contributing to its decoration by Picasso, Miro, Moore, Calder, Tamayo, Noguchi and other modern, traditional and anonymous masters.

Conclusion

In all likelihood, Unesco's share in the total flow of international cooperation in its domains has shrunk considerably, especially over the past 20 years or so. Which is in a sense to be welcomed since it implies that Unesco is now part of a much broader movement than existed at its foundation, a movement that has become strongly organized and routinized through governments under bilateral arrangements, other IGOs and regional integration, the more active participation of universities, research bodies, NGOs, the media and many other actors. New communication technologies, both for mass audiences and for highly specialized ones, have become the tools to promote globalization, the systemic characteristics of which are being increasingly investigated. Thus, it is emerging that internationalism in the sense of seeking solutions to discrete problems through agreement between nations and cooperative efforts is giving way to transnationalism, i.e., the striving after concerted management in matters which transcend the controlling capacities of any single nation, or even groups of nations, such as the global economy, global ecology or population movements across borders. One can anticipate that, in the 21st century, the concept of international cooperation will give way to a much more inclusive understanding of world solidarity and citizenship.

The perplexities and complexities to be faced are no doubt daunting but signals abound that consensus is growing apace. An end to acute East–West confrontation is in sight. The stubborn North–South contradictions can then be attacked with greater vigour. Circumstances to date have often proved highly frustrating to Unesco's endeavours but now it is no longer necessary to be of Unesco in order to be with it. The idea that education, science, culture and communication cannot be left to their own devices in the hope that an 'invisible hand' will somehow reconcile their spontaneous unfolding with the general good has taken firm root. It has become widely acknowledged that they stand in need of far-sighted policies to steer and stimulate them as well as of open debate and criticism lest they become distortedly self-serving and introspective. Exactly where responsibilities for such policies and steering lie is currently the subject of lively debate as their interdependence and cross-national implications become ever clearer. The role of IGOs might thus be decisively enhanced, opening up entirely fresh vistas. Whatever opportunities Unesco could then grasp ought to be in some relation to its record, which shows it to have been, at best considerably in advance of the time, at worst in close step with them. That holds true of its commitment on any number of important issues, from practical aims like the standardization of informational norms or the return of cultural property to its rightful owners to wider ones such as lifelong education, respect for minorities and tolerance under democratic pluralism, the advancement of women's rights, strong emphasis on the socio-cultural dimensions of development, disarmament and the re-allocation of resources thus freed, or ethics in scientific research. Such a record proves that the distillation of collective wisdom through Unesco has indeed been successful in furthering international cooperation and trust, whatever the shortcomings along the way and despite the magnitude of the challenges remaining.

At the eve of a new millenium, it appears clearer than ever that imagination and knowledge are essential if we are to achieve the conceptual and practical breakthroughs needed to ensure full respect for the rights and dignity of all human beings. New approaches, new attitudes and new behaviour patterns are required if we are properly to transmit to future generations our common heritage; and these are possible only in a context of unrestricted freedom, which is the key to a brighter future. A system in which every citizen counts is the only one that will allow participation, sharing and the full expression – in work and leisure – of creativity, which is the distinctive capacity of humanity. The name of such a system is democracy.

Select bibliography of works published by Unesco

The Dynamics of Peace (1986).
Partners in Promoting Education for International Understanding, (1986).
Teaching for International Understanding, Peace and Human Rights, (1984).
International Dimensions of Human Rights, 3rd edn, two volumes, (1983).
Scientists, the Arms Race and Disarmament, (1982).
Science and Racism, (1982).
The Seville Statement on Violence, (1991).
World Problems in the Classroom, 3rd impression, (1981).
Many Voices, One World: Communication and Society Today and Tomorrow, (published with Kogan Page, London, and Unipub, New York, 1980).
World Directory of Human Rights Teaching and Research Institutions.
World Directory of Peace Research and Training Institutions.

18

U.S.–Soviet cooperation against terrorism: Common ground

IGOR BELIAEV AND JOHN MARKS[1]

For Soviets and Americans alike, terrorism is a particularly unfortunate fact of modern life. Terrorism threatens the very fabric of civilization. Unfortunately, even the most effective countermeasures fail to eliminate it. From the point of view of the terrorists – however misguided – the potential gains outweigh the risks and costs. To terrorists, whether individuals, groups, or nations, terrorism offers a way to impose their will and gain access to the news media in a world where real and imagined grievances are not easily heard or satisfied.

Today, the U.S. and Soviet approaches toward terrorism are increasingly coinciding, and important voices in both countries are urging their governments to work together – to turn the fight against terrorism into a joint struggle. Such an effort would represent for the Soviet Union and the United States a worthy undertaking that makes hard-headed political sense – *if* the two countries are really serious about building a cooperative relationship.

To date, both governments have taken the first steps toward cooperation against terrorism. They have made clear that terrorism is not an acceptable means to achieve any end, and they have stopped accusing the other of supporting international terrorism. While disagreements remain, we believe that the best interests of our two nations will be served by a high level of cooperation in the fight against terrorism.

[1] Igor Beliaev is a Political Observer (columnist) at the weekly newspaper *Literaturnaya Gazeta* in Moscow. John Marks is Executive Director of Search for Common Ground in Washington, DC. Beliaev and Marks are the co-chairs of the Soviet–American Task Force to Prevent Terrorism, which is sponsored by their two organizations.

316

In our view, increased U.S.–Soviet action against terrorism would produce benefits on three different levels:

(1) *Preventing specific acts of terrorism.* Through collaboration, the superpowers would, on occasion, be able to stop actual terrorist attacks. This chapter will outline a framework within which this collaboration has started to occur and could greatly expand.

(2) *Shifting the global climate in which terrorists operate.* Joint Soviet–American action, even on a limited scale, would send a signal to the rest of the world that, despite their remaining differences, the U.S.A. and U.S.S.R. are united in opposing terrorism. The fact that the two countries are cooperating would probably have as much impact as the specifics of the cooperation. Although superpower collaboration obviously would not end the problem of terrorism, it could well reduce the legitimacy of terrorist violence. Both countries would be saying, in effect, that political ends cannot be achieved through terrorist tactics and that perpetrators will be treated as criminals. In recent years as regional conflicts have been winding down, there has started to exist a noticeable – but by no means overwhelming – trend toward 'de-legitimization' of terrorism. By increased collaboration, the superpowers would be reinforcing this trend.

(3) *Strengthening the U.S.–Soviet relationship.* Joint Soviet–American efforts to prevent terrorism would contribute to improving U.S.–Soviet relations, given other favorable conditions. Successful collaboration would be an important confidence-building measure which would demonstrate that U.S.–Soviet cooperation is possible, even in extremely sensitive areas.

We make our case at a time when the Cold War is coming to an end, and when the U.S.A. and U.S.S.R. have succeeded in draining most of the poison from their relationship. Leaders in both countries are faced with a dramatically changed world, and they are developing new ways to deal with issues that once aggravated the East–West struggle. While the post-World War II security system is disappearing and the replacement system is still in flux, neither country, sensibly, is letting down its guard or disarming unilaterally.

But where the emphasis in the superpower relationship once was on facing the other as the enemy, the emphasis has now shifted toward standing together facing the common danger. Terrorism is a critical part of that common danger. We believe that both our nations should deal with terrorism within a framework similar to that established by Presidents Gorbachev and Reagan on nuclear arms: Namely, to treat terrorism as a problem shared by both superpowers and to cooperate, wherever possible, to eliminate the threat. This is in keeping with Mikhail Gorbachev's call in 1987 for 'a radical strengthening and expansion of cooperation among states in eradicating international terrorism' and with similar statements from the United States Government.

Until recently, Soviet experts usually looked at international terrorism through the prism of national liberation movements. Although this attitude has now changed, Soviets often felt that specific acts of terror were 'correct'. For example, virtually all Soviets feel that any tactic whatsoever was completely justified in the fight against the Nazi occupation; that few people in the West can even faintly imagine the extent of suffering, privation, and humiliation that Soviets endured during World War II; and that when Soviet partisans, paratroopers, or secret agents carried out sabotage or murdered German soldiers, these were fitting acts of revenge and causes for rejoicing.

During World War II, the United States backed resistance groups that used similar tactics against the Nazis, and few Americans would disagree with the correctness of this policy. In the post-war years, both East and West continued to support liberation movements that at times used terror tactics.

Looking back on the Cold War, we do not believe that our two nations were equivalent or acted for similar reasons. Nor do we think that Soviets and Americans will ever agree on what happened during the Cold War – on who did what to whom; on who was right and who was wrong. Nevertheless, we feel that both countries are less likely to repeat old mistakes if they learn from the past, even when the memories are not pleasant. While friendship and cooperation are essential to the new relationship, true partnership requires moving beyond politeness and dealing honestly with the world as it is, and was.

Prior to the Gorbachev years, the Soviet government condemned terrorism as a whole but focused little public attention on terrorism committed against Soviets outside the U.S.S.R. Unlike the U.S. government which could be counted to raise a huge commotion when even a single American citizen was attacked, authorities in Moscow preferred to look the other way and 'not to overreact' when Soviet citizens were targets. Soviet authorities usually 'advised' their media not to publicize such assaults. To this day authoritative rumors circulate in Moscow about Soviet geologists and other non-political types who after many years still are held hostage in Angola and Mozambique.

In 1985 four Soviet officials were seized in Beirut. After the incident attracted widespread international publicity, the Soviet government, for the first time ever, publicly condemned an act of terrorism aimed at Soviet citizens abroad. A special envoy was dispatched to Lebanon and Syria. Undeterred, the terrorists murdered one of the hostages, raising fears in Moscow that all Soviets overseas would become targets. At this point, according to rumors widely circulated in the West and denied in the Soviet Union, KGB operatives supposedly seized an individual in Lebanon related to the suspected kidnappers and castrated him. Whether or not there is any truth to this story, the three remaining hostages were soon released. Afterwards Western terrorist experts would talk somewhat enviously of the Soviet 'no nonsense' approach to terrorism.

Before the 1985 events in Beirut, Soviet officials lived under the illusion that international terrorism was not their concern. While a few Soviet journalists and academics carefully studied the terrorism issue and the conditions that gave rise to it, the government devoted little specific attention to it. In fact, the Soviet Foreign Ministry, unlike the U.S. State Department, did not, and still does not, have a separate section devoted to terrorism. The Foreign Ministry has long regarded the subject as an international law problem and assigned responsibility to the Treaty Law Board.

In December 1988, Soviet terrorists seized a bus full of school children in the city of Ordjonikidze and demanded a $3.5 million ransom. While the KGB closely monitored the situation, there was no coordinated, pre-planned response from Soviet authorities. Soviet authorities were not able to draw from data on past incidents that would be routinely available to anti-terrorist efforts in the West. While the Aeroflot personnel involved did their utmost to cope with the terrorists, they acted intuitively, without ever having been trained in how to react in such circumstances.

In this particular case, perhaps because children were involved, Soviet authorities decided to pay the ransom and provide a plane for the terrorists to fly to Israel. During the flight, Soviet officials maintained contact with U.S. personnel who facilitated communications between Moscow and Tel Aviv which did not have diplomatic ties. When the plane landed, Israeli forces disarmed the hijackers and arrested them. Then, with Israeli approval, a high-level KGB officer flew to Israel and returned the hijackers for prosecution in the U.S.S.R. No lives were lost. The plane and money were recovered. The terrorists went to jail. The incident was a model showing how cooperation can work between countries that historically have had very different attitudes toward terrorism.

The United States has long held 'first place' as the preferred target of international terrorism, usually followed by Israel, France, and Britain. By 1989, the Soviet Union had climbed into fifth place, according to RAND Corporation figures, and sources in Moscow revealed that terrorist acts had caused the deaths of 60 Soviets in the preceding years. In short, terrorism had become a Soviet problem, and a problem that would probably increase given the increasing vulnerability of modern society to terrorist attack. Soviet leaders recognized the threat from 'techno-terrorist' attacks on computer systems, electric power grids, and nuclear power plants; from chemical, biological, and nuclear terrorism (which might involve terrorists seizing a missile from U.S. or Soviet forces and threatening to push the button); and from ethnic and 'narco-terrorism.'

In addition, under *glasnost*, the Soviet media have given much more coverage to terrorism, which, in turn, has raised general concern. In some Soviet circles, anxiety about terrorism has been heightened by the possibility that ethnic conflict – particularly Islamic extremism – will lead to increased terrorism. Today, both superpowers consider terrorism to be an important problem, and both can agree that no matter what happened in the past, terrorism should in all cases be illegal.

Despite their converging views, the U.S.A. and U.S.S.R. are by no means ready to concur on crucial aspects of the terrorism question. The difficulty in the past has usually occurred when the two nations found themselves on opposing sides of the ideological or political divide. In such cases, the superpowers or their friends supported 'freedom fighters' or 'national liberation fighters' whose struggle was opposed by the other. Examples included the Afghan Mujahedin, the Palestine Liberation Organization, the Nicaraguan Contras, the Popular Front for the Liberation of Palestine, the South African National Congress, and UNITA in Angola. Although the U.S.A. and the U.S.S.R. today are increasingly working together to curb regional conflicts, neither superpower is yet willing to cut its ties with every organization that uses tactics which its foes describe as terrorist.

Given this reality, we still believe that the Soviet and American governments can cooperate against terrorism, *if* they reject the cliché that one man's terrorist is the other's freedom fighter. We feel that one man's terrorist is, in fact, the other's terrorist. Terrorism is terrorism, when it involves the slaughter of innocents away from combat zones – no matter what the justification. We believe that, without exception, any individual, organization, or state which is guilty of terrorism should be brought to justice and severely punished in accordance with national and international law.

There have been serious breaches of this standard. Soviets recall the Brazinskas case in 1970 when a Lithuanian father and son killed an Aeroflot hostess while hijacking a Soviet plane to Turkey. Although convicted by a Turkish court, the Brazinskas were soon freed under a Turkish amnesty law. In 1976, they illegally entered the United States and were eventually allowed to stay by a U.S. court. American officials, for their part, maintain that the Soviet Union supports and arms nations like Libya and Syria which harbor known terrorists; these American officials state that if the Soviets are serious about cooperating with the United States against terrorism, they will have to put at risk their relationships with such nations.

In the foreseeable future, neither the U.S.A. nor the U.S.S.R. is going to stop fighting terrorism on its own and in collaboration with its allies. We believe that cooperation between the superpowers should be implemented *in addition to* and not *instead of* any existing counter-terrorism efforts. This cooperation would build an additional layer of protection against terrorism, and *not* interfere with existing defenses.

This chapter reflects the approach taken by Soviet and American participants in the U.S.–Soviet Task Force to Prevent Terrorism, of which we are the co-chair. Only a few years ago, a bi-national group of this sort could not have existed in any meaningful way. Today, our successful collaboration, in itself, represents a model for superpower cooperation.

Although many of the Soviet and American participants in our Task Force enjoy close relations with their governments, the Task Force is in no way official. In fact, the Task Force would not exist if we had listened to U.S. State Department officials who, during much of 1988, actively tried to discourage the first meetings. Our accomplishments demonstrate how unofficial 'citizen diplomats' can be out in front of their own governments and find agreements that their governments can then adopt.

The U.S.–Soviet Task Force to Prevent Terrorism had its origins at a conference, the 'Citizens Summit,' sponsored in February, 1988, by the Center for Soviet–American Dialogue and the Soviet Peace Committee. That gathering brought together several hundred Soviets and Americans for five days in order to generate a whole array of new U.S.–Soviet projects. The 'Summit' was in essence, a giant brainstorming session, an unparalleled opportunity for Soviets and Americans to try to be innovative together.

The two of us, Igor Belyaev and John Marks, were given perhaps the most difficult assignment. We were named co-chairs of the committee whose job was to find solutions to regional conflict. As much as the

Soviet–American relationship has improved, we recognized that our nations still have substantial differences in the Third World, from Afghanistan to Cuba to the Middle East. We immediately decided that our best chance to make a difference was to identify a single issue on which effective U.S.–Soviet collaboration might be possible. Regional conflicts seemed either intractable or firmly on the official agendas of our two governments. What could unofficial 'citizen diplomats' do? After two days of deliberations with the 15 or so Americans and Soviets who made up our committee, we agreed that terrorism would be the issue.

Somewhat grandly, we named ourselves the Soviet–American Task Force to Prevent Terrorism and participants signed an agreement to encourage U.S.–Soviet cooperation on terrorism. Among other things, we decided that the Task Force should include former officials, scholars, journalists, lawyers who were experts on terrorism; that it would meet at regular intervals in Moscow and Washington; and that its members would write a book together (from which this chapter is taken, and called *Common Ground on Terrorism: U.S.–Soviet Perspectives on Terrorism* (New York: W. W. Norton, 1991).

After inevitable delays, the Task Force had its first meetings in Moscow in January, 1989, under the sponsorship of our two organizations *Literaturnaya Gazeta* of Moscow and Search for Common Ground of Washington with the support of the Soviet Peace Committee. Far removed from the trusting cooperative atmosphere in which it had been born, the Task Force now included one Englishmen and twenty Americans and Soviets who were among their countries' leading experts on terrorism. While we still maintained we were 'private,' the Soviet delegation included officials from the Ministry of Foreign Affairs, the Ministry of Internal Affairs, and key institutes within the Academy of Sciences. On the American side, more than half of the delegation were current consultants on counter-terrorism to the U.S. government.

Yet, these 'non-official officials' were able to speak, and probe, without having to represent fixed national positions. By January 1989, both the U.S. and Soviet governments were providing at least tacit support to the Task Force as a channel in which positions could be ascertained and new initiatives tried out – at virtually no political cost. The Task Force provided both a sounding board for various approaches and a means of judging the seriousness of the other side's interest.

In the month before the first meeting, the incoming Bush administration and Soviet authorities gave their tacit blessings and asked for full reports. The week before we convened, the KGB's deputy director, Lt.

General Vitaly Ponomarev declared on Moscow Radio, 'We realize we have to coordinate efforts to prevent terrorist acts, including hijackings of planes ... We are willing, if there is a need, to cooperate even with the CIA, the British Intelligence service, the Israeli Mossad, and other services in the West.' A high U.S. State Department official opined that this statement from the KGB was timed to have an impact on our upcoming meetings. Within days, James Baker, the new American Secretary of State testified before Congress, 'We ought to find out whether Moscow can be [helpful] on terrorism and if not, why not.'

While this high level attention gave cause for optimism, Soviet and American participants in Moscow were still skeptical that anything useful would be accomplished. Cold War attitudes die hard, and the idea that the other superpower had something constructive to say about terrorism was new and beyond everyone's personal experience. Both the American and Soviet participants were taking risks. The Americans feared they might appear naive and foolish or that they might be walking into a Soviet media circus. Soviet participants feared American harangues about alleged Soviet involvement in international terrorism. They suspected that even talking with Americans about terrorism would be interpreted by some of their friends in the Third World as abandonment or as a hostile act.

In fact, no one's worst fears were realized, and the meetings turned out much better than even the most optimistic among us had hoped. Polemics were minimized. Contentious statements were regularly listened to and noted, without leading to major arguments. One member of the American delegation was Marguerite Millhauser, an expert not in terrorism but in conflict resolution. Ms. Millhauser was given a mandate by both Soviet and American participants to break deadlocks, and she contributed greatly to the atmosphere of collaborative problem-solving that resulted.

Essential to the meetings was the joint understanding that traditional ways of discussing terrorism had led nowhere and that for these talks to be successful new ways would have to be found to frame the issue. In fact, earlier talks on terrorism never seemed to get past the inability of Soviets and Americans to define terrorism. Given this reality, the Task Force chose not to define who was a terrorist but instead identified certain acts which constituted terrorism. These included:

(1) Hijacking or bombing of airplanes
(2) Taking hostages
(3) Attacks on children or internationally protected persons (diplomats, international organization employees, etc.)

The Soviet and American participants agreed that these were always crimes – never political acts. While far from exhaustive and perhaps unsatisfactory to international law experts, this mutually acceptable approach still represented a basis for moving forward. Once the two sides concurred in what they jointly opposed, they could make long lists of ways that the two countries could take to prevent it. Both Americans and Soviets were, in effect, acknowledging that conceptual differences should not stand in the way of their two nations working together to stop what both considered to be heinous acts.

This approach involved separating out, or slicing off, areas that were ripe for agreement, while agreeing to disagree on the rest. The Task Force largely avoided the twin pitfalls of assigning blame for past sins or of making vague statements about the abstract future. Soviet and American participants were able to find substantial agreement. But progress was *not* made contingent on one government being required to change long-held policies which the other claims supports terrorism. The participants accepted that neither country was likely to cease and desist all activities the other finds objectionable as a price for cooperation. Cooperation was to be based on the two nations recognizing that their own national security is best served by collaborating to reduce the common danger.

Soviet and American participants alike were aware that cooperation on terrorism – even among allies, let alone between rivals – would be quite difficult and that recommendations would be meaningless if they were not ultimately acceptable to both governments. With all that in mind, the Soviet–American Task Force recommended the following:

> Creation by the two governments of a standing bilateral channel of communications for exchange of information on terrorism; in effect, a designated link for conveying requests and relaying information during a crisis
>
> Provision of mutual assistance (informational, diplomatic, technical, etc.) in the investigation of terrorist incidents
>
> Prohibition of the sale or transfer of military explosives and certain classes of weapons (such as surface-to-air missiles) to non-government organizations, and increased controls on the sale or transfer to governments
>
> Initiation of bilateral discussions on requiring chemical or other types of 'tags' in commercial and military explosives to make them more easily detectable and as an aid in investigation of terrorist bombings

Initiation of joint efforts to prevent terrorists from acquiring chemical, biological, nuclear or other means of mass destruction

Exchange of anti-terrorist technology, consistent with the national security interests as defined by each nation

Conduct of joint exercises and simulations in order to develop further means of Soviet–American cooperation during terrorist threats or incidents

Joint action to fill the gaps that exist in current international law and institutions

These recommendations were reported directly to the White House and the Kremlin, and they attracted considerable media attention in both countries. Within two months of our first meetings, in March 1989, Foreign Minister Eduard Shevardnadze and Secretary of State James Baker agreed to put anti-terrorist cooperation on the superpower agenda. By June, the two governments had opened up official discussions at the working level and had reached their first agreements over superpower cooperation to prevent terrorism.

Both Soviet and American participants in the Task Force believed they had contributed to this new collaboration between their governments. All of us were gratified during the summer of 1989 when the Soviet and American governments worked together in a highly successful way to prevent the executions of Joseph Cicipio, an American held hostage in Lebanon. Indeed, one of our participants, Brian Jenkins of the RAND Corporation, told a BBC interviewer that, without the efforts of our Task Force, the Soviet–American cooperation that, in his view, saved Mr. Cicipio's life would not have taken place.

The Task Force was scheduled to reconvene in September, 1989 at the RAND Corporation in Santa Monica, California. Both sides recognized that missing from our Moscow meetings had been people with 'hands-on,' operational experience in counter-terrorism, and such people usually were found in the world of intelligence. Thus, it was agreed that the Task Force should expand to include individuals with experience in their country's secret services. The American side secured the first acceptances from former CIA Director William Colby and former Deputy Director Ray Cline to join the group. A telex was sent to Moscow announcing the new participants and requesting that the Soviets add former KGB officials of comparable protocol rank. The Soviet side brought in retired KGB Lt. Gen. Feodor Sherbak, former Deputy Chairman of the KGB's Second Directorate, and retired KGB Major General Valentin Zvezdenkov, former chief of KGB counter-terrorism. Never before had high-level former KGB officials come to the West and met with their counterparts. The participation of such people indicates the importance the KGB places on cooperating to curb international terrorism.

When the Task Force met in Santa Monica, Sherbak, Zvezdenkov, Colby, and Cline became the nucleus of the subcommittee on information. Their meetings were cordial. Together, the CIA and KGB veterans agreed that, while protecting 'sources and methods,' their old services should exchange information to counter terrorism. The ex-KGB men made clear that KGB–CIA cooperation would probably result in an increase in terrorist groups targeting Soviets, but they said the added risk was necessary to curb terrorism. They also agreed that neither the U.S.A. nor U.S.S.R. should provide weapons useful to terrorists (e.g., surface-to-air missiles or plastic explosives).

Ex-CIA Deputy Director Ray Cline, known for his very conservative views, wrote afterwards in the *Washington Post* that before our meetings he had only dealt with the KGB in 'an essentially adversarial context.' He continued:

> The KGB came to the United States to assure some of those who would understand that, whatever happened in the past, it really wants to exchange information with U.S. intelligence agencies to suppress terrorists now. What they can and will deliver remains to be explored in official channels. But Gorbachev's seriousness of intent was crystal clear. Our private scholars' delegation was getting an official message.

Needless to say, the bringing together of top retired CIA and KGB officials attracted considerable attention in both the Soviet and American media, including an ABC-TV 'Nightline' program which focused on our meetings and, particularly, the possibility of CIA–KGB cooperation.

While the media stressed the intelligence old boys, the other participants were, if anything, more active. In all, the Task Force made over 30 recommendations. These included suggestions that the superpowers directly cooperate to free hostages; that the already established U.S.–Soviet nuclear crisis control centers be expanded to deal with potential terrorist use of biological, chemical, and nuclear weapons; and that the U.S.A. and U.S.S.R. cooperate against narco-terrorism by attacking the laundering of drug money, exchanging data on production, smuggling, and distribution of drugs in Latin America and Southwest Asia, and providing aid to front-line countries in the drug war, particularly Peru and Colombia.

Both the Soviet and American members of the Task Force agreed to deliver these recommendations to their governments at the highest levels. Meetings were held in the U.S.A. with State Department, Justice Department, and White House policy-makers. In the U.S.S.R., a parallel effort brought the recommendations to the attention of the Central Committee and the Foreign Ministry. Special attention was given the intelligence recommendations. Generals Sherbak and Zvezdenkov reported them directly to KGB Director Vladimir Kruyuchkov and his senior staff. In December, 1989 the KGB formally accepted the recommendations regarding information-sharing. Two months earlier, William Colby and Ray Cline personally presented these same recommendations to CIA Director William Webster. Webster was interested, yet skeptical of Soviet intentions. As this chapter went to press, the CIA remained reluctant to enter into a cooperative relationship with the KGB, a service which it had so long opposed. The CIA took the position that contacts on counter-terrorism should take place on the diplomatic level between the U.S. State Department and the Soviet Foreign Ministry.

As successful as our Task Force's meetings have been, it should be noted that there was an asymmetry between Soviet and American participants. For more than 20 years, terrorism has been a subject of great concern to Americans, and U.S. experts have developed considerable knowledge on the subject. In comparison, terrorism is a comparatively new field for the Soviets. At the meetings in Moscow and Santa Monica, the Soviets listened and learned, and the Americans, to a large extent, set the agenda. While the Soviets acknowledged they have a great deal to learn, it is obvious that they will not maintain for long their relatively passive posture. For this partnership to work in the long run, it will need to be perceived on both sides as equal and mutually beneficial.

In November 1989, Renee Muawwad, President of Lebanon, was assassinated. Our Task Force, while uneasy with the prospect of denouncing all terrorist acts, decided to use this unfortunate event as an opportunity to illustrate our approach and make a further recommendation to our two governments. The joint statement read:

> We the undersigned participants in the U.S.–Soviet Task Force to Prevent Terrorism unequivocally condemn the assassination of Renee Muawwad, President of Lebanon, and we condemn the use of any and all terrorist means, no matter what the justification;

> We recognize that the U.S. and Soviet Governments share a

common interest in denouncing such heinous acts and in preventing future occurrences of all types of terrorism.

In order to demonstrate that the United States and the Soviet Union stand together in opposition to terrorism of which the murder of the Lebanese President is only the most recent example, we recommend that the U.S. and Soviet Governments jointly condemn this act and that, in addition, they coordinate their response to future acts of terrorism.

The very fact that our Task Force of Americans and Soviets could issue such a statement, regarding a country where the superpowers have long had such different priorities, points up the potential – that exists alongside the danger – of the international terrorism issue. By coming together to curb terrorism, the superpowers may develop common ground which can be expanded into other disputed areas. For example, Soviet members of the Task Force believe that while American participants loudly condemn Palestinian terrorism, Americans are mainly silent on Israeli state terrorism such as the June 1989 kidnapping from Lebanon of Shaikh Abdul Obeid, a radical Shiite leader. Leaving aside the substance of this particular issue, it would seem entirely possible that as Soviet–American cooperation on terrorism increases, Soviets and Americans will be able to bridge the remaining gaps on such questions.

In any case, opposing terrorism is much more than a superpower problem. It is an effort in which all countries should join. The United Nations and other international organizations are already actively involved. We regard our efforts to encourage U.S.–Soviet cooperation as complementary to and supportive of other international efforts. U.S.–Soviet collaboration should in no way be seen as an attempt to have the superpowers dictate to other countries how, when, and where to oppose terrorism. In fact, we have already taken steps to broaden our unofficial 'citizen diplomacy' efforts to include Western European countries.

The process we have started has already had an impact well beyond anything we foresaw when we started. The promise of our approach is enormous. We are committed to moving forward and expanding.

19

U.S. policy towards the Soviet Union from Carter to Bush

ERNST-OTTO CZEMPIEL

Looking back from 1990 the policy of the United States towards the Soviet Union seems to have come full circle. In 1972 President Nixon started the process of détente with Moscow: this, in the view of President Nixon, produced an 'Emerging Structure of Peace'[1]. Ten years later, in 1982, American–Soviet relations had deteriorated to such a degree that many analysts anticipated the danger of a Third World War. In 1989 the Pentagon acknowledged that the likelihood of conflict between the two countries was 'perhaps as low as it has been at any time in the post war era'[2]. A member of President Bush's team evaluated the situation not only as peace but as the 'end of history'[3]. In the short time-span of 16 years American-Soviet policy has lived through two periods of cooperation and confrontation. Each period took on average five to seven years. From 1972 to 1979 détente and cooperation prevailed. From 1980 to 1984 confrontation dominated. Since 1985 cooperation again has been prevalent, producing the first arms reduction treaty in history and moving on to two more treaties of this kind.

It is not easy to explain this circle and its phases. The political discussion points towards the new Soviet leader Gorbachev as a moving force. In historical perspective, he has played a much more minor rôle as an auxiliary force. The turnaround in American–Soviet relations occurred in

---

[1] Richard Nixon. U.S. Foreign Policy for the 1970's. The Emerging Structure of Peace. A Report to the Congress, Washington, D.C., 9.2.1972.
[2] Department of Defense: Soviet Military Power. Prospects for Change 1989, Washington, 1989, p. 140.
[3] Francis Fukuyama. The End of History? in: *The National Interest*, **16**, Summer 1989, p. 3ff.

329

1984–1985, well before Gorbachev came into power. The change occurred although the basic relationship remained unchanged. From 1972 to 1985 nothing happened in the Soviet Union which could have justified the shift from cooperation to confrontation and back to cooperation. What, then, explains the dramatic changes?

Looking briefly into the scientific literature the analyst does not find much consolation. Waltz's Realism cannot apply since neither the international system nor the subsystem of the East–West conflict changed in this period. The Neo-Realism had softened, but not abandoned, the strict analytical roles of Realism. The 'transformational' morality of Realism might contain an interesting change;[4] it is too early to judge its utility[5]. It is a common property of all these theories and models that they do not give any clue how they should be used to analyze a particular international relationship. There are many and good reasons for this deficit. Theory-building in international relations is probably the most complicated task facing the scientific community, certainly much more complicated than the problems caused by physical or chemical phenomena. Political Science, on the other hand, cannot wait until the theoretical puzzle is solved and the gap between theory-building and empirical analysis of distinct processes can be bridged.

Trying to escape this critical dilemma I restrict myself to the analysis of foreign policy. The analysis of action reduces the difficulty somewhat because we do have the methodological tools to deal with this level. We lack, in my view, all the instruments necessary to deal with interactions or relations. They must not, and will not, be neglected, however. At the level of the action one can recognize the influence of the international system, of interactions, and relations. It is possible to distinguish between processes and structures, to differentiate between unique and patterned events.

In addition, the level of foreign policy permits a more comprehensive view of the causes and conditions of action. In my view, there are five of these. First: the international system with its anarchical order stimulates and conditions the action. Second: the system of government which in a political unit selects actors and their freedom of choice. Third: the

---

[4] David Dessler. What's at State in the Agent Structure Debate? in: *International Organization*, **43 (3)**, Summer 1989, p. 467.

[5] I have dealt at some lengths with those methodological questions in my books: *Internationale Politik. Ein Konfliktmodell*. Paderborn: Schöningh, 1981; and *Friedensstrategien. Systemwandel durch internationale Organisationen*. Demokratisierung und Wirtschaft, ibid. 1986.

interests of those actors as defined by the system of government. Fourth: the strategic and tactical competence of those actors which contributes tremendously to their successes and/or losses. And, finally: the interactions between these actors as stimuli and conditions for actions. With the exception of the international system all the other causes and conditions can be analyzed as processes or as structures.

If this approach will never lead to a theory of international relations, it could in due course lead to a theory of foreign policy behavior. And it permits already a rather detailed analysis of the processes of foreign policy decision making[6]. The usefulness of this approach, particularly the emphasis laid upon the system of government as a condition of foreign policy behavior, is underlined convincingly these days by the revolutions in Eastern Europe, the GDR, and the Soviet Union. With all other variables of the international system intact these revolutions will change the foreign policy behavior of those states and, therefore, the system of relations which we called the East–West conflict dramatically.

For our purpose here this approach demands answers to the following questions. First: who are the actors in American foreign policy decisions and what role is being given to them by the system of government of the United States? Second: in the view of those actors, is the conflict with the Soviet Union caused by the structure of the international system or by the intentions and interests of Soviet actors? Third: what are the interests of those actors themselves and how do they pursue those interests? Fourth: how does the strategic–tactical competence of those actors influence the formulation and implementation of their interests? Fifth: did the interaction with their Soviet counterparts influence these interests and their implementation?

It will not be possible in this chapter to touch upon all those problems and to give the necessary answers. I try to do so in my book (see Notes). I feel the necessity, however, to state explicitly the analytical approach which is behind the book and this paper. Analysis and interpretation have been built around the above questions. They are put within the functional structure model of policy-making developed by David Easton, which I have expanded so that it includes the realm of foreign policy and international politics. What Easton has developed is certainly not a theory but a very elaborate model of policy-making which permits the formulation of distinct hypotheses and their testing. The model obliges its user not to forget any actor or relationship which might influence the process of policy-making.

The main answer to the five questions put by this model is that

American foreign policy towards the Soviet Union was influenced mainly by the interests of the actors to use it as a tool for improving their position within the domestic political setting of the U.S.A. Of course, the conflict with the Soviet Union was real, the Soviet challenge obvious. The Soviet Union was the main enemy of the United States, trying to expand her influence in the world and to catch up as the second superpower. There is no doubt about the fact that, principally speaking, the policies of the Soviet Union were the main cause of the East–West conflict. Thus, American policy towards the Soviet Union had the goals of keeping the Soviet Union from winning, containing the Soviet empire and of limiting its political successes in the Third World. In the same vein the United States always tried to remain strong militarily and to have, if possible, a certain margin of military superiority. This general American attitude towards, and interpretation of, the conflict with the Soviet Union did not prescribe a detailed set of ends and means. The range of strategies and tactics has been very wide indeed, and could include cooperation with Moscow as well as confrontation, détente or provocation, disarmament and arms control as well as arms build-up. Washington could try to attack or to appease Moscow, it could use power or peace. Since the U.S.A. was the major superpower she was free to select the means she would use in this conflict.

It is the main thesis of this paper (and the book) that American actors chose their strategies towards the Soviet Union not with regard to the conflict but to their domestic interests. The conflict, objective as it was, offered the possibility to select quite a number of different goals, interpretations, and strategies which could be used for the domestic power struggle. The oscillation of American politics towards Moscow, the waves of cooperation and confrontation are thus explained as the result of different groups and actors trying different strategies and tactics towards Moscow in order to improve their domestic political power. Soviet political behavior as such played only a minor role. It was rather constant from Khrushchev to Brezhnev. In this period the Soviet Union modernized and augmented her conventional and strategic arms at a slow but constant pace. She tried to expand Soviet influence in Africa and in Central America as well as in the Middle East. This conflict behavior was stable and predictable. This was not true for the American behavior, which was in constant flux and changed more or less every two years. This fluctuation, therefore, cannot be explained by the Soviet threat or by the structure of the international system. Both of them had demanded at least a continuing and coherent analysis of the conflict, but even in this field we find an astonishing range of interpretations.

To explain the phenomenon my two main theses are: first, the conflict with the Soviet Union was never primary but only of secondary importance for American policy making. In the center of American politics stood the economy: economic goals and interest rates, inflation and unemployment, social security and welfare. They drew the dividing line between Republicans and Democrats, conservatives and liberals. The elections centered upon those issues which affected also the distribution of power between the different élites and persons. They used the conflict with the Soviet Union as a store from which to draw arguments and proposals which could be used in the domestic power struggle. As will be demonstrated below, Ronald Reagan is a case in point. He started as a cold warrior and ended as a peace president, although the conflict with the Soviet Union remained more or less the same.

The second thesis is that the range of options available to American politicians was limited by the range of consensus within American society. This consensus could be manipulated and influenced. It could shift its emphasis accordingly. But the majority of societal demands during the seventies and the eighties were constant upon two points: keep the U.S.A. strong militarily and go for arms control and cooperation with the Soviet Union. Since in the United States' system of government the society plays the ultimate and decisive role in defining the basic goals of American politics, all American actors whatever their particular interests had to remain within this range of consensus. For the same reason Congress played a greater role in policy making than the executive. Acting as gatekeeper and converter Congress reacted towards societal demands, steered them into the political system and influenced the outputs of the system in such a way that they did not pass over the two threshold values. This function of Congress was all the more important since in the constitutionally prescribed division of power between the legislature and the executive, Congress prevailed to some extent. Of course, in the realm of foreign policy there was mainly presidential rather than congressional government. Nevertheless, Congress wielded considerable and substantial power, financial as well as political, in the field of foreign policy-making.

When the United States withdrew from the Vietnam War they had to realize that the Soviets had changed their military and political position. Having been up to then a Eurasian conventional power the Soviet Union had, by her rearmaments since the mid-sixties, acquired strategic and global capacities. She was on her way to achieve strategic parity with the United States and had developed a maritime capacity which, in addition to the strong air force she had always had, put the Soviet Union into a position to project power globally. Thus, the United States in the mid-seventies faced a new enemy challenging them militarily and politically worldwide. Having reached military parity with the U.S.A., the Soviet Union obviously aspired to political parity as well. The East–West conflict with its traditional center in Europe and the Atlantic region had spilled over into a global competition.

Jimmy Carter was the first American President confronted with this new situation. Nixon and Kissinger, occupied with winding up the Vietnam War, had given only a provisional answer: they had accepted the Soviet Union as the second military power in the world and had given in to military parity. On this basis Nixon concluded the first SALT Treaty and surrounded it with a number of cooperative agreements. The most important one certainly was on the 'Basic Principles' of 1972 which awarded the Soviet Union an element of political parity, too. In political reality, Nixon and Kissinger tried to bind the Soviet Union into a web of agreements of common interest in order to persuade her to abstain from further inroads into the Third World. This concept of 'linkage' in reality meant self-containment for the Soviet Union. For obvious reasons it did not work. It broke down in Angola in 1974, and finally in the Horn of Africa in 1978. Brzezinski was correct in stating that arms control and détente lied 'buried in the sands of the Ogaden'[6].

Therefore, Carter had to try a different approach. He, too, accepted the military parity of the Soviet Union and emphasized the mutual interest in arms control, particularly in the strategic realm. On the other hand, he tried to deny the Soviet Union political parity by starting an offensive for human rights. Carter intended to separate the political conflict with the Soviet Union from the military one and to face Moscow broadly in the center of this political conflict, that is, in the ideology. Human rights being the liberal equivalent to the Communist ideology, Carter tried to demonstrate that the Soviet Union was profoundly illegitimate and that Communism as well as the authoritarian variety of illegitimacy had to be fought by the U.S.A.

Carter failed with his approach for several reasons, but mainly because his approach was not sophisticated enough and collided with the strategic interests of the United States. Brzezinski used this collision to grasp the handle of foreign policy from Vance and, starting in 1978, to steer Carter

---

[6] Zbigniew Brzezinski. *Power and Principle. Memoirs of the National Security Adviser 1977–1981*, New York: Farrar, Strauss, Giroux, 1983, p. 189.

in a more traditional direction, emphasizing again military power as the dominant means to constrain the Soviet Union politically.

The main obstacle to Carter's approach was a different one. If Carter had succeeded with his strategy, American foreign policy would have experienced a profound change. With the Vietnam War ended and arms control on a promising track the necessity for a strong defence and, accordingly, for a big defense budget weakened. Of course, the United States had to remain strong. But without the war and with an understanding with the Soviet Union about a limit to nuclear weapons the defense budget could be lowered considerably. Carter had started his election campaign with a promise to cut the defense budget. This opened the perspective of a redistribution of money, influence and power in the United States. Under Nixon and Kissinger another perspective had shown up. With the U.S.A. and the Soviet Union cooperating in the fields of arms control and politics the relationships shifted towards some kind of 'partial duopol' of the two superpowers which could affect the international system. The chance of cooperation between the two in the Middle East in 1973 alarmed not only the Conservatives in the U.S.A. but also the Israelis. These forces have been united since then and stopped the process of an evolving détente and cooperation by pushing the Jackson–Vanik Amendment through Congress[7].

When Carter came to power the spectre of American–Soviet cooperation had vanished and Carter's human rights campaign succeeded in banishing it completely. But the cooperation in arms control remained and with it the opportunity for the redistribution of values in American politics. The Carter administration consisted mostly of members of the so-called 'equality school', the opponents of the currently dominant 'security school'. The Republican party and the politicians of the post-cold war internationalist brand were out of power and they faced the possibility of remaining out for a long time. If the 'national priorities' of the U.S.A. were changed according to the wishes of those who asked for such a change after the Vietnam War, a real new distribution of power could have taken place.

Thus, on the side of the Conservative opposition political considerations as well as personal interests combined with the goal of keeping this from happening. In 1976 the Committee on the Present Danger was founded. It consisted mainly of members of the old élite who were now

---

[7] Paula Stern. *Water's Edge. Domestic Politics and the Making of American Foreign Policy.* Westport Connecticut: Greenwood Press, 1979, *passim*.

'looking into [the administration] from without'[8]. With the CPD taking the lead, all conservative and neo-conservative interest groups started a campaign to destroy the society's consensus for Carter's foreign policy. The campaign focused upon the arms control issue, claiming that Carter was about to sell out American military strength and to accept the military superiority of the Soviet Union.

This, certainly, was not true. Carter did not make good on his promise to lower the defense budget considerably. In his second press conference, on February 24th, 1977, he stated that he would cut three billion U.S. dollars from this budget. In reality, he slowed the pace of rearmament. His predecessor Ford had asked for 11.8% more for the defense budget. Carter asked only for 11.4%. Comparing the defense planning of Carter with that of Ford it turns out that the Ford budget in 1982 would have amounted to 156 billion dollars and Carter's budget to 147.9 billion. There was a difference, but not a very important one. Neither the Pentagon nor any responsible official argued that Carter's defense budget would undercut American security, or that it would be insufficient even in the face of higher defense spending in the Soviet Union. Nevertheless, the conservative opposition started a campaign of criticism and discrimination. They did what all interest groups do in the United States when they try to change the course of American politics: they influenced the attentive public. And they were successful. The Chicago Council on Foreign Relations found in 1974 that 85% of the general public and 95% of the political élite were in favour of peace and arms control. Three years later, in 1978, the situation remained unchanged. The majority of the political élite and the general public still held the opinion that Carter's politics were sufficient and successful. However, there was one significant change. The attentive public altered its assessment. In 1978 one-third of this attentive public was in favor of stepping up the defence budget, 20% more than in 1974. The number of those members of the attentive public who spoke in favor of a reduction of defense spending was cut in half, from 32% in 1974 to 16% in 1978.

While the élite and the general public stuck to their assessment that American defense spending was sufficient and arms control necessary, the attentive public began to deviate. This is astonishing. The political élite consists of government officials, members of Congress, business leaders and intellectuals. They, above all the government officials, usually have the best information about the international situation. Why did the

---

[8] Charles Tyroler, II (ed). Alerting America. The Papers of the Committee on the Present Danger, McLean, 1984, p. X.

attentive public which is usually well informed, but not so well informed as the élite, take a different course? It was the Chicago Council on Foreign Relations itself which put the question, 'Where do the informed get the information that the United States have fallen back behind the Soviet Union and where do the élites get the information that this is not the case?'[9]. Although the Council could not answer it, the Council raised the question whether this different assessment was caused by a different ideology or by different information. Of course, nobody could answer this question. But the Council found out that the one-third of the élite which did not share the positive assessment of the other two-thirds was predominantly made out of special interest groups (100%), special foreign policy interest groups (87%) and members of Congress (79%). In Congress itself, 45% of the members belonged to the diverging group.

If a clear-cut conclusion is impossible, the assumption is safe that this one-third of the élite, with the general and special interest groups taking the lead, was successful in influencing the attentive public, who received and accepted the critical information spread by this part of the élite. To put it more bluntly: the Committee on the Present Danger and the other pertinent interest groups succeeded in shifting the attitude of the attentive public from arms control and détente back to rearmament and confrontation.

The growing opposition of this group made it more and more difficult for Carter to stick to his moderate change in the defense budget and to arms control. Although the defense spending of the Soviet Union had not changed in this period – as a matter of fact the CIA expected that the Soviet Union in 1976–1977 would have concluded a cycle of rearmament and would slow its defense spending somewhat[10] – the administration was forced by growing opposition to step up the rearmament of the United States. The 'B Team' challenged the analyses of the CIA.

In the administration Zbigniew Brzezinski sided with the conservative opposition, played the 'China-Card' and persuaded Carter in his famous speech at the Notre Dame University in June 1978 to demand that the Soviet Union should choose between confrontation and cooperation. In resuming diplomatic relations with the People's Republic of China at the end of 1978 the Carter administration consciously damaged relations with

---

[9] Chicago Council on Foreign Relations. American Public Opinion and U.S. Foreign Policy 1979, p. 30.
[10] United States Congress, Joint Economic Committee: Allocation of Resources in the Soviet Union and China – 1977, Part III. Washington, D.C., 1977, p. 17.

the Soviet Union. As a consequence, Moscow hardened its position, slowed down the pace of negotiation for the SALT-II Treaty and weakened its confidence in the sincerity of the Carter administration.

In the United States the conservative opposition did not react towards this new Soviet position. It reacted towards the intentions of Jimmy Carter to sign the Arms Control Treaty. The opposition stepped up its public activities. On the eve of the Vienna summit Senator Jackson blamed the Carter administration for its 'appeasement' of the Soviet Union; General Edward L. Rowny, the representative of the Joint Chiefs of Staff at the American Arms Control Delegation, resigned demonstratively. Carter had to give in. ACDA Chief Warnke had to resign in October 1978. In January 1979 President Carter submitted to Congress an additional defense budget which neutralized all the cuts and brought the real growth of defense spending to 3.1%. Before leaving for Vienna Carter ordered the construction of the MX-missile which he had postponed until then. In December 1979 the Carter administration went even further. It submitted at this early stage its defense budget for 1981, and this was 5.6% higher than the previous one. In other words, in order to save the Arms Control Treaty with the Soviet Union President Carter had to acknowledge the contrasting views of the Conservative opposition.

It is possible that these numerous concessions could have permitted the ratification of the SALT-II Treaty. But it was Ayatollah Khomeini who took 52 American diplomats hostage and Secretary-General Brezhnev who intervened four weeks later in Afghanistan, who gave the final blow to the SALT-II Treaty. Carter cancelled more or less all economic and political cooperation with the Soviet Union and pronounced in January 1980 the Carter Doctrine and the abandonment of the Nixon Doctrine.

The conservative opposition in the U.S.A. felt justified by the acts of the Ayatollah and Brezhnev. Everything it had envisaged had come true. The Soviet Union had demonstrated its aggression and its readiness even to trespass beyond the geographical boundaries which it had kept so far. The Ayatollah had demonstrated that the United States had fallen from its hitherto dominant power position. President Carter was blamed for this deterioration in the standing of the United States. He had sold out American power, had fallen into the Soviet trap of détente and had neglected to rearm the United States sufficiently.

The arguments sounded very convincing and they certainly convinced the political strata in the United States. A more differentiated analysis, however, puts these conventional assumptions in doubt. First: compared with Iran the United States certainly was a superpower commanding a vast array of military means which could have destroyed Iran completely. The power of the Ayatollah did not derive from American military weakness, but from the fact that the meaning of power had changed its content. In dealing with societal revolutions in a period of growing democratization military power had lost its usefulness. The American humiliation was not the consequence of military weakness but of an outmoded foreign policy. It was oriented towards the world as a world of states not as a world of societies.

Second: the invasion of Afghanistan by the Soviet Union also had nothing to do with American power. By no stretch of the imagination could the United States have kept the Soviet Union from invading Afghanistan. Afghanistan had belonged for decades to the Soviet sphere of influence and the decision to intervene was a purely Soviet one. We do know better today than the U.S.A. could have known in 1979 that the decision was made within a small circle of old-fashioned Communists. It was a mistake from the outset, as the U.S.A. could have known from its own experiences in Vietnam. Whatever the Soviet reasons for the intervention had been they certainly had nothing to do with the U.S. defense budget.

It is not unhistorical to suggest that those insights could have prevailed in the United States in 1979–1980. Secretary of State Cyrus Vance had tried very hard, but unsuccessfully, to spread interpretations of this kind. He advocated the ratification of the SALT-II Treaty because he saw it as independent from the Soviet invasion in Afghanistan. He criticized the ill-fated rescue mission in Iran because it would not solve the real problem. The Secretary resigned after that event, but his analyses were correct and prove that it was possible at that time to make them. In other words, the changing course of the American foreign policy after 1980 was not forced upon the U.S.A. by events in the international environment. The course was changed because of deliberate decisions made by the conservative opposition and then followed by the Carter Administration. Under the influence of Brzezinski, the Carter Administration probably had no alternative. American public opinion was stirred up from day to day by the media which pointedly commented upon the humiliating hostage crisis. The Conservative opposition continued its campaign for a stronger rearmament and for a return to more confrontational politics. It consciously destroyed whatever consensus had remained with regard to Carter's foreign policy.

There were many reasons to criticize the Carter Administration for its lack of accomplishment in the field of the economy, particularly of unemployment and inflation. But there was no reason to criticize its foreign policy. Even in 1980 the political élite, including the military branch, were of the opinion that the American defense posture was sufficient and in no way inferior to that of the Soviet Union. They all agreed that the SALT-II Treaty was in the overall interest also of the United States. But the attentive public, and in 1980 and 1981 public opinion in general, accepted the conservative interpretation. Both fell victim to the campaign for rearmament. The Committee on the Present Danger was shrewd enough to concentrate its campaign upon the arms control approach in general and upon the necessity to enlarge the defense budget considerably. It did not relate this claim to Carter's foreign policy in general or with regard to Iran and Afghanistan in particular. It tried to exploit the societal demands for military strength to the detriment of the second demand for détente and arms control. The campaign was successful in so far as more than half of the American population in the election winter of 1980 opted for more defense spending. This attitude, to some extent, contributed to the victory of Ronald Reagan. But the conservative opposition was not successful in eliminating the second societal demand for ongoing arms control cooperation with the Soviet Union and détente. This goal was very strong and proved to be the driving force which caused the Reagan Administration after 1983 to change the course of American Soviet policy again.

The shift of public opinion in 1979–1981 towards more arms and less détente was more or less homemade. It was not required by the behavior of the Soviet Union or the Ayatollah; the Europeans demonstrated that it was possible, and perhaps useful, to react differently. In the U.S.A. the conservative opposition and its foreign policy interest groups drummed up public opinion in order to benefit their own political interest. The conditions for this success can be found in the American system of government. With practically no parties effective on the federal level and with a high degree of freedom for every potent actor, political groupings and interest groups could use the vast array of media to stir up American society. It changed its demands accordingly, strengthening the Conservatives within the two houses of Congress. The arms build-up accelerated and candidate Ronald Reagan was able to base the foreign policy part of his campaign upon the demand for more military strength.

When the Reagan team took over in January 1981 and 60 leading members of the Committee on the Present Danger entered the Reagan Administration, the public campaign came to an abrupt end. Public opinion shifted back to its original distribution. While in 1981 65% of all

Americans asked for more arms, in 1982 it was only about 30%, the traditional percentage[11]. The shift had been the result of a special campaign by the conservative opposition, not the consequence of a different assessment of Soviet arms policy. The campaign proved the effectiveness of a pointed information strategy which, via the attentive public, can reach larger segments of American society. The lack of political parties which could disseminate a more adequate and constant information about the exact situation made it possible for interest groups and Political Action Committees to take over. They succeeded temporarily, as long as their campaign lasted. When this additional and artificial input ceased the demands of the American society returned to their traditional distribution.

As American President, Ronald Reagan proved once more that his call to arms was not caused by a change in Soviet behavior. As a matter of fact, it had nothing to do with the Soviet Union. Immediately after entering the White House Reagan asked for 6.8 billion dollars more for the defense budget of 1981 and 25.8 billion dollars more than Carter had asked for 1982[12]. Interestingly enough, Secretary of Defense Weinberger was unable to give any justification for his demand. He could not even tell what should be done with this additional money. The Reagan Administration had no organized concept whatever for its policy towards the Soviet Union. Secretary of State Alexander Haig developed a 'strategic concept'. But it was valid only for the Middle East and totally unsuccessful there. It took the Reagan Administration more than two years to develop a reasonably coherent strategy towards the Soviet Union. Secretary of State George Shultz presented this concept in June 1983 to the Congress, at a time when the Reagan Administration was on the point of revising it[13]. Of course, there had been some new strategic concepts which could be understood as parts of a different attitude. Secretary of Defense Weinberger developed the idea of a 'horizontal escalation', and President Reagan himself developed in 1983 the perspectives of the Strategic Defense Initiative. From the outset Reagan went for a 600-ship-navy. The possibilities inherent in 'horizontal escalation' as well as in SDI alarmed the West Europeans without enlightening them about the meaning of

[11] William Schneider. Conservatism, not Interventionism: Trends in Foreign Policy Opinion, 1974–1982, in: K. A. Oye *et al.* (eds). *Eagle Defiant. United States Foreign Policy in the 1980s.* Boston: Little Brown and Company, 1983, p. 55, 36.

[12] Wireless Bulletin 44, 5.3.1981, p. 19ff.

[13] Wireless Bulletin 108, 15.6.1983, p. 1ff.

either concept. Horizontal escalation was finally dropped in 1984, and in 1985 Reagan delivered some details about his SDI idea. Only the 600-ship-navy could be related to a strategic concept. With the Soviet Union obviously trying to expand her influence into the Third World it could be worthwhile to augment the American capability for power projection.

Neither the absence of a detailed analysis of Soviet military and foreign policy nor the lack of a coherent foreign policy concept kept the Reagan Administration from asking for enhanced defense spending. From 1981 to 1983 the Pentagon got 150 billion dollars more than in the last year of the Carter Administration. Within the first Reagan Administration the defense budget was stepped up by more than 50%, climbing from $178 billion 1981 to $286 billion in 1985. The increase was tremendous because it came on top of the increases President Carter had already been forced to order. Compared with the previous Carter defense budget of 1978 which amounted to 123.5 billion dollars the Reagan budget of 1985 with its 280 billion dollars more than doubled defense spending of the U.S.A.

If Ronald Reagan had failed to give any substantial reason for this rise in the defense budget – were there any meaningful results of augmenting American military strength to such a degree? It is difficult to tell. Of course, Ronald Reagan ordered the B-1 bomber. He ordered the production of the controversial MX-missile. He invented SDI, as mentioned; and he increased the pay for the soldiers considerably. He asked for two more aircraft carriers. But that was it. What else happened under the Reagan Administration was the result of previous decisions and production plans.

Thus, the question is pertinent why President Reagan intended to spend so much more money for purposes which could not be defined and with results the additional value of which could not be measured[14]. The question is all the more important since the United States paid heavily for Ronald Reagan's defense spending. Cutting taxes with the Economy Recovery Tax Act of 1981 and doubling the defense expenditures necessarily led to a staggering budget deficit. In 1980 the budget deficit was 40.2 billion dollars, in 1985 it was 212.3 billion. The gross debt doubled from 914.3 billion dollars to 1827.5 dollars, and in 1989 it was 2112 billion. This was a peculiar achievement for a president of the Republican party, for which the lowering of federal debt and a balanced budget were traditional sacred cows.

[14] United States Congress, Congressional Budget Office. Defense Spending: What has been accomplished? Staff Working Paper, April 1985.

It is not easy, but not impossible, to explain Reagan's fiscal and defense policies. As could be seen from his political education Ronald Reagan was interested in the redistribution of power in the U.S.A., in cutting the welfare-state and restoring the traditional conservative values and their distribution. The 'bushfire' which led his convictions burned in the domestic, not in the foreign, policy domain. Reducing the welfare state, undoing the Great Society and, if possible, weakening the New Deal demanded a big stick. Probably for this reason Ronald Reagan tried to square the circle. By lowering the revenues of the federal government and expanding its defense expenditures Ronald Reagan created a fiscal gap which, in his view, could be filled only by giving money from the welfare to the defense budget. In the first two years Congress went along to some extent with this plan, cutting some parts of the Great Society Program. But the vast majority of this program remained intact and so did social security. Reagan's revolution was stopped midway, but Reagan did not give in. Until his last budget proposal in 1988 he pressed Congress to give more money to defense and less to social security. The President preferred to incur a huge budget deficit to changing his political goals. And since Congress did the same, the deficit grew. In his policies towards the Soviet Union Ronald Reagan had sacrificed more or less everything he had believed in in 1980. But he stuck doggedly to his conviction that social security had to pay for military security.

Of course, in his foreign policy Ronald Reagan always had been a hardliner. He was, on the other hand, worried about nuclear weapons and their destructive capabilities. He was also worried about the human aspects of war and warfare. But he believed in military strength, having already in 1976 forced his political opponent Gerald Ford to replace the term 'détente' with 'peace by strength'. Ronald Reagan was profoundly skeptical with regard to arms control, not to speak of disarmament. He had to be pushed in this direction, first by the society and the freeze movements of 1983, then by the perspective that he could lose the elections of 1984 if he did not react towards the societal demand for arms control and some kind of cooperation with the Soviet Union. Having spent so much on armaments Reagan certainly could not but convert this societal demand at least into declared policies. A change in political strategies was also recommended by the fact that American business objected more and more to the strong and comprehensive controls of exports to the Soviet Union. The 'Commission on Industrial Competitiveness' claimed that those controls, enhanced considerably by the Reagan Administration, amounted to a loss of more than 11 billion dollars for the American economy. In the Jewish community of the U.S.A. the insight grew that the turn against détente probably had not been in their particular interest.

All this amounted to an emerging new constellation of societal forces and demands which were not in tune any more with the confrontational politics of Ronald Reagan. Therefore, he changed his policies towards the Soviet Union, slowly and mostly verbally, but publicly. The change had started already in mid-1983, was interrupted by the shooting down of the KAL airliner by Soviet fighters, but was fully implemented by Reagan's speech on January 16, 1984. He ended the political isolation of the Soviet Union, reopened the dialogue with Moscow and described the new political strategy of the United States as 'realism, strength, and dialog'. It is important to note that this change was made a couple of weeks after the Soviet Union had broken up the arms control deliberations in Geneva and hardened her position. Ronald Reagan's change of strategy was not caused by changed Soviet behavior. On the contrary, the reasons lay exclusively in the domestic political scene of the United States.

The change was considerable. Secretary of State Shultz, in a lecture in Los Angeles, October 1984, described Washington's New Thinking. It abandoned Carter's punishment of the Soviet Union for the Afghanistan intervention as well as even Kissinger's linkage politics. If cooperation with the Soviet Union was in the American interest, it should not be sacrificed for some ethical or political reason. This brand of 'Realpolitik' was a far cry from the politics of confrontation and isolation which Ronald Reagan had pursued during the first two years of his administration. Where the New Thinking in American policy would lead to remained to be seen. But it was obvious that this New Thinking had broken profoundly with the strategic approaches of the first three years. As a consequence, American–Soviet cooperative interaction grew considerably during 1984. The 'hotline' between the two capitals was modernized in July 1984, followed by the American lifting of the 1980 boycott of Soviet fishing rights. The agreement on economic, industrial, and technical cooperation was prolonged for ten years, and the agreement on avoiding accidents at sea for three years. The President himself delivered those and other proposals to a Washington conference on U.S.–Soviet exchange[15]. In September 1984 President Reagan met with the Soviet Foreign Minister Gromyko for the first time in four years. In November of that year the United States and the Soviet Union agreed to

---

[15] Wireless Bulletin 120, 28.6.1984, p. 1ff.

resume their arms control deliberations with a meeting of the two foreign ministers on January 7 and 8 1985 in Geneva.

In other words, the change in Reagan's policy towards the Soviet Union was a consequence of the New Thinking in Washington, and not of the appearance of Gorbachev in Moscow. When Gorbachev came to power on 11 March 1985 the changing American policy towards the Soviet Union was already in full swing. It was the product of the demands of American society, reflected in Congress, which had returned to the traditional duality of military strength and arms control and cooperation. In order not to lose contact with American society – and the election as the consequence thereof – Reagan had to adapt his administration to societal demands in the U.S.A.

It is not until this point that Secretary-General Michail Gorbachev has to be considered. When he took over in Moscow the change in the U.S.A. had already occurred. Even if somebody else had gained power in the Soviet Union the United States would have pursued the new (and old) policy mix of military strength and arms control. Up to March 1985 the American analysis of the Soviet Union was stable and invariant. Moscow was rearming slowly and continuing its political expansion into the Third World. It was the American strategy which changed and for obvious domestic reasons. It was possible that Reagan had changed only his tactics, avoiding the harsh rhetoric and the blunt criticism of arms control. The point can be made that Ronald Reagan in 1984 to 1985 had only changed the outlook of his Soviet policy, not its contents. He had resumed arms control and some kind of cooperation with Moscow because American society and American business were so much in favor of this policy. There was, in addition, in the Reagan Administration a wing of doves sitting mainly in the State Department and in the military branch of the Pentagon. It is also true that many actors after having entered the Reagan Administration became sober and analytical about the real American interest in arms control. Paul Nitze certainly is a case in point[16]. It is safe to assume, therefore, that the Reagan Administration would have continued their policies formulated in 1984–1985 even if Gorbachev had not taken over in Moscow.

It is also safe to state that Gorbachev substantially changed the international environment of the United States. He did two things and in two stages. First: by making one concession after another Gorbachev trans-

---

[16] Strobe Talbott. *The Master of the Game. Paul Nitze and the Nuclear Peace.* New York: Alfred A. Knopf, 1988.

ferred nuclear arms control into disarmament without any chance for the United States to escape this process. Second, and more important: Gorbachev's policies of *Glasnost* and *Perestroika* changed not only the Soviet Union but led to the unfolding of a slow-motion revolution in Poland, Hungary and the GDR which in 1989 practically dissolved the Warsaw Pact. It is President George Bush who has to deal with this completely new situation which will have profound impacts on American policy-making in general. Ronald Reagan only had to deal with the drive for arms control and disarmament which Gorbachev initiated. This drive interacted with the societal interests of the United States and led to the Rejkjavik summit and to the INF Treaty.

In 1986 President Ronald Reagan was by no means ready for any kind of arms control treaty. On the contrary, he still tried to pacify the societal – and congressional – demands for arms control on the declaratory level and at the same time to go ahead on the operational level with arms build-up. In January 1985 Secretary of Defense Weinberger asked, albeit unofficially, for another doubling of the defense budget during the coming five years. The Congress blocked this initiative and froze the defense budget at the 1985 level, at roughly 285 billion dollars. Reagan asked for 3.7 billion dollars for his SDI. On 6 October 1985 National Security Adviser McFarlane disclosed for the first time the so-called wide interpretation of the ABM Treaty. In 1985 Reagan was successful in pushing Congress towards the production of binary chemical weapons. In January 1986 the New Maritime Strategy was published indicating a bold offensive strategy against the Soviet Navy and homeland if war should occur. The capacity of the American navy for power projection was strengthened. Whatever the reasons given by the Reagan Administration for this massive arms build-up, the real intention obviously was to create a long-term commitment which could not be abandoned in the near future or by any successor of Ronald Reagan.

This massive rearmament clashed with the arms control rhetoric of the administration. If Reagan was put to the test, as was the case with sending the Threshold Testban Treaty (TTBT) and/or the Peaceful Nuclear Explosion Treaty (PNET) to the Senate for ratification, he baulked. But now, in 1986, he met with the resistance of Congress. The Legislature appropriated the money for 50 MX-missiles and 3.2 billion dollars for SDI in 1986 only on the condition that arms control would go ahead. In the summer of 1986 Congress passed four amendments intended to force President Reagan to adopt arms control measures in the field of ASAT, chemical weapons and nuclear tests, and keep the United States within the limits of SALT-II[17]. Reagan had made it known on 27 May 1986 that he would trespass beyond these limits and deliberately did so. This was more than Congress was ready to tolerate. The four amendments put so much pressure upon the President that he – under additional pressure from the evolving Iran–Contra scandal – had to move. Obviously under the influence of his wife Nancy, he decided for an audacious step forward. Within ten days he arranged with Gorbachev the 'non-summit summit' in Rejkjavik. He overtook, so to speak, Congress on the left and regained the initiative.

If the decision to go to Rejkjavik was a purely American one, the outcome of this summit was more or less due to Gorbachev's initiatives. They met an unprepared President who proved nevertheless to be ready to go along with disarmament in the field of INF and with the abolition of offensive strategic weapons. After a short period of confusion Gorbachev opened the way to the INF Treaty by making several concessions. The treaty was completed in December 1987 in Washington and enacted in June 1988 in Moscow. The idea of eliminating all strategic weapons within the coming ten years was swept under the carpet. The goal to cut the strategic arsenal by half, however, is being pursued by both parties in Geneva. Many people expected the completion of a treaty already in 1988. Now it should be the Spring of 1991.

The details of these accomplishments and non-accomplishments are not of interest here[18]. What is important to note is that since 1986 Secretary-General Gorbachev and his new leadership has been affecting American foreign policy decision-making. Up to then the United States was more or less independent from Soviet decisions. They were clumsy, invariable and could be counted upon. In other words, they could be taken for granted. This gave a lot of leeway to the American actors who could pursue their domestic and personal interests without taking the outside world into consideration. Of course, there were the European Allies, but they were friendly and did not go public. Under those circumstances the United States enjoyed a degree of independence which added to the impression that the United States was the master of her destiny. What she decided was exclusively up to her with nobody else interfering. The Soviet Union acted as the enemy challenging the U.S.A. and thereby

---

[17] Dante B. Fascell. Congress and Arms Control. In *Foreign Affairs*, **65 (4)**, Spring 1987, p. 737.

[18] Michael Gordon. Dateline Washington, INF. A Hollow Victory?, in: *Foreign Policy*, **68**, Fall 1987, p. 159ff.
James P. Rubin. START Finish. In *Foreign Policy*, 76, Fall 1989, p. 96ff.

justifying (or at least giving legitimacy toward) any measure which American actors believed to be in their interest. Whether Brezhnev or Andropov reigned in Moscow was of no particular interest since it did not affect the function of the Soviet Union as the enemy of the United States. Hardliners in the United States always could count upon their counterparts in the Soviet Union who, behaving in a similar fashion, made any compromise in the field of arms control difficult and improbable. As Undersecretary of State Eagleburger said in 1989 the situation up to the mid-eighties was stable and foreseeable.

All this changed with Gorbachev coming to power. The unexpected offers he made in the field of arms control indicated already the profound change he was bringing to the Soviet Union. For reasons unknown to the West in detail he substantially changed the course of Soviet world politics. Gorbachev withdrew from Afghanistan and from Angola. He lowered his help for Nicaragua and urged Vietnam to withdraw from Cambodia. He diminished the number of Soviet troops along the border with China and improved general relations with the People's Republic. He accepted the principle of asymmetrical arms reductions in Europe and began to implement the wide ranging proposals he submitted to the United Nations in October 1988. What Brezhnev had done was undone by Gorbachev. The Soviet Union really retreated from power, offering Ronald Reagan the victory of his doctrine. Reagan's goal to drive the Soviet Union out of the Third World, announced on March 14, 1986,[19] was enacted by Gorbachev who drove out by himself. In 1988 five wars, including that between Iran and Iraq, came to an end putting the world on the brink of peace. At the end of his second administration Ronald Reagan could argue convincingly that he had made good his two basic campaign goals of 1980: to accomplish arms reduction and disarmament in the field of nuclear weapons and to free the world from Communism.

As a politician Ronald Reagan was entitled to claim this victory. The analyst has to state that this victory was not gained by Reagan but given by Gorbachev. As a consequence of 'New Thinking' in Moscow Gorbachev diminished the challenge towards the United States. He did what no other enemy had done in history: he gave in, at least apparently.

It is safe to state that without Gorbachev in Moscow there would have been neither an INF Treaty nor the optimistic perspectives of conventional disarmament in Vienna and of the reduction of strategic weapons in Geneva. In other words, for the first time an outside actor intervened in

[19] United States Policy Information and Texts 39, 17.3.1986, p. 9ff.

the foreign policy decision making process in the United States in such a way that the impact was substantial. The United States could not but accept what Gorbachev was so ready to offer.

What is more, if Gorbachev continues to act more as a cooperative competitor than as an enemy or a rival the impact upon American world politics will be even greater. For 40 years policies were directed against the challenge by the Soviet Union, were oriented towards the goal of remaining the leading world power by denying militarily, at least political, parity to the enemy. If now the enemy stops behaving as such, if he does not ask for parity in the field of arms or politics, the United States faces a challenge different in kind, but great in magnitude. She must redefine all facets of her world politics, must find a new basis upon which to base her position as the world's leading country.

With the Cold War won by the West, George Bush experienced a rather easy victory in 1988. As far as the balance sheet of foreign policy was concerned it was a very positive one. Ronald Reagan delivered to his successor a world which was more or less at peace and full of hope for further disarmament treaties. But in 1989 Gorbachev's New Thinking in Moscow produced consequences which probably had been unintended and unexpected. *Glasnost* and *Perestroika* spread over to Eastern Europe and led to revolutions in Poland, Hungary, the GDR, and Czecho-slovakia. The Communist governments fell from power. The outcome differed in the different countries, but the direction was always the same: towards some kind of democracy and market-oriented economies. It is tempting to speculate about the reasons for those events. At least they give evidence for Seeley's law according to which the degree of freedom in one country is inverse to the degree of external pressure upon its frontiers. At the moment the Soviet Union alleviated the conflict with the United States, the iron brackets around the Warsaw Pact countries broke down, permitting reform and revolution to take place in Eastern Europe.

The consequence for the United States is that, militarily speaking, the Warsaw Pact does not exist any more. Accordingly, the East–West conflict has vanished. Of course, there still is a conflict with the Soviet Union. But in the case of the Eastern European states conflict has been replaced by cooperation with – more or less – western-oriented countries.

This is of tremendous consequence for the world politics of the United States and her pertinent decision making processes. With the East–West conflict there disappears the basis upon which the U.S.A. had built her position as the leader of the free world. With the conflict vanished the glue holding together Nato and the Atlantic Community is being softened. We have to be very specific here. It is not the communality of interest that is touched upon but the organization of that communality. For too many years the United States and Western Europe have stuck to the military alliance of Nato as the organizational form within which to discuss and decide political questions. When the Warsaw Pact disappears the necessity for Nato will be reduced immensely. It can be kept in reserve, of course, because there still is the Soviet Union. But not many people will value such a reserve if, by Soviet invitation, the 'Common European Home' has been constructed. The Atlantic Community urgently needs a new organization securing the relations in the field of politics, economics, and security for the coming decades.

What is more important here is that the United States needs new world policies. As has been demonstrated, the impression that the United States is still independent enough to make foreign policy decisions according to the domestic constellation of forces was dependent on the constant function of the Soviet Union as an enemy. Within the context of the East–West conflict the United States could always balance its contribution towards the security of the West (and the world) with its deficits in the field of the economy. With the view of the Soviet Union as an enemy the United States remained the dominant superpower indispensable for containing Communism and the Soviet Union. If the relevance of this contribution is diminished because of the dissolution of the Warsaw Pact and the retreat of the Soviet Union from expansion, the economic deficits of the United States will come to light. They will demonstrate that also economically, the United States is by no means as independent as she pretended to be. Foreign trade, in the sixties producing only 8% of the GNP, now amounts to 20%. Economic relations with the outside world have become much more important for the United States, having an impact upon unemployment and the labor market. Protectionism is on the rise in the U.S.A., although it is very well known that the trade deficit which amounted to 119 billion dollars in 1988 is mainly the product of imbalance between U.S. domestic savings and investment. The United States long ago lost her position as the main investor in the world; she now receives most of the investments of the developed world, with Japan as the most important investor. Foreign direct investment in the United States climbed from 25.2 billion dollars in 1981 to 58.4 billions in 1988[20].

[20] James K. Jackson: U.S. Trade Restraints: Effects on Foreign Investment, CRS-Report for Congress 89–447 E, Washington, D.C., 4.8.1989, p. 4.

Foreigners now own 6% of all investments in the U.S.A[21]. If foreign capital suddenly withdrew from the U.S.A. the economy would suffer dramatically. It is no exaggeration to state that the budget deficit of the United States is financed mainly by Japanese capital, followed by British and Dutch money.

There are many more indicators of growing American dependence. In the manufacturing sector the share of assets controlled by foreign firms reached 12% in 1986. In the chemical sector foreigners own 33%, in the sector of stone and clay their share is 22.8%[22]. Before the mid-eighties all these dependencies and interdependencies were veiled by the unique security umbrella with which the United States protected their partners and allies against the Soviet threat. In the same vein the main contribution of the United States towards world order was in the military field, with American ships and bombers safeguarding the sea lanes and containing any intruders. With the threat now gone (Iraq will be no substitute) the United States will be measured in all fields of politics separately. Their contribution to security will still be valued highly, but this cannot compensate for deficits in other fields. In the Atlantic Community, for instance, the economic competition between the U.S.A. and the European Community was dampened and restricted by the American contribution to the military security of Western Europe. With this contribution weakened the economic competition will be enhanced. At the same time, American political influence in Western Europe will be diminished. It was implemented more or less via the military alliance and the American military presence. When both have been weakened, so will the American influence.

The worldwide cooperation against Iraq's invasion of Kuwait is the exception which confirms the rule. The United States reacted quickly and firmly but she was well advised to wait for the UN Security Council to define the threat and the sanctions. And Washington had to ask its Western allies to pay for the American service. There may be more conflicts of this type to come which will justify a strong American and Western intervention force, although it is open to question whether a gunboat diplomacy will solve any problems these days. What is more important in the context of this article is that conflicts of this type will not replace the East–West conflict with its global ordering function.

Of course, the end of the East–West conflict, the future 'beyond

[21] James K. Jackson: Foreign Ownership of U.S. Assets: Past Present and Prospects, ibid. 89–458 E 1.7.1989, p. 19.
[22] Ibid., p. 5.

containment', frees the United States from many burdens. Washington is free now over other issues, especially economic ones, and in other world regions, predominantly Asia and the Pacific Basin. Washington will see a vast array of tasks waiting there without waiting for the United Nations. In Asia it will be Japan, and after that India and China competing for influence, not to forget the Soviet Union. In Africa and even in Latin America it will be the European Community striving for preponderance. The world 'beyond containment' will be a better, a safer, but a much more complex world.

For the United States this world will not bring the 'great boredom' which Fukuyama envisaged. On the contrary, it will bring new and great challenges. The United States must develop a new system of leadership relying more on political and economic competence than on military strings. This demand does not follow solely from the changed international environment, from the world without the East–West conflict. It will also be raised by American society which will not tolerate huge defense budgets if they are not necessary. Societal demands will ask for the 'peace dividend', for a new orientation of American foreign policy, even a new system of decision making. Is it necessary and/or pertinent to have huge armed forces and a big defense industry,if there is no huge and big enemy left? Is it, on the other hand, sufficient to stick to the old national security decision-making system with its dominance by the military and its complete lack of economists and sociologists? Shall the old security school which regained its power with Reagan continue to rule the country or shall the equality school get its turn again in the new world?

The whole texture of American politics is at stake. It is not protected any more by the existence of a longstanding conflict and a persistent enemy. The shelter has gone, the international environment is open and in constant flux. The political system of the U.S.A. must demonstrate to society that it is capable of handling the new environment and to fulfil societal demands under these new circumstances. George Bush is the first American president to face this new situation, his administration the first which has to produce a new equation between the different demands of American society and the divergent possibilities and constraints of the new world emerging.

Note

This paper condenses some aspects of American foreign policy towards the Soviet Union which I have extensively analyzed in my book: *Machtprobe. Die USA und die Sowjetunion in den achtziger Jahren*. München: C. H. Beck, 1989. Since the book gives the necessary empirical evidence I restrict myself here to a few notes only. It is a pleasure to appreciate once more the support given to me by the Stiftung Volkswagenwerk, the Deutsche Forschungsgemeinschaft and the Land Hessen during the academic year 1987/88. Without this help the book could not have been written.

Name index

Numbers in **bold type** *indicate bibliographical references*

Abrams, 285
Adams, J. S. **184**
Adorno, T. 298
Alexander, R. D. **25**
Alstatt, L. **146**
Anne, Princess Royal xv
Aristotle **171**
Arps, K. **145–6, 158**
Asai, M. **88**
Attlee, Clement 301
Austin, W. G. **300**
Axelrod, R. 22, **25, 48**, 279, **280**

Baker, James 323, 325
Baker, N. J. 233, **237**
Batson, C. D. **134**, 143–4, **145, 146**, 151–2, 155, **158**
Baumann, D. J. **145**
Baumrind, D. 67–8, 70, 71, **76**
Beaman, A. L. **145–6, 158**
Beck, A. T. 220, **223**
Begin, Menachem 276
Beliaev, I. x, 267, 316–28
Bercovitch, F. 24
Berkowitz, L. 141, **145, 158**
Bernzweig, J. x, 51, 54–76
Berscheid, E. **184**
Billig, M. G. 240, **261**
Blum, Leon 301
Bolen, M. H. **134**

Bond, M. H. 93, **104**
Bontempo, R. **88**
Boon, S. D. x, 160, 187, 188, 190–211, **211**
Bowlby, J. 195, **211**
Boyd, R. x, 4, 14, 27–48, **48**
Brewer, M. B. 240, 252–3, **261, 300**
Brezhnev, Leonid 338
Brickman, P. **184**
Briggs, J. L. 119, 120–2, **127, 128**
Bronfenbrenner, U. 89, **104**
Brown, J. L. 18, 26
Brown, R. 285, **300**
Brown, S. 279, **280**
Brzezinski, Zbigniew 334–5, **337**, 339
Buchanan, W. **146**
Bundy, R. P. 240, **261**
Bush, George 173, 346, 349, 352
Bygott, D. 17

Campbell, 299
Campbell, D. T. 41–2, **48**
Caro, T. 24–5
Carter, Jimmy 276–7, 334–40
Cartwright, D. 259, **261**
Cavalli-Sforza, L. L. **48**
Chammah, A. M. 232, **237**
Chapais, B. 19, 23
Chapman, M. **77**, 94, **104**
Chen, K.-S. 97, **104**

Cheney, D. 23, 25
Christie, R. 227, **236**
Cialdini, R. B. x, 132, 135–45, **145–6**,
   147, 149, 152, **158**, 160, 162,
   164
Cicipio, Joseph 325
Clack, F. L. **88**
Clark, R. D., III **146**
Clayton, S. x, 133, 173–84
Cline, Ray 325–7
Cohen, R. **184**
Colby, William 325–7
Collins, M. E. 269, **280**
Colman, A. M. 228, **236**, 241, 244,
   247, **261**
Colvin, J. 24
Crosby, F. J. **184**
Cross, J. A. **134**
Cruickshank, J. 271–2(n), **280**
Cummings, E. M. **77**
Czempiel, E.–O. xi, 267, 329–52

Darley, J. M. **134**, 199, **211**
Davies, N. B. 21, 22, **26**
Davis, K. E. 149–50, 151, 153, **158**
Dawes, R. M. 48, 252, **261**
Denton, R. K. **128**
Dessler, D. 330(n)
Deutsch, M. 191, 196–7, 198, **211**,
   229–30, **236**, 243, 245–6, **261**,
   269, 270, **280**, 296
de Waal, F. B. M. 22, **26**
Doise, W. 52–3, **53**
Doolin, D. J. 95, **104**
Doumanis, M. **87**
Dovidio, J. F. 146, 152, **158**
Draper, P. **128**
Driver, M. S. 155, **158**
Dunn, J. 5, **7**

Eagleburger, Lawrence 348
Easton, D. 331
Edwards, C. P. 109, **128**
Eisenberg, N. xi, 14, 51, 54–76, **76**, **77**,
   89, **104**
Eiser, J. R. **236**
Emlen, S. T. 18, 20, 21, **26**
Erikson, E. H. 195, **211**
Essock-Vitale, S. M. 13, **14**

Fabes, R. A. xi, 51, 54–76
Fascell, D. B. 347

Fazio, R. H. 199, **211**
Feger, H. xi, 171, 266, 281–300, **300**
Feldman, M. W. **48**
Fesbach, N. D. 2, **8**
Fischoff, B. 234
Fisher, R. 270, 276, 279, **280**
Fitzpatrick, J. 20
Flament, C. I. 240, **261**
Folger, R. **184**
Follett, M. P. 270, **280**
Ford, Gerald 336, 343
Fortenbach, V. A. **146**
Fukuyama, F. 329, 352
Fultz, J. xi, 132, 135–45, **145–6**, 147,
   158, 160, 162, 164

Gaertner, S. S. **146**
Gambetta, D. 3, 5, **7**, 235, **236**
Gartrell, C. D. **184**
Gatewood, R. 155, **158**
Geis, F. 227, **236**
Gergen, K. J. 153, **158**
Gibbons, F. X. 142, **146**
Gibson, T. 113, **127**
Giles, H. **300**
Godwin, P. H. B. 95, **104**
Goldstein, A. P. 2, **7**
Good, D. A. xi, 188, 224–36, 241, 250,
   281
Goode, E. **184**
Goody, E. xi, 13, 52–3, 106–27, 266
Gorbachev, Mikhail 317–18, 345–9
Gordon, M. 347(n)
Groebel, J. xi
Gromyko, Andrei 344

Haig, Alexander, 341, 346
Hamilton, W. D. 11, **14**, 22, **25**, **48**
Harcourt, A. H. xii, 11, 12, 14, 15–25,
   26, 159
Harris, V. A. 153, **158**
Hazan, C. 195, **211**
Heal, J. xii, 12, 132, 149, 159–71
Heider, F. 147, 150, **158**
Hendrick, C. **223**, **300**
Hendry, J. **127**
Henrick, S. **223**
Hill, K. **48**
Hinde, R. A. xii, 7, 8, 214, **223**
Ho, D. Y. F. 91, **104**
Hoffman, M. L. 62, **76**, 137, **146**
Hofstede, G. **87**

Hogg, M. A. 239, **262**
Holmes, J. G. xii, 160, 187, 188, 190–211, **211**
Hoogland, J. 23
Hornstein, H. 148, **158**
Horwitz, H. 238, **261**
Houlihan, D. **145–6**, **158**
Houston, A. I. 21, 22, **26**
Howell, S. 112, 113, 115, **127**
Hsu, F. L. K. **87**

Iannotti, R. **77**

Jackson, J. K. 350, 351
Jackson, Senator 338
Jenkins, Brian 325
Jones, E. E. 149–50, 151, 153, **158**
Jones, W. H. 193

Kahneman, D. **184**
Kanter, R. M. 214, 218, 221, **223**
Kaplan, H. **48**
Kelley, H. H. 191, 194, 197, **211**, **223**, 249–50, **261**, 276, **280**
Kelman, H. C. 268, **280**
Kenrick, D. T. **145**
Khomeini, Ayatollah 338–9
Kimmel, M. J. 231, 236, 241, 246, 247–8, 250, **261**
Kissinger, Henry 334 and (n)
Kohlberg, L. 57
Krauss, R. M. 229–30, 236, 296
Kray brothers 231
Krebs, D. L. 147, 148, **158**
Kriesberg, L. 277, **280**
Kruyuchkov, Vladimir 327

Latane, B. **134**
Lax, D. A. 271–2, **280**
Lee, S.-Y, 52, **53**
Lepper, M. 67
Lerner, M. J. xii, 133, 173–84, **184**
Leung, K. **88**
Levin, H. 71
LeVine, R. A. 41–2, **48**, **127**
Levinger, G. 220, **223**
Lewin, K. 238, 239, 240, **261**
Lewis, C. C. 100–1, **104**
Lizot, J. 117–19, **128**
Lodewijkx, H. F. M. 246, 247, 249, 250, 251, **261**
Lombardo, J. P. **146**

London, P. 156, **158**
Lucca, N. **88**
Lumsden, C. J. **48**
Lund, M. xii, 160, 188, 212–23
Luterman, K. G. 141, **145**
Lutz, C. A. **128**

Macaulay, J. **158**
McCarthy, P. M. **146**
Maccoby, E. E. 71
McFarlane, 346
McGrath, J. E. 241, 247, 251, 260, **261**
McGuire, M. T. 13, **14**
MacIntyre, A. **172**
Mackil, D. M. **300**
Mao Zedong 94
Markovsky, B. **184**
Marks, J. xii, 267, 316–28
Marwell, G. **300**
Mayor, F. xii, 267, 301–15
Mead, M. 89, **104**, **128**
Messick, D. M. 240, 252–3, **261**, **300**
Milgram, S. 233
Miller, 147
Miller, D. T. 199, **211**
Miller, N. **300**
Miller, P. A. xiii, 12, 51, 54–76
Milhauser, Marguerite 323
Mischel, W. 266–7
Mitrany, D. 304
Montagu, A. 121, **128**
Moore, B. S. **77**
Muawwad, Renee 327
Munro, D. J. 94, **104**
Mussen, P. **14, 76**, 89, **104**

Naroll, R. **87**
Neuringer-Benefiel, H. E. **134**
Nitze, Paul 345
Nixon, Richard M. 329, 334
Noe, R. 24
Noriega, Gen. Manuel 173
Norton, W. W. 322
Nussbaum, M. C. **172**

Oakes, P. J. 239, **262**
Obeid, Shaikh Abdul 328
O'Connor, J. 233, 237
Oliner, S. P. and P. M. **77**
Overing, J. 115, **127**

Peak, L. 100, **104**

Peoples, J. E. **48**
Perlman, D. 193
Piaget, J. 57
Piliavin, J. A. 137, **146**
Ponomarev, Lt Gen. Vitaly 322
Pruitt, D. G. 231, **236**, 241, 246,
    247–8, 250, **261**, 279, **280**

Rabbie, J. M. xiii, 188–9, 238–60, **261**,
    266, 281
Rachels, 149
Radke-Yarrow, M. **77**, 94, **104**
Raiffa, H. 275, **280**
Rapoport, A. 232, **237**
Rausch, H. L. **223**
Reagan, Ronald 317, 333, 340, 341–8
Reicher, S. D. 239, **262**
Reis, H. T. 206, **211**
Reyer, U. 17, 20
Richerson, P. J. xiii, 4, 14, 27–48, **48**
Ridley, C. P. 95, **104**
Robarchek, C. A. 111–12, **127**
Rokeach, M. 153, **158**
Rosaldo, M. Z. **128**
Rotter, J. B. 228
Rowny, Gen. Edward L. 338
Rubin, J. P. 347(n)
Rubin, J. Z. xiii, 159, 266, 268–80, **280**
Rubin, Z. 214, 215, 220, **223**
Rusbult, C. 218
Rushton, J. P. 149, **158**
Rutter, D. 233
Ryle, G. 171

Sadat, Anwar 276
Sands, Bobby 27
Saunders, H. H. 274, **280**
Schaller, M. **145–6**, **158**
Schmitt, D. R. **300**
Schneider, W. 341(n)
Schopler, J. 154, **158**
Schot, J. C. 238, 257, 258, 259, **261**
Schul, Y. **184**
Schwartz, S. H. 141–2, **146**
Sears, R. R. 71
Sebenius, J. 271–2(n), 280
Severy, L. J. 156, **158**
Seyfarth, R. 24
Shaver, P. R. 195, 206, **211**
Shaw, G. B. 217
Sherbak, Lt Gen. Feodor 325–7
Sherif, M. 288–90

Shevardnadze, Eduard 325
Shultz, George 341, 344
Silk, J. 23
Simmel, G. 213, **223**
Smuts, B. B. 24, **26**
Snyder, M. 140, **146**
Sorenson, E. R. 114–15, **128**
Stahelski, A. J. 197, **211**, **261**
Staub, E. **77**, 148–50, 153, 156, **158**
Stephan, W. G. **300**
Stern, P. 335(n)
Stevenson, H. W. xiii, 52, **53**, 89–104,
    169, 266
Stewart, K. 19, 20
Strayer, J. **76**
Strickland, L. H. 197, **211**
Stroebe, W. 154, **158**
Susskind, L. 271–2, **280**
Swap, W. C. xiii, 132, 147–57, **158**,
    160

Taborsky, M. 21
Tajfel, H. 239, 240, 243, 250, 255, 256,
    257, 259, **261**, 290
Talbott, S. 345
Tesser, A. 155, **158**
Thompson, V. 154, **158**
Triandis, H. C. xiii, 52, 78–87, **87–8**,
    169, 266
Trivers, R. I. 11, 14, 22, 26, **48**
Turnbull, C. M. 110, **128**
Turnbull, W. 199, **211**
Turner, J. C. 239, 240, 243, 250, 255,
    256, 257, 258, 259, 260, **261–2**,
    300
Tyler, T. **184**
Tyroler, C., III 336(n)

Ury, W. L. 270, 276, **280**

Vance, Cyrus 334, 339
van Oostram, J. 250
Varey, C. **184**
Varney. L. L. **146**
Vehrencamp, S. 20, 21
Villareal, M. **88**
Visser, L. 238, 250, **261**

Wade, M. J. **48**
Walster, E. and G. **184**
Webb, W. M. **211**
Webster, J. B. 96, **104**

Webster, William 327
Weinberger, Caspar 341, 346
Weiss, R. F. 137, **146**
Weitzman, L. J. 216, **223**
Westerterp, K. 17
Wetherell, M. S. 239, **262**
Weyant, J. M. 139, **146**
Whiting, B. B. 89, **105**, 109, **128**
Whiting, J. W. M. 89, **105**
Wichman, 233
Wicklund, R. A. 142, **146**
Wilkinson, G. 18
Williams, B. A. O. **172**
Willis, R. 112, 113, 115, **127**
Wilson, 96

Wilson, E. O. **48**
Woolfenden, G. 20
Worchel, P. **211**
Worchel, S. 289, **300**
Wrangham, R. 25
Wrightsman, L. S. 233, **237**

Yamaguchi, Y. 96, **105**

Zahn-Waxler, C. **77**, 94, **104**
Zander, A. 259, **261**
Ziller, R. C. **88**
Zimbardo, P. G. 231
Zvezdenkov, Major Gen. Valentin
    325–7

Subject index

academic tracking 294
ad hoc organizations 283
Afghanistan, Soviet invasion of 338–9
age, and prosocial behaviour 62–4
aggression 1–3
  perceptions of 155–6
  prosocial 98–9
alliances 6
allocentrism 82
altruism 12–13, 132, 159–71
  benefit to others 150–1, 162
  correspondent inferences 149–56
  definition 5, 131, 149, 156
  development of 54–5
  and disinterestedness 162–6
  a good thing? 166–71
  intentionality 154–5
  and justice 132–3, 157
  and natural selection 12–13
  and norms 152–4
  perceptions of 147–57
  and pleasure 161–2
  and self-benefit 151–2, 162–3
  situation/context 149–57
  spontaneity *vs.* premeditation 155–6
animals, cooperation 11, 15–25, 28
  partnerships 23–4
  reasons for cooperation 16–19
  and structure of society 24–5
  timing of cooperation 19–22
anthropology, learning of prosocial
    behaviour 106–27
antisocial behaviour 2–3, 140
Arapesh people, New Guinea 107

arms control 334–5, 346–7
Asia, philosophy of prosocial
    behaviour 90–4
attitudes 284
  change of 268–9

Behavioral Interaction Model (BIM)
    238, 242–4
behavioural assimilation 197
Birifor people, Ghana 119, 123, 125,
    126
Buid people, Philippines 113, 124, 126,
    127

Camp David Accord 276–7
categories, social, and groups 239–40
Chewong people, Malaysia 112–13, 124
China, development of prosocial
    behaviour 52, 89–104
class, and individualism *vs.*
    collectivism 81–2
co-acting 287
coalitions 287
cognition
  and commitment 219–21
  and justice 175–6
  and prosocial development 51,
    55–9, 61–2
Cold War 318
  see also policy, USA towards USSR
collectivism 4, 52, 79–87, 169
commitment 212–23
  decisions and cognitions 219–21
  definition 5–6, 213–14

and love 214–15
and maintaining relationships
   212–23
new 217–23
old 215–17, 221–3
and trust 187–8
and work 212, 222
communes 218
communication
   and cooperation 188
   games 232–4
   between groups 266
competition, between groups 285
competitive strategy 224, 228–9, 250
compliance 66
Con Game 227
conflict 266, 285
   realistic conflict theory 299
conflict resolution 6, 266, 268–70
conflict settlement 268–70
   common processes 271–3
   and enlightened self-interest 270–1
   and intervention 269
   negotiation from inside out 276–7
   readiness for negotiation 277–8
   relationships in negotiation 273–4
   residue 278–9
   time scale 274–6
conformist tradition of cooperation
   14, 30, 35
   evolution of 36–8
Confucianism 93–4
contact hypothesis 291–5
context *see* situation/context
cooperation 11–14, 284–8
   in animals 11, 15–25, 28
   between groups 265–7, 281–300
   and common interests 278, 286–7,
      288–91
   and communication 188
   conformist tradition of 14, 30, 35–8
   and culture 14, 27–43
   definition 4
   effects on groups 297–8
   in ethnic groups 41–3
   experience of, group 295–6
   experimental induction of 285–6,
      288–95
   games 224–36
   international 266–7, 271, 303
   intra-group 188–9, 238–60
   and justice 173–84

as norm 298
situation/context 131–2, 135–45
studies of 3
theories of 298–9
cooperative strategy 225, 228–9, 250
co-orientation 287
correspondent inferences 149–50
costs of behaviour 12, 133, 187–8
cuelessness 233
culture
   conformist cultural transmission 30,
      35–8
   and cooperation 14, 27–43
   and development of prosocial
      behaviour 78–87
   emigration 33, 40
   endogamy 38–9
   ethnic group cooperation 41–3
   extinction 33, 39–40
   and group selection 29–36, 40–3
   influence on behaviour 52–3

definitions 4–7
democracy 315
development of prosocial behaviour
   51–2, 54–7
   anthropological view 106–27
   in China and Japan 52, 89–104
   and cognition 55–9, 61–2
   cross-cultural differences 78–87
   and emotion 60–2
dilemmas, social, tasks 240–2, 252–4
discipline, and prosocial development
   69–71
disinterestedness, and altruism 162–6
distress reduction model of helping
   137, 138–9
distrust 210
divorce 212, 216, 222

egoism, and altruism 151–2
emigration 33, 40
emotions
   and justice 176–7
   and prosocial development 51, 60–2
empathy 60–1, 143
empathy-altruism hypothesis of
   helping 143–4, 164
endogamy, cultural 38–9
ethnic group cooperation 41–3
evolution 11
exchange theory 12, 133, 187–8

face-saving 273
faithfulness 213
family, prosocial behaviour within 13,
    106–27
Fore people, New Guinea 107,
    114–15, 124–5, 126, 127

games 188, 224–36
    coercion 234
    cognitive constraints 235–6
    communication 232–4
    payoffs 225, 229–31
    personality 226–9
    time 231–2
gender
    and competitive *vs.* cooperative
        strategy 228, 250
    and helping 134
    and prosocial behaviour 64–5
goals, interdependence of 278, 286–7,
    288–91
Gonja people, Ghana 106, 114, 126
group selection 29–36, 40–3
groups 4, 238–9
    and *ad hoc* organizations 283
    and altruism 171
    Behavioral Interaction Model
        (BIM) 238, 242–4
    communication between 266
    competition between 285
    conflict and negotiation 268–80
    cooperation between 265–7,
        281–300
    cooperation within 188–9, 238–60
    definition of 282
    determinants of cooperation 238–60
    education in Japan 99–101
    ethnic 41–3
    individual vs. intergroup behaviour
        281–2
    institutions within 290
    leadership 265
    loyalty to 6
    Minimal Group Paradigm 255–60
    multi-person dilemma tasks 252–4
    origins of 282–3
    Prisoner's Dilemma game 241,
        244–52
    and social categories 239–40, 266,
        282–3
    social dilemma tasks 240–2
    socio-cultural structure of 6

UNESCO role in international
    cooperation 301–15
US policy toward USSR 329–52
US/Soviet cooperation against
    terrorism 316–28

hedonism 161–2
helping 4, 284
    distress reduction model of 137,
        138–9
    empathy-altruism hypothesis of
        143–4, 164
    negative state relief model of 137–9
    normative explanations 141–3
    qualities of 135–6

idiocentrism 82
Ifaluk people 107
Ilongot people, Philippines 119, 123,
    126
individualism 52, 80–7, 169, 270
individuals, interaction between 2
inductions 69–70
innocence, stage of 91
institutions, within groups 290
intelligence
    and justice 175–6
    and prosocial behaviour 22, 59
interaction 5
    between groups 6
    between individuals 2
    influences on 6–7
interdemic group selection 33–4
interdependence 191, 256–7
intergovernmental organizations
    (IGOs) 303–5
internalization of prosocial values
    66–7
intervention, and conflict settlement
    269
Inuit, 106–7, 119–22
investment, in relationships 218–19
Iran, hostages 269–70, 338–9
Iraq, invasion of Kuwait 351

Japan, development of prosocial
    behaviour 52, 89–104
justice 173–84
    and altruism 132–3, 157
    implications for intergroup
        cooperation 183–4
    individual 175–7

relationship 177–9
social issues 180–3

kin, and prosocial behaviour 13
kin selection 13, 18–19, 23, 25
kindergartens
cross-cultural study 101–3
Japan 99–101
Kuwait, Iraqi invasion of 351

leadership of groups 265
learning cooperation 32
China and Japan 94–8
Confucian tradition 94
love 188, 201–2, 215
loyalty, to groups 6

Machiavellianism 227
marriage 215–23
Mbuti people, Congo 110–11, 124–5, 126
media 266–7
and aggression 3
Minimal Group Paradigm 255–60
modelling 72–3, 109
moral reasoning 51, 57–9
motivation for helping 135–45, 148
mutualism 4
in animals 17–18

NATO 350
natural selection 11
and altruism 12–13
negative state relief model of helping 137–9
negotiation 266, 269
between groups 268–80
from inside out 276–7
readiness for 277–8
relationships in 273–4
time scale 274–6
norm, cooperation as 298
norms for helping 141–3, 152–4

obedience study 233
organizations, *ad hoc* 283

parenting, and socialization of prosocial behaviour 66–75
patriotism 6, 266
pay-offs 225, 229–31
perception, effects of cooperation on 297–8

personality
determinants of helping 135–45
games 226–9
and prosocial behaviour 76, 134
and trust 195–6
perspective-taking 55–7
Piaroa people, Venezuela 115–16, 126
pleasure, and altruism 161–2
policy, USA towards USSR 329–52
arms control 334–5, 346–7
Bush 349, 352
Carter and conservative arms demands 335–41
changes in relations 329–30, 332
confrontation 332
and domestic economic interests 332–3
functional structure model of policy-making 331–2
and Gorbachev 345–9
new approach 350–2
and public opinion 333, 336–7, 344
Reagan and public demands for cooperation 341–5
Prisoner's Dilemma game 225, 228, 232
cooperation in groups 241, 244–52
process-orientated model of trust 194
prosocial behaviour
and age 62–4
and antisocial behaviour 2–3
and cognition 51, 55–9, 61–2
development (*q.v.*) of 51–3, 54–76
and emotion 51, 60–2
and gender 64–5
influences on 131–4
situational and personality determinants 131–4, 135–45
socialization of 66–75
studies of 3

quasi-groups 241–2

realistic conflict theory 299
reason, stage of 91
reciprocity 28–9, 187
in animals 17–18
reinforcement 12, 131–3
relationships
commitment and trust 5–6, 212–23
definition 5
investment in 218–19

in negotiation 273–4
old and new commitment 212–23
rewards in 217–18, 221
social support for 221
types of 215
reproductive success 11–12, 16–19
reputation 187
rewards 12, 131–3
in relationships 217–18, 221
risk 187, 191–4, 295
aversion to 209
rivalry 285

SALT 334–5, 338–40
satisfaction 163
schools
cooperation in 294
cross-cultural study 101–3
Self-Categorization Theory (SCT) 239,
255
self-fulfilling prophecy 200
Semai people, Malaysia 111–12, 124,
126, 126
shaping 109
situation/context
and altruism 149–57
and cooperation 131–2, 135–45
and trust 196–8
social categories, and groups 239–40,
266, 282–3
social exchange theory 133, 187–8
Social Identity Theory 239, 240, 290–1
socialization 31–2, 51–2, 66–75, 283–4
China and Japan 89–104
socio-cultural structure, of groups 6
Soviet Union *see* Union of Soviet
Socialist Republics
symbiosis 4
sympathy 60–1

Taiwan, development of prosocial
behaviour 52, 89–104
terminology 4–7
terrorism, US–Soviet cooperation
against 316–28
benefits 316–17
common danger 317–18
definition of acts of terrorism 323
recommendations 324–5, 326–8
Soviet attitude 318–20
US–Soviet Task Force 321–8
textbooks, China and Japan 94–8

tit-for-tat strategy 245
training 109
Trucking Game 229–30, 234
trust 187–9, 190–211
in animals 15–25
in close relationships 201–10
and commitment 187–8
commitment to relationships 212–23
definition 5
and history of relationship 198–9
interpersonal 190–211
and perceptual distortion 199–201
and personality 195–6
process-orientated model of 194
and risk 187, 191–4
and situation/context 196–8
trustworthiness 228

UNESCO 301–15
administrative innovation 307–11
Constitution 301
functionalism and search for
partners 304–7
intergovernmental organizations
(IGOs) 303–5
international cooperation process
303
and national statistics 309
and non-governmental
organizations 305–6
output 311–14
Participation Programme 311
Tensions Project 306–7
Union of Soviet Socialist Republics
(USSR)
and arms control 334–5, 346–7
common interests with USA 278,
317–18
Gorbachev's policies 345–9
invasion of Afghanistan 338–9
terrorism, cooperation with USA
against 316–28
and Warsaw Pact 349–50
United States of America (USA)
Chicago Council on Foreign
Relations 336, 337
Committee on the Present Danger
335–6, 340–1
common interests with USSR 278,
317–18
economic changes 350–1
and NATO 350

need for new world policies
350–2
policy (*q.v.*) toward USSR
329–52
terrorism (*q.v.*), cooperation with
USSR against 316–28
Utku Inuit 106, 107, 119–23, 123–4,
126

violence, media 3

war 1–2
Warsaw Pact 349–50
work, and commitment 212, 222

Yanomami people, Venezuela 116–19,
123, 124, 126, 127